How We Remember

How We Remember

Brain Mechanisms of Episodic Memory

Michael E. Hasselmo

The MIT Press
Cambridge, Massachusetts
London, England

First MIT Press paperback edition, 2013

This book was set in Stone Sans and Stone Serif by Toppan Best-set Premedia Limited.

Library of Congress Cataloging-in-Publication Data

Hasselmo, Michael E.
How we remember : brain mechanisms of episodic memory / Michael E. Hasselmo.
p.; cm.
Includes bibliographical references and index.
ISBN 978-0-262-01635-3 (hardcover : alk. paper), 978-0-262-52533-6 (pb)
1. Memory—Physiological aspects. 2. Recollection (Psychology)—Physiological aspects.
3. Brain—Physiology. I. Title.
[DNLM: 1. Mental Recall—physiology. 2. Brain—physiology. WL 102]
QP406.H37 2012
612.8′2—dc22

2011008719

150116994

To Chantal, Simone, Nicholas, and the rest of my family

Contents

Preface

I remember the episode when I first started to do research on memory function. By episode, I mean an experience covering a specific time interval in a specific location. I was working on a term paper one afternoon in the fall of 1981. I was searching the Psychology Department library on the 6th floor of William James Hall at 33 Kirkland Street in Cambridge, Massachusetts. This was about 10 years before I first had access to literature searches on the web, so I remember my path walking back and forth between a wooden card catalog and the shelves of books and journals. Based on my experience taking physics that same term, I expected to find a textbook on memory analogous to the treatment of mechanics in a physics textbook. I expected to find a book describing the brain mechanisms for encoding and remembering the times and locations of my path through the library, analogous to the description of the spatio-temporal trajectory of a falling object in physics. I did not.

Even now, almost 30 years later, there is not a standard, quantitative description of the brain mechanisms of memory analogous to the theory available in most areas of physics and engineering. However, exciting recent physiological data and modeling have laid out a framework for modeling the memory of episodes as spatiotemporal trajectories. These recent developments include the discovery of grid cells and their relationship to oscillatory dynamics in the entorhinal cortex. This recent work builds on an extensive body of older work on memory function.

In this book, I try to present what I was searching for in 1981, a quantitative model of the brain mechanisms for encoding and remembering an episode as a spatiotemporal trajectory. In the main text, I will present the model in a more narrative form. In the appendix, I will present the material in the quantitative style of a textbook. Because of limits on space and time, there are unfortunately many areas of research that I have not been able to include.

When describing episodic memory, researchers often give the example of remembering where you parked the car or remembering where you left your reading glasses or keys. Here, I will directly address real-world examples of episodic memory, describing how your brain might encode and retrieve the location where you parked your car. A familiar solution to problems such as finding your glasses or keys is to "retrace your steps," that is, to retrace your trajectory through space and time. Related to this, mnemonic strategies for enhancing encoding include the method of loci in which one imagines placing items in a sequence of familiar locations.

Though other models will be discussed, the book will focus on one specific model of episodic memory as the encoding and retrieval of spatiotemporal trajectories. We build models to evaluate our theories and to generate predictions that can then be tested. Testing the models provides us with information about where the models need to be revised, and this information can then be used to improve the models. The current models can address a range of physiological and behavioral data but do not accurately address the full range of data. However, despite the incomplete nature of current models, it is useful to describe how existing models address components of the data and to compare the strengths and weaknesses of different models. This process may be helpful to others interested in these questions and has certainly been helpful to me. A primary focus of this work will be data on the activity of neurons during memory-guided behavior. Some of this data has been shown in humans and primates but most data is available from research on rodents.

This book is intended to be accessible to the general educated reader. Based on my teaching experience, I find that neuroscience is so broad a field that experienced researchers in one area commonly do not know the basic principles in another area. For this reason, I present the data on a basic level that I feel will be useful to faculty and graduate students in many fields as well as advanced undergraduates. The Appendix provides mathematical descriptions appropriate for coursework or independent study by advanced undergraduates and graduate students in neuroscience.

Presenting the model of episodic memory requires some background. In the first chapter, I will present some of the behavioral data on episodic memory and give a general overview of the model. In the second chapter, I will provide an overview of the anatomical data on structures involved in episodic memory and an overview of the physiological data obtained from these structures. In the third chapter, I will present an overview of the mechanisms in the entorhinal cortex and hippocampus for encoding space and time in episodic memories, including models of grid cells. In the fourth chapter, I will present the full model of the mechanisms of episodic memory. In the fifth chapter, I will address the role of the hippocampus in forming associations between spatiotemporal trajectories and individual items or events. In the sixth chapter, I will address the role of neuromodulators such as acetylcholine in episodic memory function, and in the seventh chapter I will address models of the use of episodic memory function for memory-guided behavior.

I appreciate the encouragement of the editors at MIT Press, especially Robert Prior for finishing the job and Michael Rutter for starting it. In addition, I appreciate the inspiration from my colleagues at the Center for Memory and Brain (CMB) at Boston

University. Special thanks to Chantal Stern for her collaboration in building our family as well as our careers and her clear-sighted perspective in our many fruitful conversations, and to Howard Eichenbaum for bringing us to Boston University and for his energetic and enthusiastic insights on memory. Thanks also to CMB members John White and Nancy Kopell. Thanks also to my colleagues in the Department of Psychology, the Graduate Program for Neuroscience, and other departments and centers at Boston University.

I would like to give my very enthusiastic praise and thanks to the students and postdocs in my laboratory who contributed to the research that I describe. In particular, very special thanks for the use of figures of data and models from my graduate students Mark Brandon, Lisa Giocomo, Brad Wyble, and Eric Zilli and postdoctoral fellows Edi Barkai, Murat Erdem, Erik Fransén, Randal Koene, Inah Lee, and Motoharu Yoshida. Thanks also for the hard work in my lab by all my graduate students including (in alphabetical order) Mark Brandon, Jason Climer, Eve De Rosa, Lisa Giocomo Anatoli Gorchetchnikov, Kishan Gupta, Jim Heys, James Hyman, Terry Kremin, Caitlin Monaghan, Chris Shay, Akaysha Tang, Brad Wyble and Eric Zilli, and all my current and former postdoctoral fellows including Edi Barkai, Ian Boardman, Clara Bodelon, Thom Cleland, Vassilis Cutsuridis, Erik Fransén, Norbert Fortin, Amy Griffin, Marc Howard, Sarah Judge, Ajay Kapur, Randal Koene, Inah Lee, Christiane Linster, Jill McGaughy, Ehren Newman, Mahdvi Patil, Nathan Schultheiss, Yusuke Tsuno, Carl Van Vreeswijk, Gene Wallenstein, Motoharu Yoshida. Thanks also to the undergraduates and research assistants who have worked with me including Andrea Abi-Karam, Chris Andrews, Ross Bergman, Andrew Bogaard, Milos Cekic, Michael Connerney, Brian Fehlau, Lisa Femia, Shea Gillett, Vikas Goyal, John Holena, Greg Horwitz, Kaiwen Kam, Chris Libby, Eugene Lubenov, Michaella Maloney, Brad Molyneaux, Amanda Paley, Christina Rossi, Eric Schnell, Vikaas Sohal, and Tyler Ware. I also greatly appreciate the administrative assistance for my research from Denise Parisi, D. J. Aylward, Scott Enos, and Psychology Chairperson Henry Marcucella.

I would like to thank the other students and postdoctoral fellows at the Boston University CMB. In my collaboration with Chantal Stern, I have worked with members of her laboratory including Ali Atri, Thackery Brown, Brenda Kirchhoff, Matthew LoPresti, Marlene Nicolas, Robert Ross, Haline Schendan, Karin Schon, and Seth Sherman. I also appreciate my interactions with members of the Howard Eichenbaum lab including Paul Dudchenko, Norbert Fortin, Ben Kraus, Rob Komorowski, Megan Libby, Paul Lipton, Joe Manns, Sam McKenzie, Seth Ramus, and Emma Wood.

I would like to thank the government agencies that funded my scientific research, including the National Institute of Mental Health (NIMH), the National Institute of Drug Abuse (NIDA), the National Science Foundation (NSF), and the Office of Naval Research (ONR) as well as the program officers who facilitated funding for my scientific work. Special thanks to Dennis Glanzman, who was the program officer on my first grant and on my two ongoing R01 grants from NIMH. Thanks also to program officers Susan Volman at NIDA, Ken Whang at NSF, Bettina Osborne at NIMH, Soo-Siang Lim at NSF, and Tom McKenna, Marc Steinberg, and Joel Davis for ONR grant support.

Thanks also to my mentors over the years, including Jim Bower at Caltech, Edmund Rolls at Oxford, and Jim Stellar and Al Galaburda at Harvard. Thanks also to many of my colleagues during my education including my co-authors Matt Wilson and Brook Anderson, as well as Upi Bhalla, Jim Knierim, John Thompson, Rich Mooney, and others at Caltech and Chantal Stern, David Thaler, Gordon Baylis, and others at Oxford.

I also give special thanks to my collaborators and colleagues working on neural mechanisms of memory in the entorhinal cortex and hippocampus. Special thanks to my collaborators working on the physiology of the entorhinal cortex, particularly my former collaborator the late Prof. Angel Alonso (who sadly died in 2005) and our collaborators Clayton Dickson, Alexei Egorov, Babak Tahvildari, Mark Shalinsky, and Motoharu Yoshida, as well as other slice physiologists in the field especially including my fellow CMB member John White, as well as Nelson Spruston, John Lisman, Dan Johnston, and Uwe Heinemann. Thanks to my colleagues and collaborators in modeling including Neil Burgess, Caswell Barry, Peter Dayan, Ila Fiete, Ole Jensen, John Lisman, Bruce McNaughton, David Redish, David Touretzky, Alessandro Treves, and many others. Thanks also to my collaborators and colleagues doing unit recording in the hippocampus and entorhinal cortex, including John O'Keefe, Caswell Barry, Andre Fenton, Kate Jeffery, Adam Johnson, Jim Knierim, John Kubie, Colin Lever, Bruce McNaughton, Sheri Mizumori, Bob Muller, Jim Ranck, David Redish, Bill Skaggs, and Matt Wilson. Special thanks as well to the colleagues working on grid cells, including Edvard and May-Britt Moser as well as Charlotte Boccara, Stefan and Jill Leutgeb, Trygve Solstad, and many others. Thanks also to the researchers on head direction cells including Jeff Taube, Pat Sharp, and Tad Blair and their students. Thanks also to my colleagues working on the mechanisms and function of theta rhythm, including Gyuri Buzsaki, Steve Berry, and Bernat Kocsis. Thanks also to my other collaborators over the years, including Mark Baxter, Hans Breiter, Robert Cannon, Neal Cohen, Joe Coyle, Steve Cramer, David Gerber, Mark Gluck, Robby Greene, Michael Greicius, H. C. H. Grunze, Jonathan Hay, Brad Hyman, Max Ilyn, Bill Kath, Yael Katz, Steve Kunec, John Leonard, Hans Liljenstrom, Bob McCarley, Jay McClelland, Earl Miller, Catherine Myers, Ken Norman, Nick Roy, Martin Sarter, Matt Shapiro, Greg Siegle, Peter Siekmeier, Susumu Tonegawa, Max Versace, and Don Wunsch. Thanks for comments on manuscript sections from my current lab members and students in my modeling course, as well as Chantal Stern, Nils Hasselmo, Nicholas Hasselmo, Edvard and May-Britt Moser, Caswell Barry and Neil Burgess. Any remaining errors are my own! Despite this long list, I have probably inadvertently left someone out that I will only notice when I pick up the published and bound book. Thank you, too!

1 Behavioral Dynamics of Episodic Memory

When we think of an individual memory, many of us think of an episodic memory. We conjure up rich recollections of sequences of events from our recent or remote past that play out in our minds as if we were reliving the experience. For example, I can remember walking our dog Ollie in the snow on the morning that I wrote this paragraph on January 20, 2010. I remember some of my actions in putting on his leash, stepping out the back door and walking along our back path to the street. I remember distinct points of view, for example as I walked up the hill and then as I turned to see our cat following us. I remember some of the twists and turns in my path as we walked around piles of plowed snow or as Ollie investigated patches of yellow. I remember the movement of our neighbor's cat as it ran out of our way, and I remember how Ollie barked and ran toward the neighbor's cat. As demonstrated by this example, the term *episodic memory* refers to the memory of specific events occurring at a specific place and time. In this book, the term *remember* is used to specifically describe the retrieval of an episodic memory. As described below, episodic memory is contrasted with many other types of memory, including memory for general facts and world knowledge (semantic memory) and memory for how to do things (procedural memory).

By no means do I remember every moment in time from my walk. I did not encode some events, and others were rapidly forgotten. However, my episodic memory contains many detailed representations of the dynamics of the episode—including my location, my point of view, the type and speed of my actions. With some thought, I can remember many segments of a detailed pathway through time and space. Like the snow under my feet, my episodic memory retained mental footprints showing where I faced, where I turned, and even how fast I moved. I can retrieve my memory on the basis of these imprints in my memory, as if retracing my path by matching my foot to each footprint. This path of footprints through the snow with its features of direction and speed, and the associated actions, will be described here as a spatio-temporal trajectory or episodic trajectory. The traces left in my memory are not just discrete footprints but the continuous trace of my full body moving through three-dimensional space, as well as the continuous traces of other moving agents including our dog and our cat. A more accurate analogy may be the streaks of movement in a long-exposure photograph, where a movement appears as an unbroken, fluid arc. I will argue that episodic memory contains segments of the spatiotemporal trajectories from our prior experience. My memories of this morning are most vivid, but I can also recollect portions of episodes from the more distant past, with actions, point of view, and segments of trajectories. I can mentally relive many segments of my past life.

This book will focus on the brain mechanisms of episodic memory. On the behavioral level, I will focus on the capacity to encode and retrieve segments of spatiotemporal trajectories from personal experience, including the time and location of individual events such as encounters with other agents or items. On the biological level, I will focus on the dynamical properties of neurons in the brain regions implicated in episodic memory, presenting a model of how the cellular properties of these neurons could underlie the mechanisms of episodic memory.

This is an exciting time of progress in our understanding of the brain dynamics of episodic memory. Recent experimental discoveries have provided new insights, elucidating new details about how our brains might encode and retrieve spatiotemporal trajectories. These recent discoveries include the discovery of "grid cells" in the entorhinal cortex of rats foraging through open spaces, the discovery of cellular rhythms in individual neurons in entorhinal cortex that could underlie these grid cells, the discovery of "splitter cells" that appear selective for different sequences through the same location, and also the discovery of patterns of neural activity that appear to reflect the "replay" of previous movements through the environment. These discoveries provide for the first time a framework for understanding the encoding and retrieval of episodic memory not just as discrete snapshots of past time but as a dynamic replay of spatiotemporal trajectories. These breakthroughs also help to resolve some puzzling paradoxes of scientific findings over the past several decades.

Definition of Episodic Memory

In the domain of psychological research, episodic memory was first defined by Endel Tulving (Tulving, 1972). At that time, a number of cognitive researchers were focused on studying memory as a single, monolithic process. In contrast, researchers today acknowledge the existence of multiple memory systems.

Before Tulving, there were many writers who described different aspects of memory function. However, Tulving had an influential effect in focusing academic research on differences between the recollection of a specific sensory experience at a specific place and time versus the retrieval of a piece of factual evidence from general knowledge. Tulving first developed the term *episodic memory* as a contrast to the concept of semantic memory (Tulving, 1972, 1983, 1984, 2001; Baddeley, 2001). The term *semantic memory* was developed by Quillian (Collins and Quillian, 1969) to describe memory for facts and general knowledge about the world. For example, remembering that Stockholm is the capital of Sweden would be considered a semantic memory, as opposed to remembering a specific trip to visit relatives in Stockholm, which would be an episodic memory. Episodic memory was later contrasted with other forms of memory such as knowing how to do things, referred to as *procedural* or *implicit memory* (Cohen and Squire, 1980; Cohen and Eichenbaum, 1993), and with actively keeping recent information in mind, known as *working memory* (Baddeley and Hitch, 1974; Baddeley, 2001).

Over the years, the definition of episodic memory has matured to include many features accepted by the vast majority of neuroscientists and psychologists (though there are still some features that are controversial). There is general acceptance of the statement that the form of the retrieval query directed at the episodic system is "What did you do at time T in place P?" (Tulving, 1984). Tulving's definition also includes the capacity for mental time travel to past episodes, to replay them from memory as if reliving them (Tulving, 1983; Eichenbaum and Cohen, 2001; Tulving, 2002). These features of episodic memory fit well with subjective experience and the particular example that I presented above.

Tulving focuses on mental time travel as the mental voyage from the current time to the past episode being remembered, and his perspective differs from the focus presented here in that he does not focus as much on the mental time travel within the past episode itself. In fact, he states that his model

> ...describes a "snapshot view" of episodic memory: It focuses on conditions that bring about a slice of experience frozen in time which we identify as "remembering." The recursive operation [...] produces many snapshots whose orderly succession can create the mnemonic illusion of the flow of past time. (Tulving, 1984, p. 231)

This description differs from how I will describe the retrieval process. In particular, it suggests a discrete rather than a continuous flow of retrieval and suggests that a retrieval act (ecphory) is required for each discrete snapshot, rather than a retrieved memory being extended in the dimension of time. This runs contrary to elements of personal experience, such as my memory of my dog lunging toward the neighbor's cat as a complete fluid movement (a continuous motion rather than a snapshot).

Thus, I argue that the process of mental time travel goes beyond forming associations of single items with a single, static behavioral context of location and time and involves encoding of continuous segments of a spatiotemporal trajectory. This requires specific machinery on a cellular and circuit level. I will present models of the cellular dynamics of episodic memory that allow us to relive a sequence of events as segments of a spatiotemporal trajectory with an explicit sense of position in continuous space and duration in continuous time and an explicit reexperience of factors such as head direction and the direction of movements.

Other features of episodic memory have been elaborated by other researchers such as Martin Conway (2009). Conway reviews data showing that episodic memories have a perspective, either consisting of the point of view of the person retrieving the memory (a "field" perspective) or a third-person observer perspective (Nigro and Neisser, 1983; Robinson and Swanson, 1993; Conway, 2009). The description below will show how the field view of episodic memory could arise from the functional role of neurons that respond on the basis of current head direction. Conway also points out that visual episodic memories represent short time slices of experience (Anderson and Conway, 1993; Williams et al., 2008) with beginning and end points often related to the achievement of specific goals. This is related to other experimental work showing some consistencies in how humans parse their experience into separate events (Zacks and Tversky, 2001; Zacks et al., 2001; Agam et al., 2005). This conception of short time slices of experience is consistent with the model presented here of episodic memories as containing segments of spatiotemporal trajectories.

Episodic Memory as a Spatiotemporal Trajectory

To further illustrate the concept of episodic memory as a spatiotemporal trajectory, I will provide another specific example. Episodic memory allows me to remember different events that I experience each day in the same work environment and to recall these in detail after a few hours or days. The example in figure 1.1A shows my memory of the start of one specific day at Boston University. At location 1 in the figure, I park my car in one of many spots in the Warren Towers parking garage. I then walk along Cummington Street and pass one of my graduate students, Jim Heys, on the street outside the Science and Engineering library (location 2). I continue down the street to my office in the Center for Memory and Brain at location 3, where I speak with my wife, Chantal Stern, and then later with another graduate student, Mark Brandon. After meeting with Mark, I go back up the street for a brief meeting in the Biomedical Engineering Building (location 4). Then I walk back down the street and enter the Life Science and Engineering Building at location 5, where I greet one of my postdoctoral fellows, Nathan Schultheiss.

As shown in the figure, my episodic memory of my day includes segments from a continuous spatiotemporal trajectory shown with a dashed line that passes between different buildings. The trajectory includes my sense of absolute location but contains more than just location. I can remember my point of view at individual moments, so there is a clear directional component. I can also remember the timing of the trajectory, such as my speed of walking or the timing of events when I stayed in one location. Thus, it is a spatiotemporal trajectory. The spatiotemporal trajectory or episodic

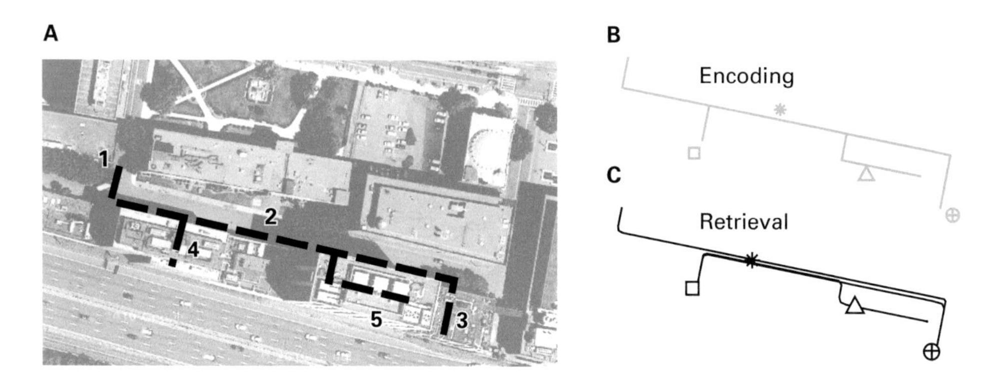

Figure 1.1
Example of episodic memory as a spatiotemporal trajectory. (A) My spatiotemporal trajectory (dashed black line) through the Boston University campus, including encounters with people at different positions. (B) The encoded trajectory is shown in gray with symbols representing the events of meeting people at different positions (asterisk, circle, plus sign, square, triangle). (C) The episodic memory model can correctly retrieve the spatiotemporal trajectory (black) and retrieve the events of meeting people encountered at specific positions (symbols).

trajectory forms a framework for the association of individual events or items. Different positions along the episodic trajectory are associated with individual events involving interacting with specific people. Figure 1.1B shows the spatiotemporal trajectory during encoding as a gray line. The events are symbolized by different symbols at different positions along the spatiotemporal trajectory (for example, the asterisk indicates meeting Jim, the circle marks the meeting with Chantal, the plus sign indicates meeting with Mark).

Consistent with Tulving's definition of episodic memory as involving mental time travel, I have the capability to remember segments of the overall episodic memory in sequential order as if reliving the experience. The complete sequential recollection of the whole episode is represented by a black line in figure 1.1C that retraces the full trajectory. My recollection includes walking speed and point of view at individual locations along the trajectory and includes retrieval of the events at salient positions along the trajectory. Recollection of the full trajectory requires considerable mental focus. However, as an alternative I can relive one isolated segment and then jump back in time to remember an earlier segment.

I have simulated the mechanisms described here in a computer program implementing a mathematical model of brain dynamics that effectively encodes and retrieves spatiotemporal trajectories. The model of brain dynamics was trained on the gray line representing the location, speed, and direction along the encoded trajectory, and then, cued by the starting position in the parking lot, the computer model generated the

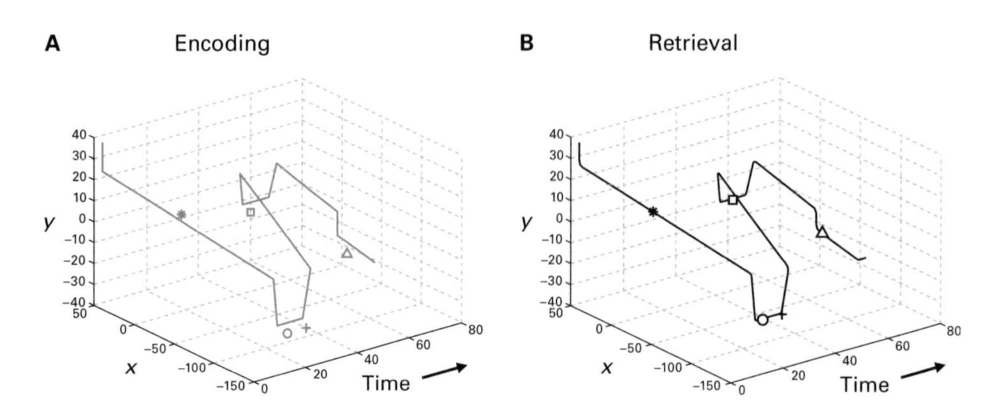

Figure 1.2
Spatiotemporal trajectory in dimensions of both space and time during encoding (A) and retrieval (B). (A) The trajectory changes in both time and space (x and y dimensions) when walking, for example, when passing Jim on the street (asterisk). The trajectory evolves in only time when I am stationary in one location and experience different events, for example, when speaking with Chantal (circle) and Mark (plus sign) at different times in my office. (B) The model generates retrieval of an extended trajectory cued from the location where I parked my car. Retrieval matches the previously experienced trajectory through time and space, including the time and location of the events of meeting people (asterisk, circle, plus, etc.). With different cues, retrieval can start from arbitrary locations or events and select specific trajectory segments.

black line in the figure, showing that the model can effectively retrieve the full spatiotemporal trajectory as well as the location and time of individual events along the trajectory.

To illustrate the role of time in a spatiotemporal trajectory, figure 1.2 plots the same spatiotemporal trajectory in a different way, showing the dimension of time as well as the coordinates of location. In the plot of the spatiotemporal trajectory, different walking speeds result in different slopes of the spatiotemporal trajectory. The model of brain dynamics effectively retrieves both the time intervals and spatial location properties, so that the line indicating retrieval matches the line indicating the original encoded trajectory.

The model of brain dynamics that retrieved the trajectories in the figures contains representations of many of the neurophysiological properties of neurons that have been discovered in recent experimental data. For example, the model includes grid cells and time cells. The elements contributing to this model of brain dynamics will be described in detail, including both the anatomical components and the physiological data. The primary focus of subsequent sections will be describing how neural circuits can encode and retrieve an episode as a spatiotemporal trajectory along with associated events. Thus, my goal is to write the book I described in the preface. The book I was seeking when first learning about memory, a book that describes how populations of neurons could encode and retrieve a complex spatiotemporal trajectory of behavior.

Note that episodic trajectories are not necessarily limited to the dimensions of physical time and space. When I recall walking between the buildings at Boston University, I also remember my thoughts about the environment. As I walked past my graduate student, I remember thinking about his experimental project on the resonance frequencies of entorhinal neurons. When I spoke to my student Mark, I remember discussing his data on drug infusions that alter grid cell firing. Thus, the episodic trajectory includes additional segments that incorporate features of sensory experience as well as trajectories through internal thought. These can all be incorporated into a multidimensional feature array (or vector) representing the state experienced at a particular time and place, and this multidimensional feature array (vector) can be encoded and retrieved with mechanisms analogous to those used for a trajectory defined only in physical space and time. This has been described as a memory space, in contrast to pure physical space (Eichenbaum et al., 1999). This point will be addressed further in the subsequent text.

A Model of Episodic Memory

As noted above, this book focuses on presenting a model of episodic memory as a spatiotemporal trajectory. The description of episodic memory as a spatiotemporal trajectory immediately puts specific demands on the model. By the definition of episodic memory, the model must address the encoding and retrieval of the place and time of the episode. In addition, the model must address the encoding and retrieval of a viewpoint and the direction of action during specific events in the episodic memory. In the model presented here, the components of time, space, and action are bound together in the manner of physics. The following chapters of the book will provide detailed descriptions of the individual components of the model, but a basic overview is useful as a guide for explaining the significance of different sets of data.

Consider my arrival in my office at the Center for Memory and Brain. Chantal came to my door to say that our paper on sequence retrieval was accepted by the *Journal of Neuroscience*. Then, after about an hour, my graduate student Mark Brandon came to the door to report on his recording data showing grid cell firing patterns altered by muscimol infusions. A short time later, I walked out of my office and back up the street to the Biomedical Engineering Department. When I came out of that building, I walked back along the exact same segment of street that I traversed in the morning, but this time entered the door into the Life Science and Engineering Building. All of these elements need to be coded in the episodic memory.

In the model, my position in time and space is coded by the relative timing of rhythmic activity in different populations of neurons. Rhythmic activity of neurons is described in chapter 2. The representation of my trajectory through space and time is determined by the nature of inputs that change the relative timing of rhythmic peaks in activity, as described in chapter 3. The model depends on the fact that rhythms at slightly different frequencies will shift in timing relative to each other. In the absence of inputs, the relative timing of rhythms with different frequencies will evolve according to time alone, as described in chapter 3, allowing items to be linked with specific points in time coded by populations of neurons, as described in chapters 4 and 5. Thus, as I sat in my office, I could encode the duration of the interval between speaking with Chantal and speaking with Mark. When I left the Center for Memory and Brain to walk to the Biomedical Engineering Department, the trajectory shifted in space. My own sense of direction and movement speed provided a velocity signal that could drive the rhythmic frequency of different populations of neurons. The shift in frequency proportional to velocity meant that the relative timing of these other rhythms could code my location in space, as described in chapter 3. Thus, the code of relative timing can be altered by different inputs that describe the sequential movement of the trajectory through space and time. In addition, the encoding of individual events requires associating the relative timing of population activity with specific items, events, or actions at individual positions along the trajectory, as described in chapters 4 and 5.

During retrieval, one component of the spatiotemporal trajectory is activated as a cue—for example, a memory of entering the Center for Memory and Brain. This activates the relevant rhythms. These rhythms then shift a small amount in relative timing to retrieve the time interval until part of the population pattern activates an association with the event of speaking with Chantal. The rhythms shift a larger amount before a different part of the population pattern retrieves the association with the event of speaking to Mark. Eventually, the recollection reaches the time of leaving the Center for Memory and Brain, and this activates a memory of the direction and speed of movement leaving the center and turning left to walk to the Biomedical Engineering Department. This shifts the frequency of rhythms coding location, causing them to update the internal representation of location in the memory. In this manner, the spatiotemporal trajectory is recreated from the starting cue, and the individual events and actions along the trajectory are reactivated. The subsequent chapters will focus on the neural mechanisms for this process.

Brain Systems for Episodic Memory

Now that I have described the behavioral properties of episodic memory, with an emphasis on encoding and retrieval of spatiotemporal trajectories, we must consider data suggesting specific brain mechanisms for episodic memory. We cannot fully understand episodic memory if we only consider data on a behavioral level. Thus, the work presented here focuses on a reductionist approach to the understanding of episodic memory. The behavior must be linked not only to the anatomical structures but to the cellular physiology within these structures and the role of particular chemicals acting as neurotransmitters and neuromodulators. Later chapters will address how the cellular properties of individual neurons and the effect of modulators such as acetylcholine may be involved in episodic memory function. However, studying the cellular mechanisms of episodic memory requires first having some sense of what brain structures are involved in episodic memory.

Where in the brain should we look for the cellular dynamics of episodic memory function? Human data provide some answers about the specific anatomical structures involved in episodic memory. Often the first source of information for localizing function in the brain arises from studies of the effects of brain damage. The brain damage can be accidental due to head trauma, or the damage can be deliberate as part of a medical treatment. The first case I will discuss involves a medical treatment that had an unexpected result.

I once met a man who had entirely lost his capacity for episodic memory. His name was Henry Molaison. To preserve his anonymity when he was alive, older journal articles refer to him only as patient HM. I met him when he came to Massachusetts General Hospital in Boston to have an anatomical scan using magnetic resonance imaging to show the specific damage to his brain that caused this loss of episodic memory. My wife was one of the postdoctoral fellows collecting the scanning data, so I accompanied her, along with our infant daughter, Simone. Henry Molaison was 68 years old at the time. He was tall and broad shouldered with a wide face and forehead. He was gregarious and talkative, though when I was first introduced to him he simply said hello, without saying "Nice to meet you." He had learned to avoid saying such things because he knew that he could not remember if he had just spoken to the individual 5 minutes previously.

In response to questions about his past, he very rapidly fell into detailed descriptions of memories from his youth, describing his work in a factory making electric motors, winding the wires around the central core of each little motor. When I met him, Henry used a walker with an attached basket filled with crossword puzzle books and issues of *Reader's Digest*. Before he went into the scanner, I saw him reading a sensationalistic story in *Reader's Digest* about doctors delivering a baby in China. It was the woman's second child, and laws in China restrict couples to one birth. We briefly discussed the details of this somewhat dramatic story before patient HM went into the scanner. After about 15 minutes, he was taken out of the scanner, and I said, "That was a sad story you were reading, wasn't it?" He looked at me blankly, clearly not remembering anything of our previous conversation. "You know, the story about the doctors in China?" I pointed at the *Reader's Digest*, where he had placed a bookmark on the story. He opened the *Reader's Digest* and, without saying anything, started reading the story again at the beginning, clearly having no recollection of any part of the previous conversation, though I had a clear episodic memory of those events.

This story gives a notion of patient HM's behavioral impairment. He could carry on a conversation, keeping track of the context of recent comments. This indicated the preservation of his working memory. He knew the meaning of most words, indicating the preservation of his semantic memory. He could rapidly unfold his collapsible walker, indicating the preservation of his procedural memory. He could remember some episodes from before his surgery, indicating he does not have loss of memory for all events from before surgery (known as retrograde amnesia) though he had difficulty remembering events shortly before his surgery, suggesting partial retrograde amnesia. Most strikingly, he demonstrated an inability to form memories for new events in his life after the surgery. This is often described as complete anterograde amnesia.

He was more patient and friendly than one normally would expect of a 68-year-old man waiting to have a scan in a hospital—I could not help feeling that his personality was more like a 27-year-old man. This was his age when a neurosurgeon in Hartford, Connecticut, performed the brain surgery that resulted in his loss of ability to form new episodic memories. He could remember events from his childhood, suggesting that episodic memories occurring some time before his surgery were preserved, but the surgery appeared to block the formation of new episodic memories.

The selective behavioral impairment of patient HM provides one piece of evidence supporting the existence of multiple memory systems (Eichenbaum and Cohen, 2001). Surprisingly, most people do not have a strong intuition about the existence of multiple different types of memory function. In colloquial conversation, we use the same word "remember" for a variety of functions. We talk about remembering the name of the capital of Sweden (semantic memory), or remembering the start of a spoken sentence (working memory), or remembering how to ski (procedural memory), or remembering our first search for articles on memory (episodic memory). However, in these examples, the same word is being used to describe very different types of memory that depend on distinct brain systems. As noted above, Endel Tulving first defined episodic memory as a contrast to semantic memory, or memory for general knowledge of the world (Tulving, 1983, 1984, 2001). As described below, short-term memory for the start of a sentence is separate from long-term episodic memory. In addition, extensive data show a distinction between episodic memory and memory for motor skills learned by the procedural memory system, such as the ability to ski or to ride a bicycle, which shows a dependence on brain systems unaffected in patient HM (Cohen and Squire, 1980; Eichenbaum and Cohen, 2001).

As a child, patient HM suffered from intractable epilepsy—regular seizures would disrupt his life daily, and drugs could not control the seizure activity. Finally, as a last resort, a neurosurgeon named William Scoville removed the regions of patient HM's brain where the seizures appeared to be starting. The surgeon entered through two openings on each side of the forehead, lifting bone flaps that gave him access to the inside and bottom surface of each temporal lobe (ventromedial surface). He then cut the tissue from the ventromedial surface, removing structures on both sides of the brain that included the anterior two-thirds of the hippocampus, the entire entorhinal cortex, and parts of the perirhinal and parahippocampal cortices as well as the amygdala and white matter (fiber) connections of these regions. The lesion that Scoville created in patient HM is shown in figure 1.3 (Scoville and Milner, 1957; Corkin et al., 1997). The anatomy is described further in chapters 2 and 5.

Patient HM was not the only case of anterograde amnesia from bilateral excision of the medial temporal lobe. The original article by Scoville and Milner described several other cases, but these other cases were schizophrenics in whom other

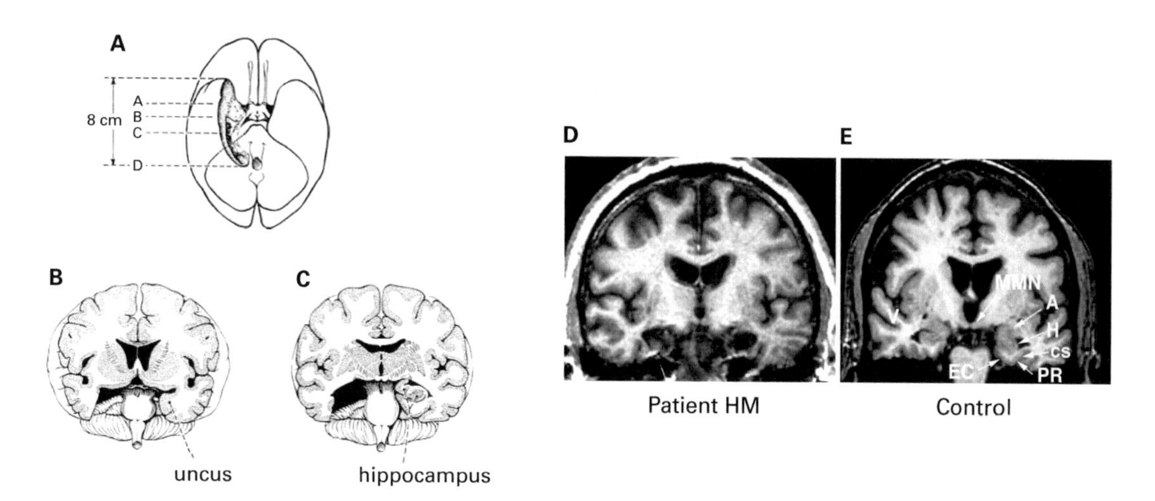

Figure 1.3
Damage to the brain of patient HM. (A) Ventral (bottom) surface of brain showing location of removal of hippocampus, entorhinal cortex and other structures. (B and C) Brain cross-sections showing areas removed on both sides in patient HM. Each figure shows the damage on the left and illustrates the area that is missing on the right. (D) Anatomical scan of the brain of patient HM, showing absence of the anterior hippocampus and entorhinal cortex on both sides of the brain. (E) Comparison scan of a different brain with intact structures, including hippocampus (H), entorhinal cortex (EC), perirhinal cortex (PR), amygdala (A), medial mammillary nucleus (MMN), collateral sulcus (cs), ventricle (v). Reprinted with permission from Scoville and Milner, 1957 (A–C) and Corkin et al., 1997 (D, E).

behavioral disorders partially obscured the nature of the memory impairment, which is strikingly specific in patient HM. The neurosurgeon Wilder Penfield at the Montreal Neurological Institute performed a similar surgery unilaterally in a patient which resulted in profound anterograde amnesia due to previous atrophy of the hippocampus on the contralateral side (Penfield and Milner, 1958).

The damage to the brain of patient HM was very extensive. However, more selective damage to the hippocampus has been shown to impair memory function also. Complications during surgery (hypoxia) has caused selective damage to subregions of the hippocampus in some subjects (Rempel-Clower et al., 1996). These subjects have amnesic effects that are somewhat less severe than those of patient HM, but they still demonstrate impairments of episodic memory in quantitative tests of human memory that allow a more sensitive measure of these differences in memory capabilities.

Tests of Episodic Memory

Whereas patient HM's deficit would be clear to anyone who spoke to him for a reasonable period of time, comparing deficits across patients cannot be done on the basis of anecdotes such as the one above. The description of his disorder gives a general sense of what memory function he retained (he could remember individual topics during a conversation, and he could remember facts about the world and memories from the distant past). These aspects of spared function can be described using the terms working memory and semantic memory, respectively, but this general description of the disorder does not provide a clear quantitative assessment of the individual dimensions of memory function.

I remember a feeling of disappointment as an undergraduate when I first read the literature on human memory function. I had expected descriptions of the mechanisms for memory recollection that had all the richness of my own episodic memories with trajectories and viewpoints and complex visual scenes. However, I primarily found quantitative studies of human verbal memory function that focused on the encoding and recall of lists of single words or groups of letters. I should mention that this was the year before Tulving's book on episodic memory was published in 1983.

However, despite their initial dryness to a student, these studies using specific human memory tasks provide an important quantification of human memory function in a consistent, standardized manner. Memory tasks include standard clinical tests used in neuropsychological evaluation of patients, such as the Wechsler Memory Scale, described below. In addition, considerable research has utilized individual memory tests, described below, that provide a larger continuous scale for measuring memory function. These latter tests also allow quantification of the effect of numerous other factors on memory function in humans, including conditions of encoding as well as effects of drugs. These tasks vary in presentation modality and timing of different components but share some features. Here I will describe some standard tests of memory that directly test episodic memory, such as free recall and cued recall, as well as tests that address aspects of other memory systems. My description is focused not on the general use of these tests but on what they say about the selectivity of effects of neural damage on different aspects of memory function. In addition, the effect of drugs on performance in these tasks will be discussed in more detail in chapter 6.

Neuropsychological Tests

Neuropsychological tests focus on assessing the behavioral symptoms of brain damage or disease. The Wechsler Memory Scale has been used in a variety of neuropsychological testing situations, including the initial studies of patient HM. This task contains a battery of different memory tasks focused on different components of memory function. Any behavioral task will put demands on multiple different memory systems, but some require episodic memory function more strongly than others. The Wechsler Memory Scale includes a test of story recall, in which a paragraph is read to participants and they are scored on how well they retrieve components of this story. This requires elements of semantic memory and working memory but clearly also tests episodic memory. Patient HM was severely impaired on paragraph recall. Patient HM also showed striking deficits in other tests on the Wechsler Memory Scale that required encoding and retrieval of items from an episode (Scoville and Milner, 1957; Corkin, 1984) such as the free recall of words from a list or cued recall of paired associates (both tasks are described in more detail below). In the list of difficult paired associates, patient HM could retrieve none of the second words in each pair. While this provides a general overview of memory impairment, the Wechsler Memory Scale has relatively small numbers of items in each component of the task. This allows it to give a broad qualitative measure of memory function but does not give much quantitative detail. For example, the test includes a list of 20 word pairs for testing paired associate learning, whereas focused paired associate learning tasks might use 64 or more (see below). Thus, the Wechsler Memory Scale can distinguish between a college undergraduate and patient HM but might not distinguish between two amnesics with slightly differing memory capacities.

Numerous other multicomponent tasks have been developed for neuropsychological research, including the California Verbal Learning Task (CVLT). This task also provides a consistent assessment which can be relatively rapidly applied across a large number of patients. However, similar to the Wechsler Memory Scale, the CVLT has a paucity of numbers within the individual components of the task. The mini-mental state examination is an even shorter test that is often used as an initial screen for dementia—for example, in studies of patients with Alzheimer's disease.

Free Recall

Better quantitative data are available from individual laboratory tests of free recall that have larger numbers of words than the Wechsler Memory Scale and can give a detailed quantitative assessment of individual components of memory function. In a free recall task, participants sit with an examiner in a testing room and hear or see a sequential presentation of a series of single words during an encoding period. Subsequently, during a separate retrieval period, they are asked to say as many words on the list as they can remember in any order, either immediately after the end of the list (immediate free recall) or after a delay period of varying length with intervening distraction (delayed free recall). Thus, the task has two separate behavioral components that occupy extended periods of time. In the encoding period, participants are presented with the individual words. In the retrieval period, they must generate the words that were on the list. (Note that these longer periods of encoding and retrieval during a behavioral task differ from possible rapid changes between dynamics of encoding and retrieval in the brain.) In an immediate free recall test, normal participants usually show a much greater probability for recall of words at the end of the list. This is called the "recency" effect and appears to involve short-term memory. In contrast, retrieval of words from the start of the list in immediate free recall and from the entire list in delayed free recall is believed to require episodic memory.

Remarkably, only in the past few decades have the fields within cognitive science accepted the existence of multiple different memory systems. One of the first distinctions was between short-term memory and long-term episodic memory, as supported by cognitive data in patients with amnesia caused by Korsakoff's syndrome. Korsakoff's syndrome results from insufficient levels of the vitamin thiamine in the diet, causing neurological damage to structures including the medial mammillary nuclei and anterior thalamus. Baddeley and Warrington (Baddeley and Warrington, 1970) tested patients with Korsakoff's syndrome as well as some with hippocampal damage in an immediate free recall task. Multiple tests allowed plotting of a serial position curve, in which the probability of recall is plotted for words at different positions in the encoded list. As shown in figure 1.4, the control participants in the task showed a normal recency effect with enhanced retrieval of words from the end of the list. The amnesic patients showed a similar recency effect with normal retrieval for words at the end of the list. However, the amnesics showed a dramatic reduction in retrieval of words from the start of the list relative to control participants, indicating a selective loss of long-term episodic memory but not short-term memory.

Thus, the serial position curve shows similar responses to recent stimuli in both controls and amnesics, with a clear decrease in performance on earlier portions of the curve in amnesics. The figure demonstrates the impairment of delayed free recall (long-term memory) in amnesics with sparing of recency (short-term memory). Patient HM and other hippocampal patients show a similar pattern of deficits (Corkin, 1984; Graf et al., 1984). For example, patient RB had a lesion selective to a subregion of the hippocampus termed CA1 caused by hypoxia during surgery. In tests of the free recall of ten words from the middle of a fifteen-word list, patient RB only recalled 10 percent of the words whereas controls recalled about 40 percent (Graf et al., 1984). This

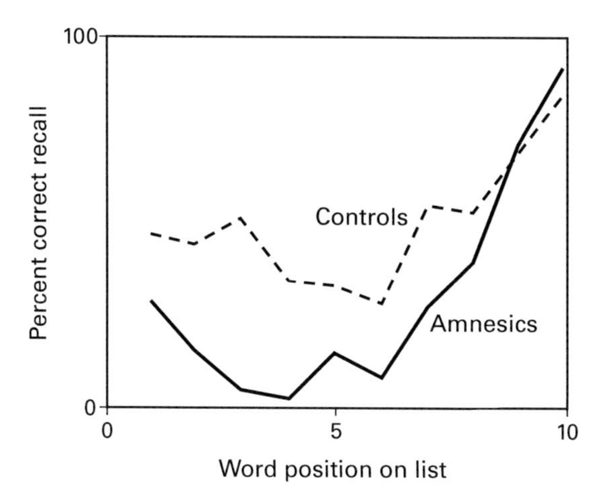

Figure 1.4
Serial position curves for control participants (dashed line) and amnesics (solid line) showing percent correct recall of words based on their presentation time during encoding at different serial positions in the list.

indicates a selective effect of hippocampal damage on long-term episodic memory. The separation of short-term memory from long-term episodic memory is further supported by the existence of a different group of patients with a different locus of damage (e.g., the parietal lobe) showing a selective loss of short-term memory relative to long-term episodic memory (Warrington and Shallice, 1969; Shallice and Warrington, 1970).

In the popular media, there is sometimes confusion about the difference between long-term and short-term memory. For example, in the film *Memento* the main character has a memory impairment similar to patient HM that is referred to as a short-term memory problem, but the character displays normal short-term memory function as defined in the psychology literature. This confusion arises because medial temporal lobe damage spares semantic memory and some long-term episodic memory from before the damage and sparing of short-term memory in patient HM. The confusion could be alleviated by using the label intermediate term episodic memory to describe the loss of encoding of new episodic memories in patient HM, despite the sparing of old episodic memories that he consolidated before the damage.

As noted above, the impairment of delayed free recall is similar in the different types of amnesia arising from either hippocampal damage or Korsakoff's syndrome, though the anatomical basis appears to be very different. In particular, Korsakoff's amnesia is associated with atrophy of the mammillary nuclei and the anterior thalamus (Vann and Aggleton, 2004). Other sources of damage to these regions have also been shown to cause amnesia, including two rather unusual cases in which damage was caused by sharp objects entering the brain through the nose, including a miniature fencing foil and a pool cue (Vann and Aggleton, 2004) that damaged the mammillary nuclei and anterior thalamus. Physiological data suggest that the role of these structures could involve the coding of speed and head direction by the activity of neurons in the mammillary nuclei (Blair et al., 1998) and anterior thalamus (Taube, 1995). The model presented here shows how the input from these regions would be essential to the encoding of spatiotemporal trajectories for episodic memory.

The verbal free recall task seems quite far from a model of spatiotemporal trajectories in episodic memory. However, it is clearly a test of episodic memory, so the relationship to spatiotemporal trajectories is worth evaluating. A free recall task focuses on a restricted subspace of the spatiotemporal trajectory, defined in the dimension of time by the start and end of testing and defined in space by the testing room. In fact, most free recall tasks divide time into testing on individual lists, and participants must limit their free recall to words presented in the most recent list.

The subjects in a free recall task experience a full spatiotemporal trajectory as they walk into the building for their experiment, find the appropriate room, and learn about the testing rules. If they were tested on their recall of the approach to the room, they could just as easily describe memories of this approach as they describe memories of individual words in the test. During the free recall task, the participants experience a wide dimensionality of individual words that could be seen as excursions into multiple dimensions of orthography and semantics. Thus, the trajectory does not just involve the dimensions of time and physical space but also involves movement through other dimensions of what has been termed a memory space (Eichenbaum et al., 1999). Multiple dimensions of sensory experience and internal thought are a constant component of episodic experience. Participants are likely to have heard some of the same words in the free recall task during a conversation earlier that day, but they are asked to ignore these previous multidimensional memories because they fall outside of the temporal segment of memories experienced within the subspace defined by the free recall task. In free recall, no temporal order is required in the response. Even though retrieval of a series of words involves an excursion along a spatiotemporal trajectory of previous experience, most experimenters will only score the individual words recalled. In a sense, this collapses the time dimension of the retrieved spatiotemporal trajectory onto a plane of verbal dimensions defining only the words. However, some free recall experiments will sample the temporal domain by keeping information about order of retrieval (Howard and Kahana, 2002) or specifically requiring the serial recall of the words (Burgess and Hitch, 1999), constraining retrieval to recreate the spatiotemporal trajectory in the same temporal order.

In addition to damage to brain regions, administration of different types of drugs can have a profound effect on memory performance in these tasks. For example, drugs that block the activity of acetylcholine receptors or enhance GABA receptor activity can strongly impair the encoding of new words in this task, while sparing the recall of previously learned words. Drug effects on episodic memory are reviewed in more detail in chapter 6.

Cued Recall

In a cued recall or paired associates task, participants sit in a testing room and hear or see sequential presentation of a series of pairs of words, for example, *barber* followed by *crumb*, then *razor* followed by *tact*, then *vulture* followed by *aspirin*. They are then given the first word of each pair and asked to say or type the second word of the pair. For example, they would be given the word *razor* and should respond with the word *tact*. This task also shows strong impairments in amnesic patients with hippocampal damage. Patient HM was not able to learn a single paired associate from the unrelated list in the Wechsler Memory Scale (Scoville and Milner, 1957). Even patients with much more selective damage affecting only subregions of the hippocampal formation such as region CA1 show significant quantitative impairments on this task (Rempel-Clower et al., 1996). For example, patient RB has damage restricted to region CA1 of the hippocampus. When tested three times sequentially on a list of ten unrelated paired associates, patient RB only recalled 3.7 pairs out of 30 while controls recalled correct associations over 20 times (Zola-Morgan et al., 1986). Cued recall does not completely ignore the order of encoding as in free recall but focuses on retrieval of very short temporal segments of an episode.

Richness of Episodic Detail

Tasks such as free recall and cued recall test the episodic memory for the specific event of seeing a particular word at a particular time and a particular place. However, they do not test the full richness of detail about an episode from real life. More recently, specific behavioral scoring methods have been developed to fully quantify the richness of detail in human episodic recollection (Levine et al., 2002) by scoring the amount of contextual detail internal to specific remembered episodes. These techniques show significant reductions in the recall of internal details from an episodic memory after lesions of the hippocampus and parahippocampal cortices (Steinvorth et al., 2005; Kirwan et al., 2008). In contrast to the impairment of richness of detail in episodic memory, hippocampal and parahippocampal lesions have less effect on long-term semantic memory (Steinvorth et al., 2005). Interestingly, recent data show that lesions of the medial temporal cortices also cause significant impairments in the description of future or imagined episodes (Hassabis et al., 2007; Schacter and Addis, 2007). These data on richness of detail support the notion that the hippocampus and parahippocampal cortices create a detailed representation of a complex spatiotemporal trajectory on which a range of detailed items and events can be fastened. The loss of this spatiotemporal trajectory impairs the ability to retrace a trajectory and access the relevant details and even impairs the ability to imagine walking along a future trajectory and experiencing events and viewpoints along that future trajectory.

Spatial Memory

Given the focus of this text on the encoding of spatiotemporal trajectories in episodic memory, perhaps the most relevant quantitative data for this model concerns the learning of visual mazes by patient HM. In the time of initial testing on patient HM, one available task used an array of metal pegs on a board and a metal stylus. Participants were required to put the stylus on the pegs and learn by trial and error the winding path between the start and finish pegs with wrong responses detected by a click. In an average of 25 trials, normal participants can learn to contact only the 28 correct pegs between start and finish, but patient HM did not reduce his number of errors in 215 trials (Milner et al., 1968). I interpret this as evidence for his difficulty in remembering the spatiotemporal sequence of correct pegs encountered on a given trial. On a dramatically shortened task with 8 pegs in an array of 20, patient HM did show a reduction in errors over several days, indicating a capacity for learning over extended training that could reflect learning in semantic memory or procedural memory. This slow semantic learning is consistent with the ability of patient HM to draw a floor map of the house he lived in after his surgery after he had lived there for eight years (Milner et al., 1968). This could indicate the slow updating of neocortical semantic memory over time in contrast to rapid learning of new spatial memories. Recent studies on patients with unilateral thermal lesions of the hippocampus performed for treatment of epilepsy show striking impairments in memory for the spatial location of objects. Participants were shown one to six objects in an arena and asked to mark the locations or replace the objects. Right hippocampal lesions caused significant impairments relative to controls even for the location of a single object (Bohbot et al., 1998; Stepankova et al., 2004).

Recognition Tasks

In a recognition task, participants are usually presented with a list of single words during the encoding period. Subsequently, their recognition of these words can be tested by presenting word pairs containing one novel word and one word presented during encoding. Participants must choose the word presented during encoding. For example, participants might see the words *horse, bone, church, snow* during encoding. Then, during recognition testing, they would be asked to indicate the familiar word in a series of word pairs, such as the pair *horse, glove*. As an alternative test of recognition, participants might be presented with single words and requested to indicate whether the word is old or new. Considerable research has focused on characterizing aspects of memory function retained in humans with hippocampal damage. This remains an active and controversial area of ongoing research. In general, amnesics with hippocampal damage perform better on recognition tasks than on free recall or paired associate tasks, but this might reflect sparing of just one mechanism for recognition.

Considerable research has focused on the possibility of two separate mechanisms involved in recognition memory. During the recognition period, participants might explicitly retrieve a detailed episodic memory of hearing a word (recollection) or they might base their response on a vague sense of the greater familiarity of one word versus another (familiarity). This was initially tested by Tulving using what is known as the "remember–know" paradigm. Participants were told to rate the confidence of their response by responding with "remember" for recollection and "know" for familiarity. More recently, these mechanisms have been tested based on participants' rating their confidence in their recognition judgment. Many researchers and models have proposed that hippocampal circuits are required for recognition based on recollection, but not for recognition based on a general sense of familiarity (Norman and O'Reilly, 2003; Fortin et al., 2004; Yonelinas et al., 2005; Eichenbaum et al., 2007). However, amnesics with hippocampal damage show impairments on recognition memory tasks including both "remember" and "know" responses (Manns et al., 2003). Full review of the extensive research and controversies in this area would require a separate book. This book primarily focuses on mechanisms that would be involved in recognition judgments based on recollection. In this task, if a subject can recollect a segment of a spatiotemporal trajectory containing the word, he or she will give a "remember" (recollection) response.

Comparison with Short-Term Memory Tests

The data on the serial position curve above show sparing of short-term memory during impairment of long-term memory. Similarly, damage to the hippocampus and adjacent cortical structures does not affect the articulatory loop for holding information in working memory. This can be seen by the preserved performance of patients with hippocampal damage in the digit span task. In this task, participants are presented with sequences of numbers and requested to repeat them back (without an interfering task) at the end of the sequence. Typically, participants are very effective at repeating lists of numbers up to seven digits long but begin to make errors with longer lists, particularly when there are ten or more digits in the list. Participants with hippocampal damage show normal memory span (they can usually repeat back six or seven digits effectively, similar to normal participants; Cave and Squire, 1992). Patient HM had a memory span of six digits after his operation (Milner et al., 1968), and patient PB with anterograde amnesia from unilateral left hippocampectomy combined with atrophy of the right hippocampus had a forward digit span of nine (Penfield and Milner, 1958). Amnesics usually show significant impairments on longer, supraspan lists, suggesting the necessity for use of hippocampal circuits once working memory capacity has been exceeded.

The Brown-Peterson task provided initial behavioral evidence for a short-term memory process with rapid decay that was distinct from long-term episodic memory. In this task, participants are presented with groups of three words or three consonants (called trigrams). For example, the experimenter might say "YQM" and the subject repeats this. After each trigram, the subject must count backward by threes from different starting numbers. Then after 3, 9, or 18 seconds, they must repeat the trigram that was presented to them. This task has traditionally been described as a test of short-term memory. However, participants such as patient HM show some reductions of performance on this task (Corkin, 1984), indicating a potential role for the hippocampus or parahippocampal cortices in this type of task despite the short retention interval. The presence of a distractor task during the delay in this task appears to be important as it may prevent holding of information by active maintenance in the neocortex and thereby require the use of hippocampal circuits for episodic memory. This type of task emphasizes the importance of developing a model that can directly address task variables and data from individual tasks rather than attempting to summarize the behavior in multiple individual tasks in terms of a small number of discrete verbal categories.

Functional Imaging Studies

As described here, the attribution of behavioral function to specific brain regions draws on a long history of neuropsychological work involving effects of localized brain damage on behavioral function. Patient HM and other patients with localized damage to the hippocampus and adjacent areas were influential in focusing memory research on these regions. Though some of this research may seem old, research on the neural mechanisms of other cognitive functions such as language have an even longer history that extends back to the work of neurologists in the mid-nineteenth century.

Given this long history, the use of functional imaging techniques in humans presents a very recent addition to the field. When the technique of positron-emission tomography (PET) became available for measuring the activity of brain regions by measuring uptake of radioactive glucose or oxygen, studies of memory focused on finding activation of the hippocampus. Initial studies analyzed PET activation during verbal memory tasks such as word stem completion based on recall (Squire et al., 1992; Buckner et al., 1995) or retrieval of episodic versus semantic paired associates (Fletcher et al., 1995), but these studies did not show robust activation of the hippocampal formation. In contrast, PET activation during encoding of faces showed greater regional cerebral blood flow in the hippocampus (Grady et al., 1995; Haxby et al., 1996).

A later imaging technique involved measuring changes in blood oxygenation and flow associated with neural activity using functional magnetic resonance imagining (fMRI). The first demonstration of hippocampal activation using fMRI was obtained with a study using encoding of complex visual scenes for subsequent postscan recognition (Stern et al., 1996). Participants in this study showed robust differences in activation in the posterior hippocampus and parahippocampal gyrus between encoding of different visual scenes and repeated presentation of a single visual scene. This effect was replicated in a later study using the same techniques (Gabrieli et al., 1997). Refinement of fMRI techniques allowed measurement of event-related activity associated with individual stimuli. These techniques showed differences in hippocampal and parahippocampal fMRI signal associated with stimuli that were later remembered versus those that were forgotten in a subsequent memory task, both for complex visual images (Brewer et al., 1998) and for word stimuli (Wagner et al., 1998). Subsequent studies showed that both types of stimuli could activate hippocampal regions, with greater left hippocampal activation for word stimuli versus bilateral coding for pictures (Kirchhoff et al., 2000). Interestingly, most imaging studies show greater activation of posterior hippocampal regions with encoding (Stern et al., 1996; Greicius et al., 2003), consistent with rat lesion studies suggesting the dorsal hippocampus may be more strongly involved in episodic memory (Moser and Moser, 1998). Dorsal hippocampus in rats is analogous to posterior hippocampus in humans. High resolution imaging studies analyzing the function of individual hippocampal subregions (Bakker et al., 2008) will be discussed further in chapter 5.

Studies have also demonstrated the involvement in encoding of episodic memory of a broader circuit of regions beyond the hippocampus and parahippocampal structures, including the inferior prefrontal cortex, the parietal cortex, and the retrosplenial cortex. One study used a virtual world in which subjects received items from individuals in different locations of the virtual world and were later tested for their memory of the items and the individuals (Burgess et al., 2001). Results support a model in which the location of events is coded by parahippocampal and hippocampal structures with translation to head centered coordinates by parietal cortex (Burgess et al., 2001; Byrne et al., 2007). The retrosplenial activity appears to correlate with the recall of specific spatial locations from memory (Epstein et al., 2007b; Epstein et al., 2007a), rather than with more general perception of spatial scenes mediated by parahippocampal cortices. The involvement in spatial processing is consistent with retrosplenial neural responses based on head direction in rats (Cho and Sharp, 2001). Imaging studies have also shown correlates of sustained entorhinal cortex activity during encoding with subsequent cued recall performance (Fernandez et al., 1999). In a delayed match to sample task, entorhinal activity during the delay period after the sample period correlated with subsequent long-term recognition memory for the sample stimulus (Schon et al., 2004; Schon et al., 2005). These studies support the proposal that cellular mechanisms of persistent spiking in entorhinal cortex may contribute to encoding of episodic memories as described later.

Episodic-like Memory in Animals

Our capacity to remember episodes in our lives provides a profound and highly valued aspect to our existence, allowing us the capacity to relive events in our lives. We value the ability to relive a proud moment such as a graduation or the birth of a child. We expect and in many cases require that we will remember a meeting with a coworker or a visit to a family member. We start to question our mental capabilities if we forget an episode, and we question the capabilities of others if they forget an episode that we remember. Thus, the memory for episodes is an essential component of human behavior.

The richness and complexity of human episodic memory have led some researchers to argue that episodic memory and mental time travel are a purely human capacity (Tulving, 1983, 1984; Suddendorf and Corballis, 1997; Tulving, 2001) whereas others argue that animals have this capacity as well (Clayton et al., 2003; Dere et al., 2006). The personal experience of animals is beyond experimental test, as they cannot provide a description of consciously reliving the richness of detail in a recollection of a prior event. However, behavioral data provide a compelling argument that many of the capacities for episodic memory shown in humans can be found in animals (Dere et al., 2006). This has led to the development of behavioral tasks testing what is referred to as "episodic-like memory" in animals (Fortin et al., 2002; Clayton et al., 2003; Fortin et al., 2004; Dere et al., 2006). Studies on episodic-like memory in animals provide an important opportunity to perform experiments for testing the brain mechanisms of episodic memory.

As described in chapter 2, electrophysiological recordings from animals show phenomena that support the existence of mental time travel along previously experienced trajectories (Skaggs and McNaughton, 1996; Johnson and Redish, 2007). In addition, many of the qualitative anatomical (Amaral and Witter, 1989) and physiological (Halliwell, 1986, 1989) properties of neural circuits observed in human cortical structures are also found in other mammals. Thus, it is reasonable to conclude that the cellular dynamics mediating episodic memory in humans are also present in animals, even if there is a difference in the quantity of experimental data on episodic memory available from other species. This section will review some of the available behavioral data indicating the presence of episodic memory in animals.

Birds

Some of the most compelling data on episodic-like memory in animals come from research on birds such as the scrub jay, which demonstrate remarkable episodic memory for hiding food in specific locations at specific times and making responses on the basis of both the location and time that they hid the food (Clayton et al., 2003). These data even indicate that scrub jays can plan for the future by hiding food in locations where they know the food will be more desirable in the future. The brains of birds contain structures considered homologous to the hippocampus, but the anatomical structure of these regions is dramatically different from the structure within mammalian species, and there has not been much physiological analysis of these structures. Thus, the behavioral data provide a convincing case for episodic memory in birds, but the anatomical homologies with episodic memory in humans provide less detail than the studies of mammalian species.

Nonhuman Primates

Work on monkeys provides an important link between the human data and work in other mammals, showing that analogous structures in animals are involved in episodic memory and allowing a more detailed analysis of the neural structures that could underlie the deficits of episodic memory in patient HM. The research on patient HM inspired extensive research in nonhuman primates testing memory effects of lesions of the hippocampus and associated cortical structures. Only a brief review is provided here with more extensive reviews available elsewhere (Mishkin and Appenzeller, 1987; Eichenbaum and Cohen, 2001; Murray and Wise, 2010). The surgeon who worked on patient HM performed the same bilateral removal of the medial temporal lobe on monkeys and tested performance using an existing memory task based on delayed match to sample. In this task, the monkey is initially presented with a sample stimulus concealing a reward. The monkey can lift the stimulus to get the reward. After a delay, the monkey is presented with two stimuli in a test trial. One stimulus matches the sample and conceals a reward; the other is a distractor stimulus. The monkey must make its response to the test trial on the basis of matching the earlier sample trial. Surprisingly, early studies using sample and distractor stimuli drawn from a small number of stimuli found only small impairments with hippocampal lesions (Correll and Scoville, 1965) that did not match the magnitude of behavioral impairment in patient HM (Murray and Wise, 2010).

Later studies altered the testing procedure in a manner that both simplified learning of the task for monkeys (Mishkin and Delacour, 1975) and made the task much more sensitive to medial temporal lesions. In some studies, training was made easier by asking the monkey to choose the test object that did not match the sample object (delayed nonmatch to sample). A crucial alteration was the use of delayed matching tasks with sample items chosen out of a large set of items (Gaffan, 1974; Mishkin, 1978). This may shift the task from testing active maintenance of a highly familiar sample stimulus to testing the episodic memory for a trial-unique event. When I worked with monkeys at Oxford University, the labs contained large boxes of small objects, toys and other cast-off junk items that were used as samples or distractors by Gaffan's group and others before the use of computers with touch screens. Because of this physical limit on total number of items, the sample and distractor items were trial unique on a given day for a given episode but would repeat over a long term so that they could not be distinguished by familiarity alone but required memory of a specific episode. Testing shows a lesser impairment if a small number of highly familiar items are used but also if the distractor items are truly novel on the sample trial, as then animals can use simple familiarity of the item to succeed rather than using a specific episodic memory of the sample (Murray and Wise, 2010).

The studies testing memory for trial-unique objects using the delayed nonmatch to sample tasks allowed detailed analysis of effects of damage to different anatomical subregions. These studies showed permanent impairments after hippocampal lesions (Zola-Morgan and Squire, 1986), transient impairments after entorhinal lesions (Gaffan and Murray, 1992; Leonard et al., 1995), and stronger, permanent impairments after perirhinal and parahippocampal lesions (Meunier et al., 1993; Zola-Morgan et al., 1993). Other studies in nonhuman primates have tested memory for associations between visual stimuli and specific spatial locations, indicating that lesions of the fornix that remove modulatory inputs to the hippocampus impair the construction of a snapshot memory for the spatial location of visual features (Gaffan and Harrison, 1989) or associations with specific responses (Gaffan et al., 1984). The fornix provides a major pathway between the hippocampus and subcortical structures. Lesions of the fornix cut fibers of cholinergic and GABAergic cells that regulate oscillatory dynamics in the hippocampus. The role of this input will be described at greater length in chapter 6.

Rodents

In describing the mechanisms of episodic memory, I will focus extensively on research in rodents. This is for two reasons. The first is that studies of rats and mice are commonly performed with the animal moving freely along complex spatiotemporal trajectories, so the behavioral and physiological data more directly correspond to our own capacity to remember spatiotemporal trajectories. In contrast, studies in monkeys usually involve the animal sitting stationary in one location. The second is that studies of rodents allow more detailed analysis of the cellular neurophysiological mechanisms that for a wide variety of practical and ethical reasons are rarely performed in humans and monkeys. The wide range of intracellular and molecular studies of neuronal properties in rodents allows more detailed analysis of the brain mechanisms of memory.

A number of behavioral tasks test episodic-like memory in rodents. These tasks provide useful data on the behavioral effects of lesions and pharmacological manipulations that allow testing of the model. Later in this book, in chapter 7, I will show how the model can be used to simulate memory-guided behavior by rodents in these tasks.

These tasks all test retrieval of trial-specific information that could include segments of spatiotemporal trajectories.

Delayed Spatial Alternation A common test of memory-guided spatial behavior that could be solved by encoding and retrieval of spatiotemporal trajectories is the delayed spatial alternation task. The spatial alternation task usually uses a T-shaped maze (figure 1.5) that is elevated or enclosed with high walls to keep the rat on the maze. Rats are placed at the base of the stem of the T and allowed to run to the intersection. At this point, they must choose whether to go right or left. In a spontaneous alternation task, a rat is allowed to follow its natural inclination to visit the arm that it did not visit on the previous trial. In a rewarded alternation task, rats are explicitly rewarded for making a choice on each trial that differs from the choice on the previous trial. Thus, if they went left on the previous trial, they must remember this choice and turn right on the subsequent trial. In a delayed spatial alternation task, a delay is interposed between each of the trials, requiring the rat to remember its choice on the previous trial across the delay interval. In a forced choice delayed alternation, the first choice in the maze is forced by the insertion of a barrier on one arm. After a delay (usually 10 seconds), the rat must then make the correct choice of the arm that was previously blocked. Rats with hippocampal lesions perform more poorly than control rats in delayed spatial alternation (Aggleton et al., 1986; Ainge et al., 2007), including the forced choice version (Aggleton et al., 1986). This suggests an inability to accurately retrieve memory of the previous spatial response after a delay due to hippocampal damage.

Lesions of the fornix or the septum also impair performance in delayed spatial alternation tasks (Stanton et al., 1984; Freeman and Stanton, 1991; Aggleton et al., 1995; Ennaceur et al., 1996; see figure 1.5). Fornix or septum lesions are an indirect way of altering hippocampal function by damaging the input to the hippocampus from neurons in the medial septum that normally release the neurochemicals acetylcholine and gamma-amino butyric acid (GABA) in the hippocampus. The role of acetylcholine is described in detail in chapter 6. Damage to both these inputs reduces oscillations in the theta frequency range in the hippocampus (Rawlins et al., 1979; Mitchell et al., 1982). As will be elaborated in the model presented in chapter 3, the loss of theta rhythm oscillations could impair memory by removing the baseline signal for determining relative phase of oscillations. Impairments in delayed spatial alternation are also caused by entorhinal lesions (Bannerman et al., 2001). Models presented later in chapter 7 show how spatial alternation could be performed by episodic retrieval of the most recent spatiotemporal trajectory from the choice point (Hasselmo and Eichenbaum, 2005; Zilli and Hasselmo, 2008b). Models also show that the task could be performed by persistent neural activity holding the most recent response in working memory (Zilli and Hasselmo, 2008a), but the delay period appears to prevent rats from using this active maintenance as a strategy.

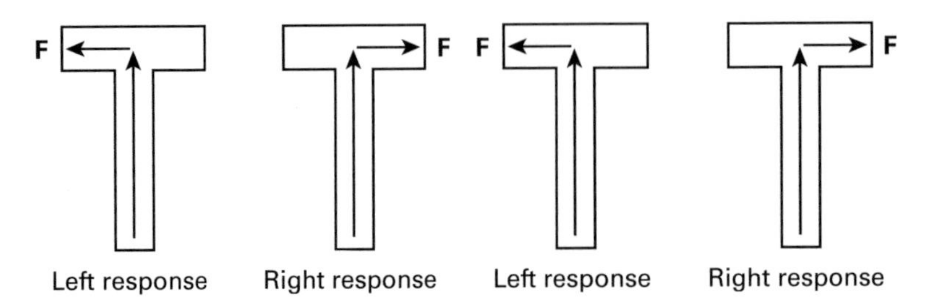

Figure 1.5
Delayed spatial alternation in a T-maze. On the initial trial, the rat runs up the stem of the T-maze and makes its own choice of left turn. It receives a food reward (F). After a delay period, the rat is allowed to run up the stem for the next trial. To receive food on this trial, the rat must run up the stem and turn right. After each delay, it must generate the opposite direction of response.

This task does not necessarily require retrieval of a full spatiotemporal trajectory. It might be sufficient to just retrieve a single event from episodic memory, such as the turn at the last choice or the location when eating the most recent reward. In addition to episodic memory, performance of the task probably involves interaction with other memory processes. For example, performance might involve retrieval or planning of a spatiotemporal trajectory followed by active maintenance of the planned response processes (Zilli and Hasselmo, 2008b). Alternately, the solution may involve active maintenance of a cue that then triggers episodic retrieval of a spatiotemporal trajectory at the necessary time.

Spatial Reversal Fornix lesions have also been shown to impair reversal learning in a T-maze (M'Harzi et al., 1987). In this task, rats are initially trained to find food on one side of a T-maze (e.g., the left side). After extensive training on one side, the food reward is moved to the opposite side. Rats with fornix lesions take longer to stop responding to the unrewarded side. This impairment of extinction could be due to lack of an episodic memory for the spatiotemporal trajectory leading to an unrewarded visit to the initially rewarded arm but could also be due to loss of normal updating based on nonreward.

Eight-Arm Radial Maze The eight-arm radial maze also tests episodic-like memory in rats (figure 1.6). This task is composed of eight different segments (arms) radiating from a central area to hidden food reward sites at the end of each arm. The maze can be enclosed with high walls and include gates at the entrance of each of the segments, or it can be elevated to a height such that rats will not jump from a maze without walls.

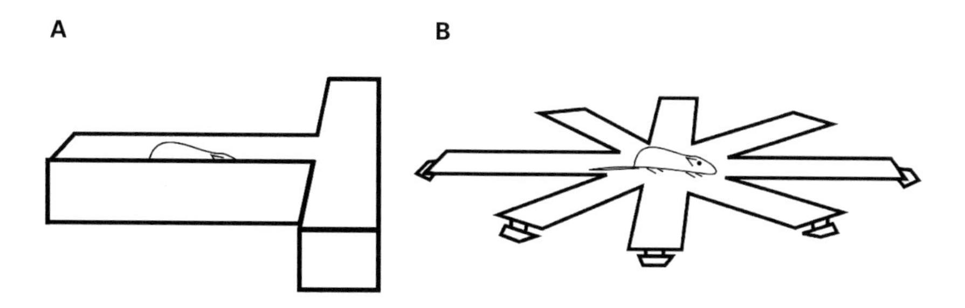

Figure 1.6
(A) Schematic side view of rat in T-maze task for delayed spatial alternation. (B) Schematic side view of rat on elevated eight-arm radial maze without walls. Note reward wells at end of each arm.

In a common task run on this maze, food rewards are placed at the end of each of arm of the eight-arm radial maze. A rat is then placed in the center of the maze. The rat is allowed to walk to the end of individual arms to obtain the food reward (which is usually hidden in a well below the level of the floor—therefore not visible from the choice point). When the rat returns to the center platform after visiting one arm, the entrances to the arms are blocked (by raising or lowering a door or a drawbridge). After a 10-second delay, the blockade is removed and the rat can choose another arm. The most efficient strategy for the rat is to visit each arm only once. Therefore, it must keep track of which arms it has visited on a given trial as the trial progresses. Repeat visits to a single arm are scored as errors since an arm no longer contains reward after a rat has removed that reward once.

Hippocampal lesions have been shown to increase the number of repeat visits to arms, suggesting an impaired memory for which spatial locations were visited on a given trial (Becker et al., 1980). The number of arm reentries is also increased by fornix lesions (Olton et al., 1979; Hudon et al., 2002) and lesions of the medial septum (Mitchell et al., 1982). In contrast, hippocampal lesions do not impair the ability of rats to avoid arms which are never rewarded on any trial. This indicates a loss of episodic memory for the trial-specific information but not a loss of longer-term semantic memory for consistent absence of reward. Unfortunately, Olton used the term working memory to refer to the memory for trial-specific information. In Olton's terminology, the term working memory essentially refers to episodic memory of previous responses in the task that is impaired by hippocampal lesions. In contrast, the term working memory as defined by Baddeley and Hitch (1974) refers to information held online with active maintenance during performance of a task (e.g., remembering a telephone number while dialing a phone or remembering intermediate numbers while mentally adding numbers). Here the term working memory will be used with the Baddeley definition. Traditional working memory tasks in humans are not impaired by hippocampal lesions.

In the eight-arm radial maze, the rat could avoid the error of repeating an arm entry by approaching each arm and testing for recall of a previous trajectory into that arm on the same day, thereby using a strategy dependent on episodic memory. However, the task could also be performed by avoiding arms with strong familiarity from the same day.

Morris Water Maze In many memory tasks, a rat could leave some olfactory or somatosensory cue which signals where it has previously been in the maze. This problem can be avoided in the Morris water maze, a large circular tank containing water that has been rendered opaque by the addition of chalk powder or other substances (figure 1.7). In my own work as an undergraduate, we used milk powder to make the water opaque, but this puts an unpleasant burden on the olfactory system of the experimenter. The water is deep enough that the rat cannot touch the bottom of the tank except in one location where a platform has been placed just under the surface of the water so that it is not visible.

A common task run in the Morris water maze consists of a series of training sessions during which the platform is kept in the same location. On each trial, the rat is placed in the water at a different location on the circumference of the tank. Initially, the rat swims aimlessly, trying to escape the tank, but eventually bumps into the underwater platform and climbs up on it. The rat is then removed until the next trial. Over individual trials, the rat gradually becomes better at finding the platform from different starting locations as quantified by either the time spent swimming or the path length of the swim. Richard Morris showed that lesions of the hippocampus impair the ability of rats to learn the platform location in this task, resulting in longer swim times and path lengths during training (Morris et al., 1982). I should mention that Richard Morris is the only person I have met who always refers to the task simply as "the water maze." Learning of platform location in this task is also slowed by drug manipulations that block the modification of synapses in the hippocampus as described later. These impairments occur with a single fixed platform location (Morris et al., 1986) and also with a platform location that changes between days (Steele and Morris, 1999). The moving location provides a more direct test of the ability of a rat to retrieve an episodic memory of new platform location on each day. Each day, the rat must discover the new platform location on the first trial and then retrieve the episodic memory of the first trial in order to find the platform on subsequent trials. Impairments in performance on the Morris water maze also appear with lesions of the dorsal entorhinal cortex (Steffenach et al., 2005) and dorsal presubiculum (postsubiculum; Taube et al., 1992) and with lesions of the fornix (Eichenbaum et al., 1990) that will reduce hippocampal theta rhythm. The study of fornix lesions showed an impairment only with

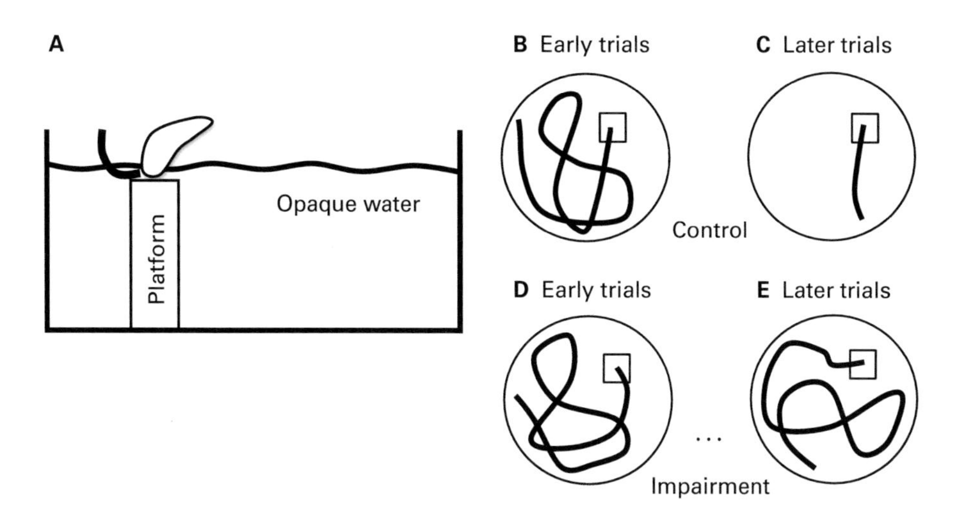

Figure 1.7
(A) Schematic side view of Morris water maze with rat sitting on its goal on each trial, a platform submerged just below the surface of opaque water. (B) On early trials, rats swim long distances before finding the platform. (C) On later trials, they swim straight to the platform. (D and E) Lesions of the hippocampus or pharmacological blockade of synaptic modification impairs behavior, preventing the decrease in swim time on later trials (E).

changes in starting locations (Eichenbaum et al., 1990) but not with a single starting location, indicating that this task puts demands on the capacity for planning a future trajectory from different starting locations to a goal location. A model of this task using hippocampal circuits for the planning of trajectories to the goal location is presented in chapter 7.

Object Investigation Time Recent experiments in rats have focused on potential episodic memory function by testing the specific requirement for memory of what, where, and when. These tasks take advantage of the strong tendency of rats to focus their investigatory behavior on novel stimuli with less investigation of more familiar stimuli. Rats are presented with novel stimuli, or stimuli in novel locations, and their object investigation time is measured to determine their capacity to differentiate between novel stimuli and familiar stimuli. For example, rats have been tested for their change in investigation time to objects that were presented at different times and moved to new locations during a recognition period (Dere et al., 2006). This task effectively tests memory for what, where, and when, corresponding to a single event in episodic memory, but could be performed based on retrieval of discrete single time snapshots of object and location.

Another test in rats suggests the retrieval of full spatiotemporal trajectories. In this task, rats learn trajectories from a central choice point to two different hidden objects in the side arms of an E-shaped maze (Eacott and Norman, 2004; Eacott et al., 2005; Easton et al., 2009). Thus, the rats learn the nature of the objects and their location at the end of two trajectories. Then the rats are kept in a holding chamber with one of the objects, so that they become highly familiarized with that object (if one anthropomorphizes, one could say they are bored with the object). After this period of familiarization, they are placed into the central arm of the task. The rats run up the central arm, and without seeing either object, they commonly make a choice to visit the arm with the less familiar object, indicating their desire to explore the less familiar object rather than the highly familiar object in the opposite arm. Thus, the trajectory followed from the choice point in this task appears to depend on actual episodic retrieval of a prior trajectory to the less familiar item associated with that trajectory, rather than depending on just the familiarity of cues or even a single snapshot memory (Eacott and Norman, 2004; Eacott et al., 2005; Easton et al., 2009). Note that this task avoids the question of familiarity of the arms at the choice point because the rat is making its choice not on the basis of the spatial locations but on the basis of the familiarity of the objects that are out of sight around the corners of the E-shaped maze. The selection of the choice toward the less familiar object is impaired by fornix lesions, which reduce theta rhythm oscillations in the hippocampus. Thus, this rat behavioral data supports a role for the hippocampus and fornix in the retrieval of episodic spatiotemporal trajectories for memory-guided behavior.

2 Neural Dynamics of Episodic Memory

What is the stuff that memories are made of? The data suggest that our ability to remember episodes depends on the physical circuits of the hippocampus, entorhinal cortex, and other parahippocampal regions located within the medial temporal lobes, as well as connected regions such as the medial septum, mammillary bodies, and anterior thalamus, as described in chapter 1. Damage to these regions causes impairments in experiments ranging from tests of word pair associations in human patients to tests of repeated visits to a location in spatial memory tasks in rats. The data on brain lesions tell us where to look first for the mechanisms of episodic memory, but we need anatomical and physiological data to provide a closer look at the specific structure of these regions and to provide a framework for understanding the dynamics of episodic memory. In this chapter, available data on the anatomy and physiology of these regions will be introduced in terms of their functional role in the specific model of episodic memory presented here.

Anatomical Circuits for Episodic Memory

The hippocampus and parahippocampal cortices have a fascinating structure that appears distinctive compared to many other regions of the brain, as shown in figure 2.1 (plate 1). The elegant scrollwork of the hippocampal formation almost immediately suggests structured dynamical interactions, the way that the shape of a violin or a cello might suggest the dynamics of vibrating strings and the corresponding pitch, and the dynamics of resonant surfaces and corresponding timbre. I specifically use this analogy with musical instruments because data suggest that the dynamics of oscillations contribute to the role of the hippocampus and entorhinal cortex in episodic memory (O'Keefe and Recce, 1993; Skaggs et al., 1996; Burgess et al., 2007; Giocomo et al., 2007). Because of their associated function, researchers have come to refer to the hippocampus and associated parahippocampal cortices collectively as the hippocampal formation. This section describes the anatomy of the hippocampal formation starting with the entorhinal cortex (see figure 2.1), which provides the bulk of cortical input to the hippocampus.

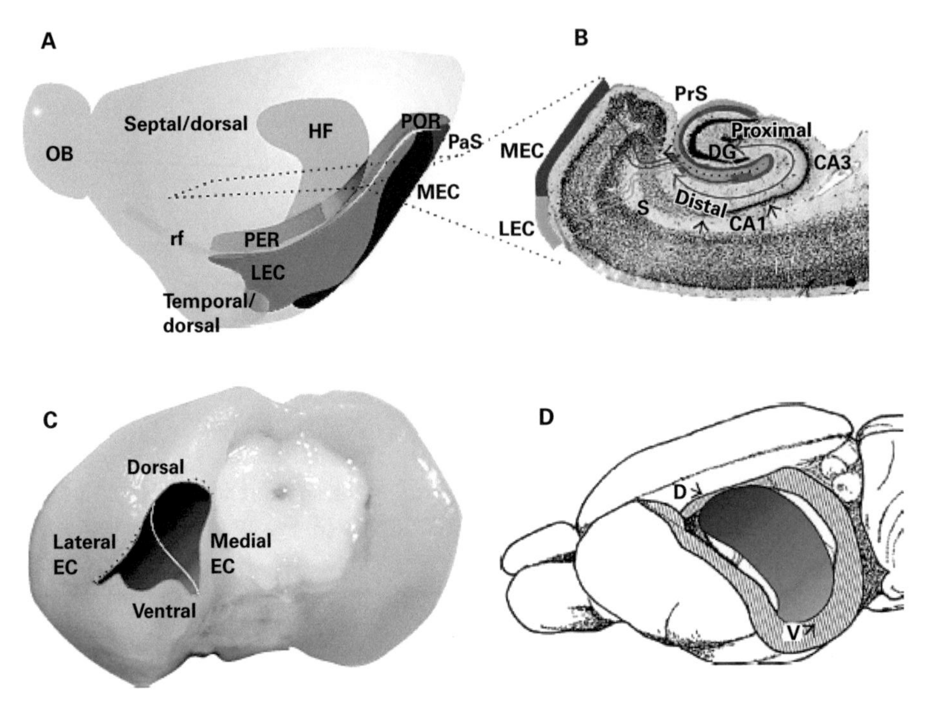

Figure 2.1 (plate 1)
Anatomy of entorhinal cortex and hippocampus in the rat. (A) Side view of left hemisphere of rat brain showing location of the rhinal fissure (rf) and the lateral entorhinal cortex (LEC) and medial entorhinal cortex (MEC) receiving input from perirhinal (PER) and postrhinal (POR) cortex and parasubiculum (PaS). Underneath these structures is the hippocampal formation (HF). To the left is the olfactory bulb (OB). Figures shown here reprinted with permission from Menno Witter (Canto et al., 2008). (B) Dotted lines in A show location of a horizontal cross-section showing projections from MEC and LEC into the dentate gyrus (DG) and hippocampal subregions CA1 and CA3 (cornu ammonis = CA). Cross section also shows location of presubiculum (PrS) and subiculum (S). (C) Picture of rat brain from behind. Dark to light shading shows the dorsal to ventral dimension of the lateral and medial entorhinal cortex (EC). (D) The same shading shows the areas of hippocampus receiving input from dorsal (D) and ventral (V) entorhinal cortex and sending back reciprocal connections.

For a variety of reasons, the analogy of an orchestra appears appropriate for introducing the structure and dynamics of the entorhinal cortex. In particular, recent data suggest that the frequency dynamics of the entorhinal cortex may provide an essential component of mechanisms for the encoding of memory, and these frequency dynamics have an orderly arrangement like the position of stringed instruments with different pitch within an orchestra. This analogy helps describe both the anatomical and functional properties of these regions.

Imagine the entorhinal cortex as an orchestra arranged upon a stage. Normally, in an orchestra, the highest pitched instruments, the violins, are on the left, and progressively lower pitches are positioned to the right, starting with the violas, then the cellos, and finally the basses. The percussion instruments are arranged behind the string section.

For the orchestra analogy of entorhinal cortex, imagine sitting in a balcony just to the right of the orchestra and looking down at the orchestra. The full array of stringed instruments on your left comprise the medial entorhinal cortex, as shown in figure 2.2. Toward the top of your view on the left, you see the violins, below them the violas, then the cellos, then the basses. As described below, this corresponds to the topography of temporal frequencies and spatial periodicities observed in the medial entorhinal cortex. These instruments will play an odd symphony conducted by the twists and turns of a spatiotemporal trajectory through space. The events and items along this trajectory will conduct bursts of percussion at specific time points, performed by the percussion on your right (figure 2.2). In this analogy, the coding of events by the percussion represents the proposed role of lateral entorhinal cortex. For the purpose of this analogy, imagine an unusually large percussion section with a wealth of tympanis, snare drums, triangles, and cymbals. The anatomical locations of the medial entorhinal cortex and the lateral entorhinal cortex are summarized in figure 2.1.

The entorhinal cortex receives input from neocortical regions, including perirhinal input to lateral entorhinal cortex and postrhinal (or parahippocampal) input to the medial entorhinal cortex. The medial entorhinal cortex also receives strong input from the postsubiculum. For the input from these regions, imagine not one conductor but an array of conductors positioned on the other side of the stage from your balcony. The individual musicians of the entorhinal cortex orchestra are paying attention to one or more conductors in this array. These are unusual musicians in that they are able to follow the hand movements of the conductors not only for tempo but also for pitch. Thus, during encoding of a memory they are not reading their music but playing music based on the signals from one or more of the conductors. The overall tempo is provided by regulation of theta rhythm by the medial septum, like a metronome that changes frequency over time.

In the model presented here, the circuits of the hippocampus play an essential role in encoding associations between elements of episodic memory. For example, the

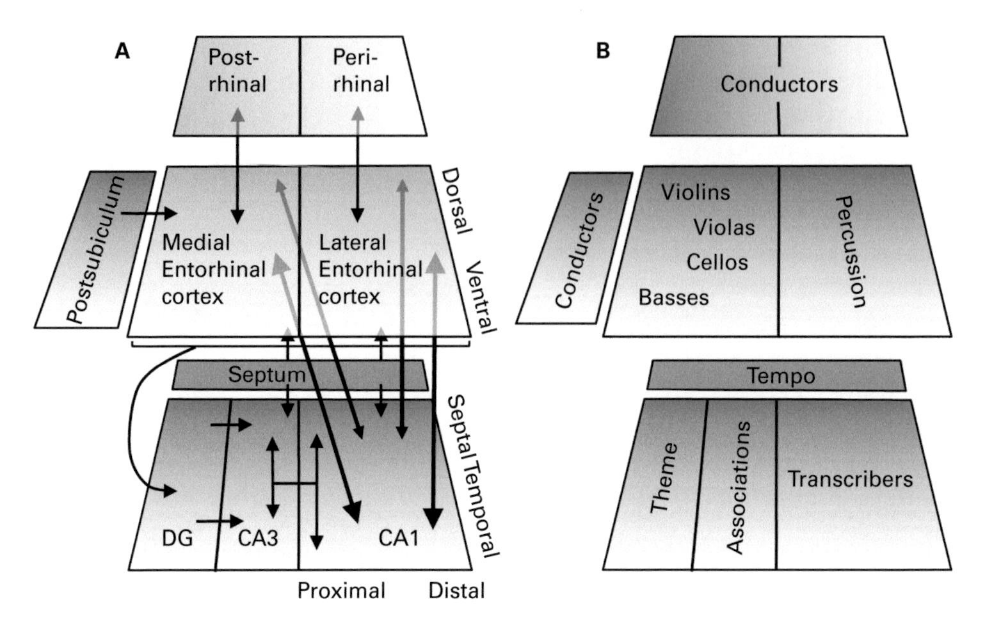

Figure 2.2

Anatomy of entorhinal cortex and hippocampus compared to analogy of the orchestra. (A) The medial entorhinal cortex (MEC) receives selective input from the postsubiculum and converging input from the neocortex via the postrhinal cortex (POR). Lateral entorhinal cortex (LEC) receives input via the perirhinal cortex (PER). Darker regions of entorhinal represent dorsal regions with higher intrinsic frequency and narrower spacing of grid cell firing fields. Lighter color represents decreasing intrinsic frequency and wider spacing of grid cell firing fields. Neurons along the dorsal to ventral extent of entorhinal regions send topographic input to neurons along the septal to temporal length of region CA1 which project back directly or indirectly to the corresponding entorhinal region (arrows). Both entorhinal regions send input to the dentate gyrus (DG) and CA3. The dentate gyrus projects to region CA3 which projects to region CA1. The medial septum paces oscillations in both hippocampus and entorhinal cortex. (B) Orchestra analogy. The resonant frequency properties of neurons in medial entorhinal cortex are like the different properties of stringed instruments. The dorsal neurons have higher resonant frequency, corresponding to violins in the analogy. The decrease in resonant frequency along the dorsal to ventral axis corresponds to lower resonant frequencies of violas, cellos and basses. The lateral entorhinal cortex is analogous to percussion. Orchestra conductors represent input from postsubiculum, postrhinal and perirhinal cortex, with tempo provided by the medial septum. The transcribers in region CA1 write the music from medial and lateral entorhinal cortex, while the larger themes of music are detected in dentate gyrus. See text for more details.

hippocampus encodes associations between the position in both space and time along a spatiotemporal trajectory and the actions and viewpoint on that trajectory, as well as the events or items encountered at that position in space and time. In the orchestra analogy, imagine region CA1 of the hippocampus as a set of superhuman musical transcribers, sitting under the stage (perhaps the stage is a wooden grate, rather than a solid floor). These are not composers, but transcribers, each of whom can listen to the musician above them, and copy down important components of the pitch and rhythm of their performance determined by the conductors. One part of the notation consists of associations in time when a set of stringed instruments and percussion play together. Then, during retrieval, the instrumentalists in the entorhinal cortex can look down through the grate and read this transcribed music from the hippocampus.

There are transcribers as well in the dentate gyrus and region CA3. However, these transcribers have a more complex task of listening to a full group of instruments and recording overall themes (e.g., they might notice recurring themes such as the four distinctive opening notes of Beethoven's *Fifth Symphony*). This corresponds to mechanisms for encoding and retrieval of the context of an episodic memory. The orchestra analogy will be used to describe many of the principles of function of the hippocampal formation.

Principles of Connectivity

A number of anatomical researchers contributed to the discovery of principles of the anatomical connectivity of the hippocampal formation and parahippocampal cortical regions (Ramon y Cajal, 1911; Swanson et al., 1978; Amaral and Witter, 1989; Insausti et al., 1997). One of the researchers who helped discover these properties was Menno Witter (Witter et al., 2000; Naber et al., 2001; Witter and Moser, 2006; Canto et al., 2008), who is now at the Norwegian Institute of Science and Technology. Prof. Witter described the properties described below that helped lead to the breakthroughs on the physiology of these regions by Edvard and May-Britt Moser. The anatomy described here will draw on his summaries of the anatomical connectivity (Witter et al., 2000; Canto et al., 2008). As summarized in figure 2.1, the anatomical data show a detailed topography of mapping between the different regions that indicate important functional principles of anatomical structure: (1) consistent scales of space and time, (2) parallel streams for what and where, (3) reciprocal richness, and (4) contextual convergence. The orchestra analogy helps give a sense of these basic functional principles.

Consistent Scales of Space and Time

Exciting recent physiological and anatomical data indicate an elegant topography that maintains consistent relations between the entorhinal cortex and hippocampus for the spatial scales of a memory. What do I mean by the spatial scales of a memory? One of the most common examples of the use of episodic memory concerns remembering where you put something, such as your car keys, or your glasses, or your pencil. Solving this problem requires answering the question on different spatial scales, including what room in the house, what furniture in the room, and even where on a given piece of furniture. Chapters 3 and 4 will describe how encoding and retracing your spatiotemporal trajectory through the house appears to involve simultaneous processing of these different scales, coding space and time on multiple scales simultaneously in different anatomical locations within the hippocampal formation.

Accurate encoding and retrieval requires that the coding of space and time be consistent within the episode. The difference in scaling must be maintained accurately throughout the circuit to prevent the spatiotemporal trajectory from being retrieved out of synchrony. If you meet your friend on a walk, you don't want the memory of the start of your walk to call up memory of your friend's nose, while the memory of the end of the walk calls up memory of your friend's eyes.

The recent physiological data described in this chapter show how the anatomical data might provide a consistent mapping of the scale of space between regions, mapping small spatial scales to small and large spatial scales to large. Further cellular data indicate that the consistent mapping of spatial scales corresponds to consistent mapping of temporal frequencies analogous to the pitch of the instruments in different parts of the orchestra.

The anatomical connectivity through the hippocampal formation maintains an exquisite topography to ensure consistent scales of space and time (Witter et al., 2000; Canto et al., 2008). As shown in figure 2.1, the dorsal portions of both medial and lateral entorhinal cortex send input to the dorsal end of the hippocampus (often referred to as the septal end, because it is adjacent to the septal nuclei). The ventral portions of both lateral and medial entorhinal cortex send input to the ventral end of the hippocampus (often referred to as the temporal end because it is continuous with the temporal lobe in primates).

In the orchestra analogy, a particular set of transcribers representing the septal (dorsal) end of the hippocampus sits directly below the violins representing dorsal entorhinal cortex and listens to the violins. A different set of transcribers representing the temporal (ventral) end of the hippocampus sits directly below the basses representing the ventral regions of entorhinal cortex. In addition, when the sheet music from the transcribers is played, the same relationship is maintained. The notes transcribed from the music of the violins are played by the violins. The synaptic projections from the septal end of region CA1 back to dorsal entorhinal cortex could be conceived as the sharp-eyed violinists reading the music below them through the grated floor. Correspondingly, the synaptic projections from the temporal end of region CA1 back to ventral entorhinal cortex can be seen as sharp-eyed bass players reading the music transcribed below them. The region CA1 circuits interact with entorhinal circuits in a systematic manner because of the systematic topographic relationship between entorhinal projections to CA1 and the return projections from CA1 back to deep layers of entorhinal cortex (Tamamaki and Nojyo, 1995; Naber et al., 2001).

This spatiotemporal topography is shown in the structured connections summarized in figure 2.1. As described later in this chapter, this anatomical topography corresponds to recently discovered systematic differences in physiological properties within these different regions.

Parallel Streams for What and Where

On an intuitive level, the memory of spatial location (where) and the memory of individual items (what) feel qualitatively different. This appears to correspond to a separation of coding of where and what into separate processing streams converging on the medial and lateral entorhinal cortex, corresponding to the strings and the percussion in the analogy presented above. The topography is maintained in a loop arising from medial and lateral entorhinal cortex. As shown in figure 2.2, the medial input goes to the part of CA1 adjacent to CA3 (referred to as proximal CA1) and the lateral input goes to the part of CA1 distal from CA3. The same separation is maintained in CA1 projections to the subiculum and back to medial and lateral entorhinal cortex (Witter et al., 2000; Canto et al., 2008).

The coding of events or items (what) is proposed to involve one stream of processing that flows from the perirhinal cortex through the lateral entorhinal cortex to region CA1 of the hippocampus and back (Hargreaves et al., 2005; Eichenbaum and Lipton, 2008). The events or items in a memory are often discrete. For example, one might experience dropping a glass in the kitchen. In the orchestra analogy, imagine that the drop of the glass triggers a conductor in the perirhinal cortex on your right to wave his arms, triggering a cymbal player in the lateral entorhinal cortex to clash his cymbals. The clash is dutifully recorded by a transcriber in the distal part of region CA1.

In contrast, the spatial coding of memory (where) and possibly the temporal code (when) appears to involve a separate stream of processing that involves medial rather than lateral entorhinal cortex (Hargreaves et al., 2005). The where stream flows from the parahippocampal cortex (known as postrhinal cortex in the rat) through the medial entorhinal cortex to region CA1 of the hippocampus and back. Another feature that makes this processing stream anatomically distinct is that the medial entorhinal cortex receives a strong, separate input from the dorsal presubiculum (postsubiculum) that does not project to lateral entorhinal cortex in the rat. As described below, these direct projections from the postsubiculum appear to provide a way for velocity information to enter the system and update spatial location independently of the cortical circuits of sensory perception (Witter et al., 2000). In the orchestra analogy, this could correspond to a separate group of conductors that code head direction and speed of movement. These could influence the stringed instruments, conducting rapid melodies in the violins and progressively slower melodies in the violas, cellos, and basses.

For example, before dropping the glass, one might have walked into the kitchen, and after dropping the glass, one might walk over to the closet to get a broom. In the orchestra analogy, walking to the closet would cause a set of conductors representing the postsubiculum to wave their arms based on the speed and direction of movement and those representing the postrhinal cortex would wave their arms coding the flow of visual stimuli. The instruments of the medial entorhinal cortex would respond with violins playing a melody of rapid movement and flow of features (e.g., each note occurring near one tile on the kitchen floor) and the basses playing a slower rhythm (e.g., one note near the refrigerator, one near the closet). Region CA1 can form associations between locations and events.

Reciprocal Richness

The richness of detail in an episodic memory covers a range of modalities. For example, you might remember the face of a neighbor that you encounter on a walk as well as the smell of budding trees and the feeling of sunshine on your face. The variety of modalities and richness of detail suggests that a large number of neurons are involved in coding the sensory features of an episodic memory.

The anatomical data suggest that this richness of detail arises from a systematic pattern of reciprocal connectivity between the entorhinal cortex and the sensory cortices involved in the initial perception of features (Van Hoesen, 1982; Witter et al., 1990; Burwell and Amaral, 1998). The bidirectional connections of neocortical association areas with the perirhinal and postrhinal cortex are extensive for all sensory modalities. For example, portions of the temporal lobe cortex might code the face of a friend that you encounter on a walk. This information could spread along feedforward connections to activate particular entorhinal cortex neurons. As described in the modeling sections below, encoding an encounter with your friend could include strengthening of connections within the hippocampus that link a position along a spatiotemporal trajectory to neurons coding the event of encountering your friend. Then, during retrieval of the episodic memory, the activation of entorhinal neurons coding a component of the spatiotemporal trajectory can activate hippocampal circuits that activate the entorhinal neurons coding the event of meeting your friend. The entorhinal neurons only code a link to features of the face represented elsewhere, but the projections from entorhinal cortex to the face-responsive areas in the temporal lobes (Hasselmo et al., 1989) could code the full details of your friend's face, so that you can remember a detailed image of seeing your friend at a particular location.

Convergence for Context

Our episodic memories include many details from different modalities, but the general context of the memory can trigger the full range of details. For example, you might be prompted to remember dropping the glass in the kitchen either by thinking about a glass or by thinking about the kitchen. This suggests some convergence for representing the overall context of the memory. The connections from lateral and medial entorhinal cortex to region CA1 and the subiculum appear to maintain very clear segregation. However, this contrasts with an anatomically distinct, convergent pattern in the connections from lateral and medial entorhinal cortex to the dentate gyrus and region CA3. This convergence could allow the dentate gyrus and region CA3 to compute the context of the overall episode. Rather than maintaining separate parallel loops as in region CA1 and the subiculum, the projections from entorhinal cortex converge together on neurons in the dentate gyrus and CA3. The lateral entorhinal cortex projects to all of the dentate gyrus and CA3, contacting the distal ends of the dendrites in these areas (Witter et al., 2000). Similarly, the medial entorhinal cortex projects to all of the dentate gyrus and CA3 but contacts slightly more proximal segments of the dendrites of these neurons.

This convergence could provide convergent information for coding the context of an episodic memory, allowing separation of different events based on an overall theme. For example, one could separate two episodes in the kitchen as "the morning a glass was broken" versus "the evening after we played soccer." In the orchestra analogy described above, the convergence of information on dentate gyrus and CA3 could be represented by separate transcribers that listen to the whole orchestra. These transcribers would listen for particular combinations of instruments, recognizing the combinations of instruments and chords and short themes the way a classical musician could immediately differentiate the start of a piece by Beethoven from the start of a piece by Stravinsky. They might also respond to short thematic segments, such as the four characteristic notes of Beethoven's *Fifth Symphony*, allowing events to be encoded accordingly. The recording of themes is analogous to remembering the context for distinguishing the different events in the kitchen. This shows another principle of the anatomical organization of the hippocampal formation. More details of the internal structure of the hippocampus will be reviewed in chapter 5.

Cellular Mechanisms and Episodic Memory

As summarized in chapter 1, lesion data support the role of the entorhinal cortex and hippocampus in episodic memory function. This raises the further question: What cellular processes in these structures provide the mechanisms for episodic memory?

The model of episodic memory function presented here uses the cellular properties of entorhinal and hippocampal neurons as a component of the oscillatory dynamics of the model. Therefore, describing the model requires at the outset an overview of the specific intrinsic properties of neurons that are used in the model, as well as a brief overview of the basic functional properties of neurons, such as spiking and synaptic potentials. Over a century of research has demonstrated these basic properties, and detailed biophysical models can replicate these properties.

Neurons provide the mechanisms for transmission of activity throughout the nervous system. This activity can be induced by input to neural circuits caused by sensory experience and can regulate output to muscles to direct voluntary responses to the environment. Changes in the spread of neuronal activity within cortical structures provide the circuit-level mechanisms for memory function, including the oscillatory dynamics that form the basis of the model presented here.

Within cortical circuits, neurons have many different anatomical forms, which can influence the functional properties of the neuron. The cortical neurons in most models contain the same basic structural components, including branching dendrites that receive input from thousands of other neurons that make synapses on the dendrite, as well as a branching axon sending output via synapses on thousands of other neurons. An electrical potential can be recorded across the membrane that forms the surface of the neuron. The spread of activity between different parts of the neuron occurs through changes in this membrane potential. Figure 2.3A shows two neurons, illustrating the basic structural components of a neuron and their interaction via a synapse.

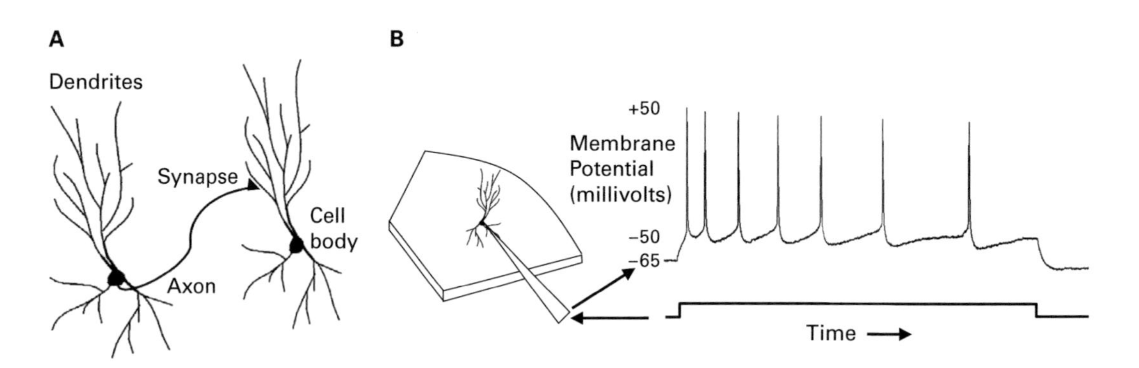

Figure 2.3
(A) Schematic of two pyramidal cells with extensive dendritic branching. The neuron on the left gives rise to an axon. A single branch of this axon is shown with a single synapse on the postsynaptic neuron. (B) Diagram showing an electrode recording of membrane potential from a neuron in a slice preparation of entorhinal cortex. A long pulse of depolarizing current injection causes the cell to generate spiking activity that slows down during the current injection. Data obtained in the Hasselmo lab by Lisa Giocomo.

The membrane potential of a single neuron can be recorded by inserting a recording electrode into the neuron and recording the difference in electrical potential between the interior of the neuron and the surrounding extracellular fluid. The membrane potential of a cortical neuron recorded in an intact brain changes constantly due to constant changes in synaptic input from other neurons. However, for intracellular recording in my laboratory we use preparations consisting of slices of entorhinal cortex. In these slices, neurons have little input and therefore show a stable resting potential. The response properties of the neuron can then be recorded by giving current injection through the recording electrode. In the absence of input, neurons have a negative resting electrical potential across their membrane of around –70 to –60 millivolts relative to the outside (more than 100 times weaker than the potential across a standard 9-volt battery).

The resting potential results from a difference in the concentration of electrically charged ions in the fluid inside and outside of the neuron. The ions include the components of dissolved table salt, sodium ions (Na^+) and chloride ions (Cl^-), that are in high concentration outside the cell, as well as potassium ions (K^+) that are in high concentration inside the cell. The membrane potential changes in proportion to the conductance of the membrane to these different ions (see appendix section on Biophysical Simulations). If the membrane conductance increases for potassium, the membrane potential will go to larger negative values (hyperpolarization). If the conductance increases for sodium, the membrane potential will go to less negative values (depolarization).

The membrane conductance can be changed by channels that allow ions to pass through the membrane. Some channels open or close in response to neurochemicals released from other cells. For example, the neurotransmitter glutamate depolarizes cells by activating receptors with channels that increase conductance to sodium and potassium, while the neurotransmitter GABA reduces depolarization by activating receptors that open channels and increase conductance to chloride or potassium ions. The neuromodulator acetylcholine changes spiking patterns by influencing channels for potassium and/or sodium. Other channels open and close based on the membrane potential voltage. The membrane potential can also be altered by currents delivered through the recording electrode.

As shown in figure 2.3 (and in appendix figure A.16C), if the membrane potential goes over a threshold value (usually around –50 millivolts), action potentials (spikes) are generated. The start of an action potential results from sodium channels in the membrane that respond to membrane voltage. These channels start to rapidly open when the membrane voltage reaches threshold, allowing sodium ions to flow into the cell. This causes a very rapid increase in membrane potential from –50 millivolts to positive values. The sodium channels then close and other channels selective for potassium respond to the membrane depolarization by opening slightly more slowly, allowing potassium ions to flow out of the cell. This causes a rapid drop from positive values to well below –50 millivolts all in less than 1 millisecond. In a plot showing the data, this rapid change in membrane potential in less than 1 millisecond looks like a sharp spike, and therefore action potentials are commonly referred to as "spikes." Because the potassium channels close more slowly than the sodium channels, the cell is hyperpolarized for one or two milliseconds after the spike, causing a dip called the fast afterhyperpolarization. Once a spike is initiated in a neuron, it activates channels sequentially along the axon, allowing the spike to spread down the axon and to cause release of chemical transmitters at synapses connecting to other neurons.

Recordings in brain slice preparations show specific properties of neurons in entorhinal cortex and hippocampus that have been used to model the mechanisms of episodic memory in these regions. Electrophysiological data from the entorhinal cortex reveals intrinsic properties of individual cells that could combine with synaptic modification to contribute to the encoding and retrieval of episodic trajectories. Understanding these properties of individual neurons will help in understanding the models presented in later chapters. The following sections will summarize the properties of persistent spiking, membrane potential oscillations, spike frequency accommodation, and synaptic modification.

Persistent Spiking

A model presented here uses the persistent spiking of entorhinal neurons for representation of space and time in episodic memory. Persistent spiking has been described using intracellular recordings from single neurons in entorhinal cortex. This cellular phenomenon can arise from intrinsic mechanisms within individual neurons, and has exciting implications for network mechanisms of episodic memory. Figure 2.4 shows the phenomenon of persistent spiking, in which a neuron fires for an extended period after a stimulus even when all synaptic transmission is blocked.

The control condition in the figure shows the typical response usually shown for a cortical neuron. The neuron generates a series of spikes during an input current injection and then stops spiking when the current injection ends. The second row shows the application of a drug (carbachol) that activates acetylcholine receptors. In this condition, the neuron shows persistent spiking. The neuron responds with spiking to current injection but then continues firing in a persistent manner after the current injection terminates. This persistent spiking could allow input stimuli to be actively maintained in working memory after the stimulus has ended.

The cellular property of persistent spiking may contribute to episodic encoding mechanisms. Even during the pharmacological blockade of all excitatory and inhibitory synaptic transmission, neurons in the entorhinal cortex demonstrate the capacity to display persistent spiking, indicating intrinsic single cell mechanisms for persistent spiking. Persistent spiking can be described as bistability, in which a neuron showing

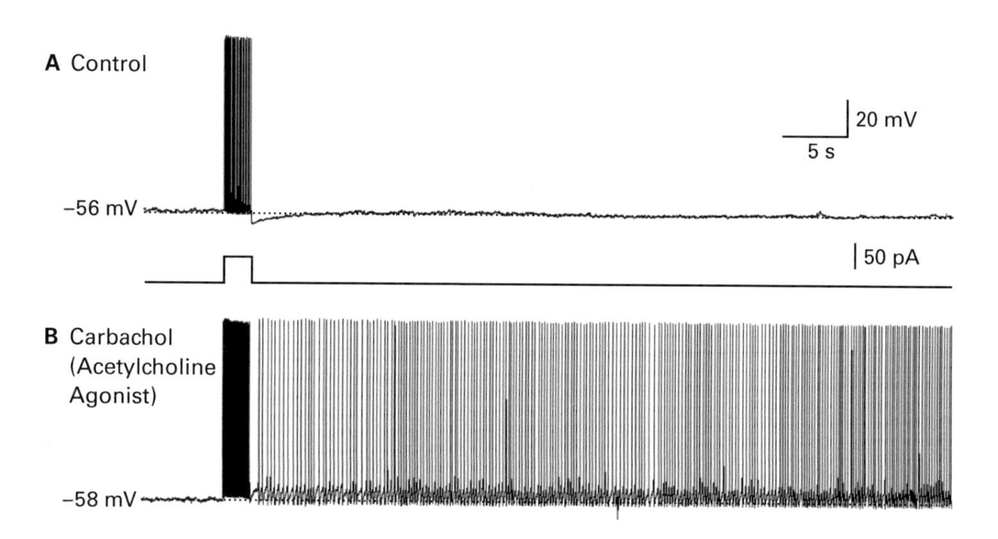

Figure 2.4
(A) In control conditions, a neuron responds to current injection by spiking during current injection and then stopping when current injection ends. Vertical black lines are individual action potentials (spikes). (B) After activation of acetylcholine receptors, the neuron spikes during current injection and then after the end of the current injection shows persistent spiking that continues for an extended period of minutes (Yoshida and Hasselmo, 2009). This trace shows firing for more than 40 seconds in a neuron in subiculum.

no spiking activity before injection transitions to a characteristic steady frequency of stable persistent spiking activity after a transient depolarizing current injection through an electrode or transient repetitive synaptic input induced by stimulating axons contacting the neuron (Klink and Alonso, 1997a; Tahvildari et al., 2007; Yoshida et al., 2008; Yoshida and Hasselmo, 2009a). The conditions are the same before and after the end of stimulation, but the neuron is silent beforehand, and spikes persistently afterward. A transient but strong hyperpolarizing input can shut off the spiking and make the cell remain quiet, generating no spikes, until another transient depolarizing current again turns on the persistent spiking with the same characteristic frequency. Persistent spiking appears to be a prominent phenomenon in entorhinal cortex. In contrast, some neurons in other cortical regions such as the hippocampus rarely show persistent spiking and will spike only during current injection but will stop after termination of the current injection.

Some pyramidal neurons in layer II of medial entorhinal cortex show stable persistent spiking whereas others show spiking that self-terminates over periods of many seconds (Klink and Alonso, 1997a). Pyramidal cells in layer III show stable persistent spiking that can last for 2 minutes or more (Yoshida et al., 2008).

Pyramidal neurons in layer V of entorhinal cortex can maintain stable persistent spiking at different graded frequencies for many minutes. Their firing frequency can be shifted to higher stable frequencies by depolarizing current injection or to lower stable frequencies with hyperpolarizing current injection (Egorov et al., 2002). This is referred to as graded persistent spiking rather than bistable persistent spiking. Graded persistent firing could allow these neurons to integrate synaptic input over extended periods. Persistent firing has also been shown in layer III of lateral entorhinal cortex (Tahvildari et al., 2007). The persistent spiking effect can persist after weak transient hyperpolarizing current injection, indicating resistance to transient distractors. The persistent spiking effect depends on activation of the muscarinic subtype of acetylcholine receptors, as the effects are blocked by muscarinic receptor blockers such as atropine. Effects of this type were also previously described in prefrontal cortical slices (Andrade, 1991; Haj-Dahmane and Andrade, 1996, 1998) and other regions (Schwindt et al., 1988).

On a molecular level, persistent spiking appears to arise from a separate channel in the membrane of entorhinal cortex neurons that allows the positively charged ions (cations) sodium and potassium ions to pass through (Shalinsky et al., 2002), so it is referred to as a nonspecific cation current. This causes depolarization similar to the excitatory synaptic glutamate receptors, which are also nonspecific cation currents. This channel appears to respond to activation of slow modulatory receptors such as the muscarinic receptor for acetylcholine or the metabotropic glutamate receptor (Yoshida et al., 2008). In addition, the channel is activated by increases in intracellular calcium. This calcium-activated, nonspecific cation current is often referred to as a CAN current (Magistretti et al., 2004).

In biophysically detailed models, the properties of the CAN current have been implemented in computer simulations of layer II nonstellate neurons (Fransén et al., 2002) and layer V pyramidal neurons (Fransén et al., 2006) using the GENESIS (GEneral NEural SImulation System) software package for biophysical simulations of the membrane potentials of populations of neurons (Bower and Beeman, 1995). The CAN current provides a mechanism for intrinsic working memory dependent on the conjunction of acetylcholine receptor activation and cellular spiking (Klink and Alonso, 1997b). Acetylcholine alone will not induce sustained spiking in neurons, and spiking in the absence of acetylcholine (or metabotropic glutamate receptor activation) does not initiate persistent spiking. However, in the presence of acetylcholine, the generation of spiking by intracellular current injection or by synaptic stimulation causes the neuron to enter an internal regenerative cycle of persistent spiking. Each new spike causes an influx of calcium through voltage-sensitive calcium channels, and the new influx of calcium activates the CAN current (Magistretti et al., 2004). The CAN current then causes additional depolarization, leading to another spike that again activates voltage-sensitive calcium channels that further perpetuate the cycle.

The cellular phenomena described above appear in entorhinal neurons even in the presence of synaptic blockers. This intrinsic mechanism contrasts with most models of working memory, in which regenerative spiking activity depends upon excitatory recurrent synapses propagating activity between different neurons within a population of active units (Lansner and Fransén, 1992; Amit and Brunel, 1997; Lisman et al., 1998). These models generate persistent spiking due to circuit level mechanisms, as described further in chapter 5. For example, network persistent spiking can be obtained due to the effect of excitatory and inhibitory synaptic feedback that drives neurons into steady activity patterns known as stable attractor states (Wilson and Cowan, 1972; Hasselmo et al., 1995a). As an alternative, the dynamics of interacting populations of neurons can result in network oscillations in the frequency range around 7 Hz known as theta (Denham and Borisyuk, 2000) or in the higher frequency range over 30 Hz known as gamma (Chow et al., 1998; White et al., 1998b; Borgers and Kopell, 2003).

This cellular data has exciting links to behavioral data on memory. The cellular persistent spiking effect in the entorhinal cortex could regulate encoding of episodes in the hippocampus by acting as a buffer for incoming information, holding behaviorally relevant information for a period of time longer than the presence of the sensory input itself. This potential role is supported by evidence that lesions of the entorhinal cortex and perirhinal cortex impair performance in delayed nonmatch to sample tasks in nonhuman primates (Gaffan and Murray, 1992; Leonard et al., 1995) and rats (Otto and Eichenbaum, 1992). In these tasks, a sample stimulus is presented at the start of the trial, and after a delay period the animal must respond to a stimulus that differs from the sample stimulus. This lesion effect suggests that entorhinal cortex may be important for active maintenance of information during the delay period. Persistent spiking in the entorhinal cortex and perirhinal cortex may also be important to allow trace conditioning, in which a conditioned stimulus (such as a tone) is presented and ends many seconds before presentation of an unconditioned stimulus such as a shock. Trace conditioning is impaired by blockade of muscarinic receptors in the entorhinal cortex (Esclassan et al., 2009) or perirhinal cortex (Bang and Brown, 2009), suggesting that blockade of persistent spiking might prevent active maintenance of the conditioned stimulus (tone) and thereby prevent association with the unconditioned stimulus (shock) and the response (freezing).

Electrophysiological recording demonstrates that some neurons in the entorhinal cortex region do maintain persistent spiking of stimulus selective activity during delay periods in a delayed nonmatch to sample task (Suzuki et al., 1997; Young et al., 1997). Extensive modeling has demonstrated how the CAN current could underlie the full range of unit activity observed in delayed nonmatch to sample tasks. The spiking caused by the CAN current could directly cause delay period activity as well as enhancement of response when the same stimulus appears again, and network interactions could cause phenomena such as match suppression (Fransén et al., 2002). Other models have also addressed how this phenomena could underlie working memory (Lansner and Fransén, 1992; Lisman and Idiart, 1995; Jensen and Lisman, 1996b; Fransén et al., 2002; Jensen and Lisman, 2005) and encoding into episodic memory (Jensen and Lisman, 1996a; Koene et al., 2003). Effects such as this were initially proposed to form the basis for short-term memory by John Lisman (Lisman and Idiart, 1995; Jensen and Lisman, 1996a).

Persistent spiking might underlie neural activity during memory encoding in humans. In humans, fMRI shows that activity for a sample stimulus that persists during the delay period of a delayed match to sample task correlates with subsequent recognition memory for the sample stimulus (Schon et al., 2004). Blockade of muscarinic acetylcholine receptors with scopolamine reduces this delay period activity (Schon et al., 2005), indicating that the delay period activity may depend upon induction of persistent spiking activity on a cellular level due to effects of acetylcholine (Klink and Alonso, 1997; Yoshida and Hasselmo, 2009). The retention of activity could be very important for providing a sustained source of afferent input for a time period sufficient to induce synaptic modification within the hippocampal formation for encoding of episodic trajectories.

Membrane Potential Oscillations and Resonance

Another exciting body of recent work concerns the possible role of resonance in episodic memory function. The anatomy section above reviews the topography relating spatial scales to temporal frequency properties of neurons, using the analogy of the pitch of different instruments in the orchestra such as violins and cellos. The resonant properties of a violin or bass can be heard by tapping on the instrument, or plucking a string. But how does one measure something analogous to pitch in a neuron? The measurement of pitch is analogous to the measurement of the favored frequency of an individual neuron, by measuring the frequency of its membrane potential oscillations and resonance. The models described in this chapter and chapter 3 show how these properties could help code continuous dimensions of space and time.

Neurons in entorhinal cortex resonate to a particular pitch. Unusually strong oscillations and resonance properties have been shown in neurons of layer II of the entorhinal cortex known as stellate cells (Alonso and Llinas, 1989; Alonso and Klink, 1993; Dickson et al., 2000; Erchova et al., 2004; Giocomo et al., 2007; Giocomo and Hasselmo, 2009). These neurons show clear membrane potential oscillations when depolarized near firing threshold, and they show resonance in response to an oscillating input that changes in frequency. Figure 2.5 shows an example of subthreshold oscillations (Giocomo and Hasselmo, 2008b) with an amplitude of a few millivolts. These oscillations can influence the timing of spikes (Acker et al., 2003; Fransén et al., 2004; Rotstein et al., 2005; Pervouchine et al., 2006; Rotstein et al., 2006) and may contribute to network theta frequency (4-10 Hz) oscillations in entorhinal cortex

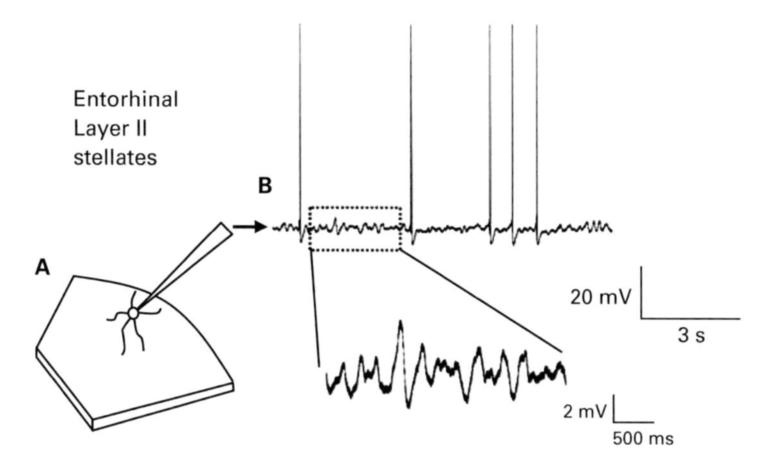

Figure 2.5
Example of intracellular recording of membrane potential oscillations in a stellate cell in layer II of a slice preparation of medial entorhinal cortex (A). Current injection pushes the cell near its firing threshold. This results in intermittent spiking activity (B). In between the spikes, the membrane potential shows subthreshold membrane potential oscillations in the enlargement that have a frequency in the range of theta rhythm (Giocomo and Hasselmo, 2008b).

(Mitchell and Ranck, 1980; Alonso and Garcia-Austt, 1987; Brandon et al., 2011b) and hippocampus (Buzsaki, 2002; Hasselmo, 2005a).

Exciting recent data support the hypothesis that there is a systematic change in the favored frequency of neurons at different dorsal to ventral positions within the entorhinal cortex and hippocampus, as if the neurons were instruments of different pitch in an orchestra. Neurons show a systematic change in the frequency of membrane potential oscillations and resonance along the dorsal to ventral axis of the medial entorhinal cortex (Giocomo et al., 2007; Giocomo and Hasselmo, 2009) in data that will be presented in more detail below. Neurons have molecular properties that make them resonate at different frequencies, just as the pitch of an instrument results from its resonant properties. For example, the reed of a clarinet just vibrates at one frequency, but as the holes on the clarinet are covered by fingers or pads, the resonant properties change to cause different frequencies. The differences in resonance of entorhinal cells may result from differences in molecular properties of membrane channels along the dorsal to ventral axis of medial entorhinal cortex (Giocomo and Hasselmo, 2008b). Oscillations and resonance do not require synaptic transmission.

Membrane potential oscillations appear less prominently in pyramidal cells of layer II (Alonso and Klink, 1993) and do not appear in layer III (Dickson et al., 1997) but are observed in layer V pyramidal cells (Yoshida and Alonso, 2007). The layer V membrane potential oscillations also show a decreasing gradient of frequency from dorsal to ventral medial entorhinal cortex (Giocomo and Hasselmo, 2008a). Membrane potential oscillations are not as prominent in neurons of the lateral entorhinal cortex (Tahvildari and Alonso, 2005).

The membrane potential oscillations have been simulated in biophysical models based on the interaction of voltage-sensitive membrane currents (Dickson et al., 2000; Fransén et al., 2004) using the GENESIS package for biophysical simulations of neurons. The oscillations in superficial layers of medial entorhinal cortex may be due to a hyperpolarization activated cation current or h current (Dickson et al., 2000; Fransén et al., 2004). The "h" refers to "hyperpolarization," which is a decrease in membrane potential to more negative values. In these neurons, presentation of a hyperpolarizing current pulse will slowly activate the h current, which increases the conductance of the membrane to the cations sodium and potassium. This causes depolarization that counteracts the hyperpolarization. For example, applying a strong negative current that causes an initial hyperpolarization will cause slow activation of the h current over a few hundred milliseconds that results in a depolarizing rebound that is described as a sag in the hyperpolarizing response to the current injection (Alonso and Klink, 1993; Giocomo et al., 2007), as shown in figure 2.6A.

At more depolarized levels, the h current can participate in oscillations, because the h current responds slowly to changes in membrane potential (on the order of tens to hundreds of milliseconds). This mechanism for membrane potential oscillations has been shown in simulations (Fransén et al., 2004) as described near the end of the

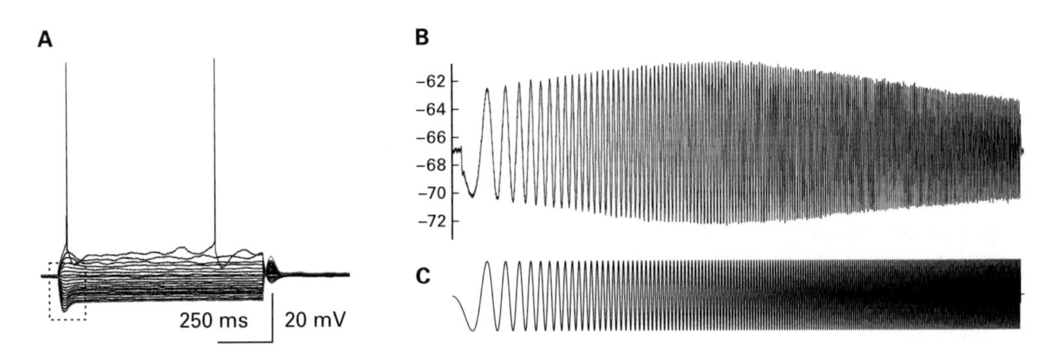

Figure 2.6
(A) Response of medial entorhinal stellate cell to different positive and negative step pulse current injections. Depolarizing injections cause spikes. Hyperpolarizing injections slowly activate the h current and cause a slow depolarizing "sag" back toward resting potential (Giocomo et al., 2007). (B) Membrane potential of a medial entorhinal stellate cell recorded by Chris Shay in my laboratory in response to a 20-second sine wave of increasing frequency shown in (C). The neuron response shows an increase in amplitude of oscillations up to the peak resonance response at about 8.8 Hz (tallest point), followed by a decrease in response at higher frequencies.

appendix. Within each oscillation in the model, a depolarization of membrane potential will slowly shut off the h current resulting in a rebound hyperpolarization. The rebound hyperpolarization will slowly turn on the h current resulting in a depolarization that then slowly turns off the h current and so on. The slow activation time constant of the h current results in theta frequency oscillations in the membrane potential.

These properties also result in resonance of the neuron. As shown in figure 2.6B, resonance can be measured by presenting the neuron with an oscillating input current injection that gradually increases in frequency. The neuron shows larger changes in membrane potential when the oscillating input current is near the resonant frequency of the neuron (Erchova et al., 2004; Giocomo et al., 2007; Heys et al., 2010), which is in the theta frequency range for entorhinal stellate cells. The resonance properties of these neurons also change along the dorsal to ventral axis of medial entorhinal cortex as described further below.

Spike Frequency Accommodation

Spike frequency accommodation is another intrinsic mechanism that could contribute to the coding of episodic memory. Figure 2.3 above shows an example of spike frequency accommodation reducing the firing frequency during a period of current injection. This neuron responds to a step pulse current injection by firing a series of action potentials. The firing starts rapidly at a high frequency but gradually slows down and stops. This slowing is referred to as adaptation or spike frequency accommodation. This adaptation due to previous firing can be considered a type of memory, as the reduction in firing depends on the duration of previous input. Spike frequency accommodation occurs in a number of cortical structures including the hippocampus (Madison and Nicoll, 1984), piriform cortex (Barkai and Hasselmo, 1994), and entorhinal cortex (Alonso and Klink, 1993).

Spike frequency accommodation appears to result from activation of membrane potassium currents (Storm, 1990) that can cause spike frequency accommodation in biophysical simulations (Barkai et al., 1994; Migliore et al., 1999) as shown in chapter 6. There are many channels regulating potassium currents (Storm, 1990) in neurons. There are fast voltage-sensitive potassium channels (known as delayed rectifier channels) that turn on and off rapidly during each spike. These cause the faster afterhyperpolarization visible after each spike in figure 2.3. Other potassium currents include the M current, which is activated more slowly by depolarization, resulting in a hyperpolarization that decreases the firing rate. The M refers to the effect of the muscarinic acetylcholine receptor in turning off this current. Accommodation also results from an additional potassium current that is turned off by acetylcholine. This is a calcium-sensitive potassium current known as the AHP current that is activated by the calcium influx that occurs through voltage-sensitive calcium channels activated by each spike.

This current also causes a slow hyperpolarization that slows down the firing rate and eventually stops the generation of spikes. After the current injection ends, the cell will show a long, slow afterhyperpolarization as the calcium gradually diffuses and the AHP current shuts off, resulting in the name AHP current. (It should really be called the slow AHP current, to contrast it with the fast AHP occurring after each spike). Spike frequency accommodation can also contribute to the generation of theta frequency oscillations in cortical structures (Liljenstrom and Hasselmo, 1995) and could contribute to experimental phenomena such as the reduction of activity with repeated presentation of a stimulus (Sohal and Hasselmo, 2000), which could allow recognition over a short term, as familiar items would have lower activity.

There is insufficient space for a full review of all the intrinsic properties of neurons here. There are other cellular properties of neurons that are relevant to memory function, including changes in burst properties and spiking after rebound from hyperpolarization (Izhikevich, 2007). All of these properties can be modeled based on the conductance of the membrane to different ions.

Synaptic Modification

Another set of physiological data supporting the role of the entorhinal cortex and hippocampus in episodic memory function concerns the ease of changing the strength of synaptic connections in these structures. Most models of memory function use changes in strength of synaptic connections as a mechanism for encoding of memory, and they are an essential component of the models presented here. Only a brief review will be presented of research on synaptic modification, which is reviewed more extensively elsewhere (Bliss and Collingridge, 1993; Bi and Poo, 2001; Lisman, 2009).

Even before the research on patient HM, the psychologist Donald Hebb had described how memories could be stored by physiological changes associated with spiking activity in a first neuron correlated with spiking activity in a second neuron. He proposed that these physiological changes could enhance the influence of spikes in the first neuron on the generation of spikes in the second neuron (Hebb, 1949). This basic idea is referred to as the Hebb rule, and sometimes summarized as "neurons that fire together, wire together." The impairment of memory in patient HM led researchers to take interest in potential physiological mechanisms of synaptic modification in the hippocampus.

Voltage-sensitive conductances on the axon allow transmission of action potentials along extensive distances to synapses connecting to other neurons. At the synapses, neurons communicate with other neurons by releasing transmitters from the presynaptic terminal that open channels that allow ions to flow in and out of other postsynaptic neurons, changing membrane conductance to specific ions. The membrane potential changes induced by this synaptic input are called synaptic potentials. Most presynaptic terminals in cortical structures including the entorhinal cortex and hippocampus release the neurotransmitter glutamate. Glutamate causes fast, excitatory potentials in the postsynaptic neuron due to activation of glutamate receptors that are named after the agonists that activate them (α-amino-3-hydroxy-5-methyl-4-isoxazolepropionic acid [AMPA] and *N*-methyl-D-aspartate [NMDA]). Synaptic potentials are much smaller than action potentials, only a few millivolts in size, but when they sum together they can trigger action potentials in other neurons.

Synaptic modification was first shown in the hippocampus many decades ago. Early experimental studies in anesthetized rabbits analyzed the effect of repetitive stimulation on synaptic potentials induced by stimulation of the perforant path fibers connecting entorhinal cortex layer II to the dentate gyrus (Bliss and Lømo, 1973). The researchers measured the size of synaptic potentials before repetitive stimulation and compared this with the size of the potentials after 100 stimulus pulses delivered in one second. They found a substantial increase in the size of the potential that was initially termed posttetanic potentiation. The size would decrease over time, but after many hours the potentials were still larger than before the repetitive stimulation, consistent with a potential role in long-term memory. This difference in size after a long period was termed long-term potentiation (LTP). Subsequently, the researchers used chronic implants in rabbits to analyze the size of potentials over longer periods and found that the change in size of potentials could last for weeks (Bliss and Collingridge, 1993). Though the repetitive stimulation was unrealistically strong, the fact that long-term synaptic changes could be induced by a single brief period of stimulation appeared consistent with a potential role in episodic memory.

Extensive further work has focused on the properties and mechanisms of LTP. While initial studies focused on induction of LTP with a single presynaptic stimulus train, additional studies analyzed the effect of combined stimulation of different pathways, to test whether the synaptic change satisfied the requirements of the Hebb rule. Experiments demonstrated that weak stimulation on its own does not cause a change in synaptic potentials, but weak presynaptic stimulation coupled with strong stimulation of a separate pathway activating the same set of neurons will cause LTP (McNaughton et al., 1978; Levy and Steward, 1979). The dependence of weak stimulation on separate postsynaptic activation indicates that LTP depends upon a combination of presynaptic transmitter release and postsynaptic action potential generation. This dependence on both presynaptic and postsynaptic spiking activity in certain hippocampal pathways corresponds to the Hebb rule. This was first shown in extracellular studies of the dentate gyrus (McNaughton et al., 1978; Levy and Steward, 1979). Subsequently, many studies of LTP have focused on the synapses of the Schaffer collaterals from CA3 terminating in stratum radiatum of region CA1 (Bliss and Collingridge, 1993). The Hebbian property of LTP of the Schaffer collateral synapses was shown with intracellular recording in region CA1 (Kelso et al., 1986; Wigstrom et al., 1986; Gustafsson

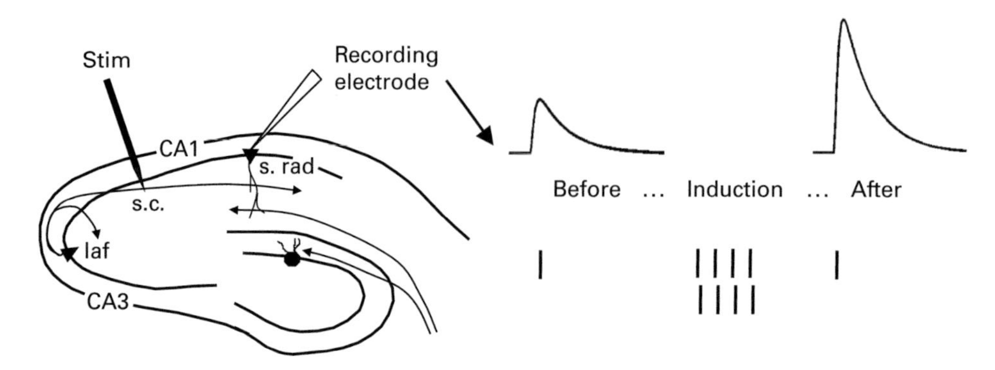

Figure 2.7
Schematic representation of the induction of Hebbian synaptic modification in the hippocampus. A presynaptic stimulation (Stim) electrode in the Schaffer collateral (s.c.) pathway provides presynaptic stimulation. A postsynaptic recording electrode in a region CA1 pyramidal cell can be used to measure the intracellular synaptic potential. Before induction of synaptic modification, a single presynaptic pulse elicits a synaptic potential of small size (Before). During induction of synaptic modification, each pulse of repetitive presynaptic activity is rapidly followed by a postsynaptic input causing a spike. After induction of synaptic modification, a single presynaptic pulse elicits a synaptic potential that has a larger rising slope and higher peak potential (After), making it more likely to elicit postsynaptic spiking activity. s. rad, stratum radiatum.

et al., 1987). A simple example of the change in synaptic strength before and after repetitive stimulation is shown in figure 2.7.

In numerous models of memory function, the formation of associations between different components of a memory depend upon the Hebb rule. Examples of these types of models are described in detail in chapter 5 and in the Appendix section on Associative Memory Function. Early models of hippocampal memory function used Hebbian modification for encoding of static associations between features of items (Marr, 1971; McNaughton and Morris, 1987; Hasselmo and Schnell, 1994; Treves and Rolls, 1994) or between items and context (Hasselmo and Wyble, 1997; Norman and O'Reilly, 2003). In models presented here, the Hebb rule is used to store associations between components of a spatiotemporal trajectory and associated items or events.

Further studies in anesthetized rats showed that LTP depended strongly on the temporal juxtaposition of presynaptic and postsynaptic activity (Levy and Steward, 1983; Holmes and Levy, 1990, 1997). In the first such study, LTP only occurred when the presynaptic spike preceded the postsynaptic spikes by less than 40 milliseconds, and this was proposed to arise from the time course of calcium concentration in the postsynaptic terminal (Levy and Steward, 1983; Holmes and Levy, 1990, 1997). More recent studies used intracellular recording to demonstrate this same time dependence of synaptic modification in cortical structures (Markram et al., 1997; Bi and Poo, 1998), and the time dependence was emphasized by using the term spike-timing-dependent (synaptic) plasticity (STDP), though this depends on many of the same mechanisms as LTP. A number of models have addressed the mechanism and role of STDP in cortical function (Gorchetchnikov et al., 2005; Rubin et al., 2005; Shouval et al., 2010).

The Hebbian properties of LTP and STDP appear to arise from the properties of one subtype of glutamate receptor, the NMDA receptor. The NMDA receptor responds to glutamate released from the presynaptic terminal but will not open unless there is a correlated depolarization of the postsynaptic membrane that pushes a magnesium ion out of the NMDA receptor channel (Bliss and Collingride, 1993). If glutamate release is combined with postsynaptic depolarization to release the magnesium block, then the NMDA channel allows influx of calcium that initiates the molecular mechanisms of LTP for strengthening synapses, which includes insertion of new AMPA receptors in the postsynaptic membrane (Lisman, 2009).

The NMDA receptor blocker 2-amino-5-phosphonovaleric acid (APV) prevents the induction of Hebbian LTP and STDP in hippocampal regions (Bliss and Collingridge, 1993). Behavioral studies support the role of NMDA-dependent, Hebbian synaptic modification in episodic memory function. Richard Morris demonstrated that infusion of APV into the ventricles adjacent to the hippocampus during encoding trials impairs the learning of a fixed location of a hidden platform in the Morris water maze (Morris et al., 1986) and strongly impairs learning of a new platform location on each day (Morris and Frey, 1997; Steele and Morris, 1999). APV also increases errors in the eight-arm radial maze (Caramanos and Shapiro, 1994). These data indicate that NMDA-dependent Hebbian synaptic modification may be necessary for encoding of spatiotemporal trajectories in episodic memory.

However, is Hebbian STDP sufficient as a cellular mechanism for episodic memory function? Many models have shown how Hebbian properties of STDP could allow the encoding and retrieval of sequences of discrete neural activity (McNaughton and Morris, 1987; Jensen and Lisman, 1996a; Tsodyks et al., 1996; Morris and Frey, 1997; Wallenstein and Hasselmo, 1997b; Koene et al., 2003). In these models, STDP can mediate chaining of associations between sequentially activated discrete populations, potentially allowing a population of neurons activated at location A to activate a population activated at location B that can then activate a population at location C. However, such a chaining mechanism is not sufficient to account for memory of sequences in humans and animals, as a number of studies have shown that recall of sequences can occur despite omissions or transpositions of individual stimuli that would prevent a mechanism based on chaining (Henson, 1998; Burgess and Hitch, 2005; Terrace, 2005; Rosenbaum et al., 2007). In addition, the very short timescale of STDP raises problems for the formation of sequential associations between behavioral items separated by intervals many times longer than the time window for STDP

(Jensen and Lisman, 1996a; Mehta et al., 2002; Koene et al., 2003; Jensen and Lisman, 2005). In addition, humans can remember different temporal intervals between events, but the chaining mechanism using STDP cannot retrieve different time intervals between events because STDP depends upon synaptic strengthening with a brief fixed time window, and the retrieval interval depends upon synaptic transmission with an even faster fixed time course.

A further problem for mechanisms dependent on STDP alone concerns the issue of an episodic representation of a continuously varying dimension, such as movement through space or the passage of time, or the expansion of a balloon or fading between different colors in a film. Continuous dimensions are difficult to represent with synaptic links between discrete neural populations. Thus, Hebbian synaptic modification is clearly important, but models based on Hebbian synaptic modification alone have difficulty encoding continuous dimensions such as time and space and have to be supplemented by mechanisms for coding changes in continuous dimensions within an episode.

Network Oscillations

The cellular properties of entorhinal cortex and hippocampus provide potential mechanisms for episodic memory, but the circuit-level processes are also important. As noted above, the connectivity of these memory-related structures intuitively suggests elegant and structured dynamical properties like those of a musical instrument. The term dynamics refers to the physical mechanics of an object or an agent moving through space, as in my example of spatiotemporal trajectories above, but the term also refers to the change in time of the state of a system, such as the oscillations of a vibrating string. To understand the episodic encoding of dynamical behavior, it is necessary to understand the neural dynamics of the change in activity within the regions encoding and retrieving episodic memories.

The hippocampal formation demonstrates salient features of oscillatory dynamics that have attracted attention for many decades. The model described here uses these oscillatory dynamics as an essential component to code the position along a spatiotemporal trajectory by the relative phase of oscillations.

Initial physiological studies of the dynamics of memory-related structures were performed with large, low-impedance electrodes outside the brain that most effectively picked up currents occurring at the same time across a wide number of neurons. If one thinks of neurons as individual people carrying on conversations, one could think of the brain as an audience of people at a sports event. Up close, each individual person can be heard, but at a distance, or behind a wall, the individual voices become a noisy hum. In this analogy, the most prominent signals are when everyone chants together in unison. The early electrodes best detected synaptic potentials occurring in unison within a region. This unified activity often appeared as oscillations in the signal recorded from individual electrodes.

During the early development of electroencephalographic (EEG) measurements from outside the skull, individual frequencies measured in humans were named with Greek letters designating specific frequency ranges. In the 1930s, the EEG pioneer Hans Berger first observed waves occurring 8–12 times per second in his resting participants and named them "alpha waves." When the participants were more actively engaged, the alpha waves were reduced and replaced by waves in the 12–30 Hertz (cycles per second) range that were termed beta waves. Subsequently, other characteristic bands were named theta (4–7 Hz), delta (below 4 Hz), and gamma (30 to 100 Hz).

Theta Rhythm Oscillations

Early electroencephalographic (EEG) recordings with electrodes inserted into the hippocampus of rabbits and cats found prominent large-amplitude oscillations around 4–6 Hz (Green and Arduini, 1954) in the local field potential which arises from synchronous synaptic interactions between populations of neurons. These oscillations were the most prominent oscillations observed in the EEG of animals and thereby attracted extensive attention over the subsequent decades. The prominence of these oscillations in the hippocampus suggests an important role in function, and in the model presented here they play an essential role in coding location in space and time. Because these oscillations fell within the 4- to 7-Hz frequency range of human "theta" rhythm, the oscillations in animals were called theta. However, the term now refers to a wider range of frequencies in animals because the same mechanisms appear to underlie hippocampal rhythms in moving rats that range from 6 to 10 Hz and in anesthetized rats that go down to 3–4 Hz (Buzsaki, 2002). Experimental data on theta rhythm oscillations in the rat hippocampus are shown in figure 2.8 (plate 2). Theta rhythm oscillations also appear in the local field potential in the entorhinal cortex (Mitchell and Ranck, 1980; Alonso and Garcia-Austt, 1987; Brandon et al., 2011b). The hippocampus and entorhinal cortex in humans are too deep to contribute to the theta rhythm recorded in a scalp EEG, but intracranial electrodes implanted to detect seizure activity have shown theta rhythm oscillations in the ventral temporal lobe in humans associated with performance of memory tasks (Kahana et al., 1999; Raghavachari et al., 2001; Rizzuto et al., 2006).

The theta rhythm appears to correlate with arousal, such as when a rabbit or rat attends to a predator (such as a cat; Sainsbury et al., 1987b) or when a cat stalks its prey (such as a rat). Theta rhythm commonly appears in rabbits and cats even when they are immobile and focusing on sensory input. In contrast, theta rhythm is strongly associated with movement in the rat. Researchers such as Vanderwolf performed studies correlating theta rhythm with specific behavioral states and argued that theta rhythm was particularly prominent in association with voluntary movement

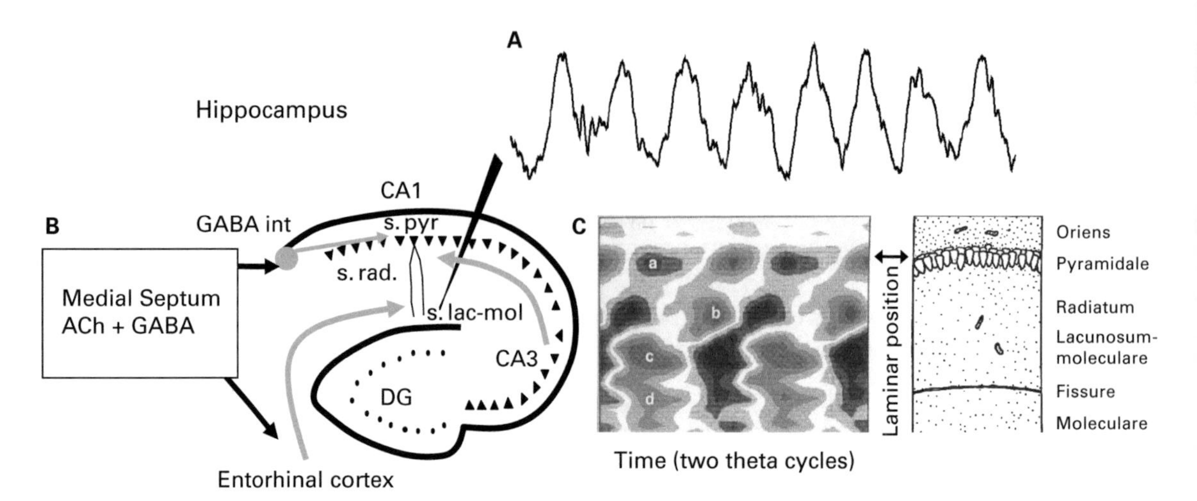

Figure 2.8 (plate 2)
Theta rhythm oscillations in the hippocampus. (A) Theta rhythm oscillations recorded in the EEG from stratum lacunosum-moleculare (s. lac-mol) of hippocampal region CA1, where oscillations are largest amplitude. (B) Oscillations are paced by cholinergic and GABAergic input from the medial septum. The GABAergic input causes rhythmic inhibition of GABAergic interneurons in region CA1, contributing to changes in current sinks in stratum pyramidale (s. pyr). (C) The current sources and sinks in different layers of region CA1 are shown over time for two cycles of theta. A current source appears in s. pyr (small a) at the same time that a current sink appears in s. lac-mol (small c). At the opposite phase a current sink (small b) appears in stratum radiatum (s. rad). (C) reprinted with permission from Brankack et al., 1993.

(Vanderwolf, 1969; Whishaw and Vanderwolf, 1973; Bland and Oddie, 2001). Theta rhythm correlates with the size of a jump in rats and with eye movements in cats (O'Keefe and Nadel, 1978). Theta rhythm appears prominently during free running (O'Keefe and Nadel, 1978; Skaggs et al., 1996) and during running in a running wheel (Buzsaki et al., 1983; Hyman et al., 2003) or on a treadmill (Fox et al., 1986; Brankack et al., 1993). This association with running is consistent with the proposed role of theta in coding velocity and location in the models presented here.

The phase of theta rhythm also appears to correlate with the phase of motor output, including the timing of sniffing (Macrides et al., 1982) and the timing of whisker movement (Semba and Komisaruk, 1984; Lerma and Garcia-Austt, 1985). Considerable data support a role of theta rhythm in the sensorimotor interface (Bland and Oddie, 2001). Theta rhythm at lower frequencies appears during immobility in rats and mice during learning of a conditioned fear response to a tone (Whishaw, 1972; Sainsbury et al., 1987a; Seidenbecher et al., 2003) and in immobile rabbits (Berry and Seager,

2001; Seager et al., 2002; Griffin et al., 2004) during both learning of conditioned eye-blink responses to an air puff and conditioned jaw movements to reward.

Evidence shows that theta rhythm in the rat increases frequency in proportion to the running speed of the rat (Whishaw and Vanderwolf, 1973; Rivas et al., 1996; Maurer et al., 2005; Jeewajee et al., 2008a) and also increases in power with speed (Sabolek et al., 2009). The relationship between running speed and frequency proves highly relevant to the firing properties of neurons in the hippocampus and entorhinal cortex as well as the behavioral mechanisms for encoding of spatiotemporal trajectories described below.

Considerable research has also focused on the correlation of theta rhythm with learning and memory (Berry and Thompson, 1978; Winson, 1978; Givens and Olton, 1990; Vertes and Kocsis, 1997; Seager et al., 2002). Lesions of the medial septum and fornix reduce hippocampal theta power in the hippocampus (Rawlins et al., 1979) and entorhinal cortex (Mitchell et al., 1982), and temporary inactivation of the medial septum also reduces theta power in the hippocampus (Mizumori et al., 1990) and entorhinal cortex (Jeffery et al., 1995). Thus, the medial septum appears to play an important role in regulating hippocampal theta rhythm, as summarized in figure 2.8B. As mentioned in chapter 1, lesions of the medial septum cause impairments in a number of memory-guided tasks, including delayed spatial alternation (Givens and Olton, 1990; Aggleton et al., 1995), delayed nonmatch to position (Markowska et al., 1989), delayed alternation of lever presses (Numan and Quaranta, 1990), spatial reversal (M'Harzi et al., 1987), the Morris water maze (Martin et al., 2007), and the eight-arm radial maze (Mitchell et al., 1982). The impairment in a spatial memory task caused by lesions of the medial septum correlates with the amount of reduction of the hippocampal theta rhythm (Winson, 1978). The impairments appear specific to recent, episodic memory as fornix lesions do not impair the initial learning of a goal location but impair the learning of reversal (M'Harzi et al., 1987), and medial septal inactivation does not impair reference memory but impairs recent episodic memory in continuous conditional discrimination (Givens and Olton, 1994).

The role of theta rhythm in learning is supported by extensive data in rabbits. The rate of learning of an eye-blink response to an air puff is faster in individual rabbits with the highest amount of theta power in the hippocampal EEG (Berry and Thompson, 1978). When delivery of the conditioned stimulus (a tone) is timed to arrive during periods of theta rhythm, the rate of conditioning to the stimulus is enhanced in both delay conditioning (Seager et al., 2002) and trace conditioning (Griffin et al., 2004).

Resetting of the phase of theta rhythm also appears to play a role in learning as discussed further in chapter 6. Theta rhythm shows phase reset during presentation of new visual stimuli for encoding in a delayed match to sample task (Givens, 1996) but not during a reference memory task, and this phase resetting allows enhanced induction of LTP (McCartney et al., 2004). Phase resetting shows specificity for item encoding versus retrieval probe phases in human memory tasks (Rizzuto et al., 2003), suggesting a role for phase reset in determining appropriate dynamics for encoding and retrieval. An early study using a runway task showed a learning-induced shift in the phase of hippocampal theta relative to entorhinal theta from hippocampus leading entorhinal cortex by 20–35 milliseconds, to the entorhinal cortex leading hippocampus by 65 milliseconds (Adey et al., 1960). As described further in chapter 6, these effects support a model of the role of theta rhythm in separating encoding and retrieval mechanisms.

Gamma and Beta Rhythm Oscillations

As described above, the first frequency bands described in the human EEG were alpha, beta, and gamma bands that are higher frequency than theta. They are more prominent in the neocortex of both rodents and humans and, therefore, more accessible for recording outside the skull. The beta and gamma frequencies in neocortex are associated with attention and arousal and are often referred to as desynchronization because they are smaller in amplitude in comparison to neocortical alpha waves and the prominent 1-to-4-Hz neocortical delta waves that are a defining feature of slow wave sleep (Buzsaki, 2006).

In animals, large-amplitude theta rhythm in the hippocampus is correlated with appearance of gamma and beta frequencies in the neocortex during arousal and attention. Traditionally, the theta rhythm in hippocampus includes the frequency band referred to as alpha in humans, but other frequency bands have been described in the local field potential data in the hippocampus. For example, there is evidence that gamma frequency oscillations occur during specific phases of the theta frequency oscillations (Bragin et al., 1995). This has been referred to as theta-nested gamma. Gamma frequencies appear in the superficial layers of the entorhinal cortex (Chrobak and Buzsaki, 1998) and have a higher gamma frequency range that appears to be coherent with higher gamma frequency in region CA1 (Colgin et al., 2009) at a specific phase of hippocampal theta when CA1 pyramidal cells are firing the most. In contrast, lower gamma frequencies in CA1 appear at earlier phases of theta and are coherent with lower gamma frequencies in region CA3 (Colgin et al., 2009). These data support the model described below for phasic interactions of region CA1 with entorhinal cortex for encoding and with region CA3 for retrieval on different phases of theta (Hasselmo et al., 2002a).

Sharp Waves and Ripples

Extensive work has also focused on the patterns of EEG activity in the absence of theta rhythm oscillations. Early researchers (Vanderwolf, 1969; Kramis et al., 1975) characterized a period of large irregular activity in the hippocampal EEG during immobility and consummatory behaviors such as eating and drinking. In this period, John O'Keefe discovered brief, large-amplitude field potential waves lasting about 50 to 100 milliseconds in stratum radiatum of CA1 referred to as sharp waves (O'Keefe and Nadel, 1978; Buzsaki et al., 1983; Buzsaki, 2002). These sharp waves are synchronized with high-frequency 300-Hz field potentials in stratum pyramidale referred to as ripples with similar duration to the corresponding sharp wave (O'Keefe and Nadel, 1978; Buzsaki et al., 1992). The sharp waves appear to arise in region CA3 and spread through region CA1 to the deep layers of the entorhinal cortex (Chrobak and Buzsaki, 1994, 1996). These sharp waves have been proposed to play a role in memory consolidation (Buzsaki, 1989) and are associated with the replay of sequences of spatiotemporal trajectories (Davidson et al., 2009) as described below.

Spiking Patterns

A series of exciting breakthroughs concerning patterns of spiking activity in the hippocampus and entorhinal cortex have elucidated potential mechanisms for episodic memory. The physiological data on spiking activity in rats and mice provide a rich source of additional support for the existence of episodic memory in animals. In particular, unit recording data indicate the encoding and retrieval of spatiotemporal trajectories. These data indicate that rat neural circuits can selectively encode the timing of spatial locations during a sequence of events within a trial and can also selectively encode and discriminate between the timing of events at different spatial locations encountered on different trials.

To describe the physiological data on episodic memory, it is first necessary to describe how the activity of neurons in the hippocampus and entorhinal cortex codes the dimensions of experience. Episodic memory is defined as occurring at a specific place. Therefore, the representation of location is essential to episodic memory. Exciting recent physiological data indicate neural mechanisms for the coding of location in episodic memory. As described above, behavioral data also indicate that episodic memories are commonly associated with a point of view that can change during the memory. For example, call up a memory of a conversation at a recent party that you have attended, and you can probably say what direction you were facing in the room and who was visible. Thus, the coding of heading direction is relevant to our ability to relive our point of view in past events. This section will describe neural activity relevant to coding of both location and direction.

Place Cells

An early breakthrough arose from development of the technology for extracellular recording of the action potentials (spikes) of single neurons in behaving rats in the late 1960s. In these recordings of spiking activity, the neurons are often referred to as "single units" to contrast them with recordings of action potentials from multiple cells. More recently, technology has allowed simultaneous recording of multiple different single units.

When single-unit recording was first developed, a number of researchers investigated the neural mechanisms of learning and memory, usually in traditional conditioning tasks. However, other researchers took a different approach of investigating neural activity during behaviors more natural to a rat. Rather than using a single task and quantifying all cells in relation to that task, John O'Keefe took the approach of recording from a single cell and quantifying the full range of possible behaviors or sensory cues that could drive that cell. This approach proved fruitful, as John O'Keefe was the first to discover place cells—neurons in the hippocampal formation that respond selectively when the rat is in a single location in the environment (O'Keefe and Dostrovsky, 1971; O'Keefe, 1976). John O'Keefe showed that individual place cells respond when a rat is in a specific location of the environment, often no larger than 20 centimeters in diameter. These cells could provide a mechanism for the representation of spatial location in episodic memory.

John O'Keefe performed a number of parametric studies. For example, he tested the dependence on place by testing the rat on a T-maze surrounded by curtains with prominent visual cues in each of four directions (O'Keefe and Conway, 1978; O'Keefe and Nadel, 1978). He showed individual place cells responding in one location (e.g., on the arm near the fan). He could then rotate the T-maze and all the cues together by a quarter circle (90 degrees). The firing field of the cell would rotate with the cues. Removing any single cue would not prevent the firing of a place cell, though removing multiple cues would sometimes cause the place cell firing field to expand. Firing would continue to be selective for the location even in darkness, indicating that no single visual cue was essential to the firing of the cell. The place cells usually would not change their firing if one portion of the maze was physically replaced with an equivalent section, thereby changing any olfactory or somatosensory cues. Many cells did not depend on the task or reward. Cells would respond in specific locations even if the rat was running away from the reward location. Thus, the cells appeared to code specific spatial locations, justifying the use of the term place cell by O'Keefe and colleagues. This physiological finding formed a central focus for a groundbreaking book by O'Keefe and Nadel on the role of the hippocampus as a cognitive map (O'Keefe and Nadel, 1978). An example of a place cell firing response is shown in figure 2.9 (plate 3). This neuron was recorded in region CA1.

As their name suggests, place cells respond on the basis of spatial location cued by multiple different types of sensory input. The firing of place cells has been modeled based on the self-organization of input to the hippocampus based on a combination of sensory cues (Sharp, 1991; Arleo and Gerstner, 2000), or based on the distance from

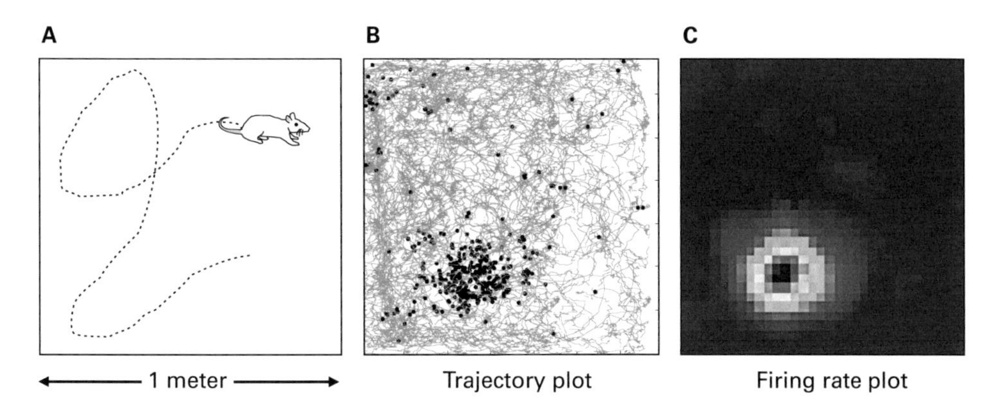

Figure 2.9 (plate 3)
Example of a place cell recorded by Mark Brandon in the Hasselmo laboratory. (A) A rat performs a foraging task in a 1-meter-square open field. (B) Lines show the trajectory of the rat during foraging, and dots show the location of the rat when a single place cell fires. Multiple cells were recorded but only a single cell is shown here. (C) Smoothed plot of firing rate in different locations (the number of spikes divided by the time spent in each location) that resembles place cells described in studies by O'Keefe (O'Keefe, 1976) and Muller (Muller et al., 1987).

environment boundaries (Barry et al., 2006). However, place cells often respond on the basis of additional dimensions of behavior relevant to episodic memory, such as the direction of movement. This was first described in a study in 1983 using recordings from rats performing the eight-arm radial maze task done by Bruce McNaughton and Carol Barnes in the John O'Keefe lab (McNaughton et al., 1983). In this task, rats run in one of two directions on the arms of the maze (outward or inward), and many hippocampal place cells showed strong directionality—responding only in one direction of running through a place field. In addition, they showed a change in firing rate with running velocity.

Perhaps the most natural behavior for a rat consists of scurrying around and foraging for bits of food. This behavior of rats and mice is a nuisance for human food storage but provides a useful behavior for scientists that was put to use by Bob Muller, John Kubie, and Jim Ranck at the State University of New York (SUNY) Health Science Center at Brooklyn. Ranck had already performed extensive recordings in hippocampus, and Muller was a classmate of O'Keefe at the Bronx High School of Science, so they knew O'Keefe's work well. Seeking to better quantify the place field activity, they developed a task that gave more uniform spatial sampling of the environment by the rat (Muller and Kubie, 1987; Muller et al., 1987). They put the rat in a 1-meter-diameter circular environment surrounded by walls and threw bits of cereal randomly into the environment. Rats took to this task and were easily induced to run around the environment picking up the pieces of food like a vacuum cleaner. This task is commonly referred to as the "Hoover" task, based on the manufacturer of vacuum cleaners. This task provided extensive sampling of spatial locations in the environment, and it allowed generation of plots showing the mean firing within individual spatial bins (taking total spiking and dividing by time spent in that location). Most plots of place cells in recent studies use this format. Spiking is usually only shown for times when the rat is moving because when the rat sits stationary and performs behaviors such as eating or drinking (or sleeping), the firing pattern becomes less specific as multiple previously inactive neurons fire during sharp wave/ripple events. This shows a prominent effect of modulatory state on the firing of neurons. As shown below, some of this activity during ripples appears to reflect the replay of previously experienced trajectories.

The work on place cells started an entire field of research focused on the neural responses of the hippocampus during spatial behavior. More detailed reviews have described the extensive data on place cell responses (Redish, 1999). However, this data raised an interesting paradox relative to previous studies on humans. Behavioral interest in the hippocampus was prompted by the striking memory impairment in patient HM, and yet the most robust electrophysiological data from the hippocampus concerned a response to specific locations that was stable across repeated visits to the same location. For the same set of cues, place cells will commonly exhibit a firing field in the same spatial location over many sessions (and even over many days).

The paradox of stable place cell firing is somewhat mitigated by the phenomenon of remapping. Place cells will fire differently in a different room or after dramatic changes in input cues (Muller and Kubie, 1987; Lever et al., 2002; Fyhn et al., 2007). Thus, they can clearly provide a separate code for memories in different locations or with different cues, but there remains the question of how they can code different episodic memories in the same location. One important study showed that performance of a different task during a different time interval could cause remapping of neuron firing within the same environment (Markus et al., 1995). In this task, the rat switched from a period of performing the Hoover task on an open tabletop to a period of performing a more structured task in the same environment in which the rat moved sequentially between four corners of a square trajectory. Many cells would completely change their firing location between tasks. Another study showed that changing the length of a linear track would shift cell firing locations dependent on whether they coded the beginning or end of the track (Gothard et al., 1996). Another solution to the paradox is suggested by the fact that place cells do not fire reliably each time the rat passes through the firing field (Fenton and Muller, 1998). This high variability could indicate the encoding or retrieval of different memories on each pass through the firing field, some of which activate the neuron and some of which do not.

Though many cells are primarily sensitive to location, hippocampal neurons do code a wide range of other cues in the environment. O'Keefe and Nadel (1978) already described displace cells that respond when an object is changed or moved. Howard Eichenbaum provided an alternative perspective to the theory of the hippocampus as a cognitive map. Research in Howard Eichenbaum's laboratory showed that during a different behavioral task individual neurons characterized as place cells during random foraging would show responses to specific cues in the environment, including different odors and the presence or expectation of reward (Wiener et al., 1989). In a conditional odor discrimination task, even neurons responding to place will become progressively more selective for the conjunction of odor and place (Komorowski et al., 2009). In addition, studies showed that hippocampal cells do not code single locations but may fire in a number of different locations in the environment (Eichenbaum et al., 1989; Fenton et al., 2008).

Context-Dependent Firing

A more recent breakthrough in the laboratory of Howard Eichenbaum in the Center for Memory and Brain at Boston University provides an important new perspective on the possible role of hippocampal neuron firing in guiding behavior in memory tasks. This finding helps resolve the paradox of stable place firing in a structure associated with episodic memory. Howard Eichenbaum proposed an alternate theory that hippocampal neurons code a memory space that includes multiple dimensions of events in an episode (Eichenbaum et al., 1999). Work by Emma Wood in the Eichenbaum lab demonstrated that firing could be strongly influenced by the context of prior or future components of a spatiotemporal trajectory (Wood et al., 2000). In their initial study, a rat was trained to run a continuous spatial alternation task on a T-maze with return arms, so that the rat was running a path with the topology of a figure-8. On the stem of the T-maze, the rat would run through the same location in the same direction, but the firing of some hippocampal cells would differ dependent upon past or future location. For example, as shown in figure 2.10, some cells fired on the stem only after the rat returned from the left side before going to the right side, and fired much less when returning from the right side before going left. Other cells showed the opposite pattern of response. Because they split their response based on prior context, these neurons are sometimes referred to as "splitter cells."

This result clearly demonstrates the capacity of hippocampal neurons to code different spatiotemporal trajectories during movement in the same direction through the same location during the same behavioral task. A subsequent study separated the retrospective and prospective components of the trajectory (Ferbinteanu and Shapiro, 2003; Shapiro et al., 2006), showing that firing in a shared part of a trajectory could depend either on the past history of the trajectory (when two different starting points converged on the same end point) or on the future trajectory (when the same starting point led to two different end points). Subsequent studies have demonstrated further context-dependent features of hippocampal neuron firing (Bower et al., 2005;

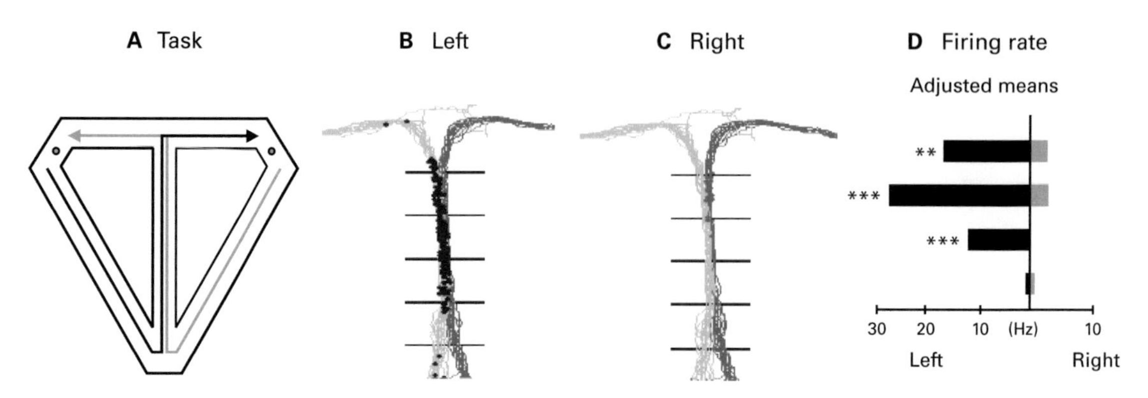

Figure 2.10
Context-dependent firing of hippocampal neurons in a continuous spatial alternation task. (A) To obtain reward, the rat must alternate between turning right and turning left. After a left turn, the rat comes down the left diagonal return arm, goes up the stem, and turns right (black trajectory). After a right turn, the rat runs down the right diagonal return arm, goes up the stem, and turns left (gray trajectory). (B) The black dots show the location of the rat when a single hippocampal cell fires after returning from a right-turn response before going left. Gray lines show rat trajectories. (C) The gray dots show firing of the same cell after returning from a left turn, showing much less activity. (D) The firing rate of the cell in different parts of the stem after left turns (black bars) and after right turns (gray bars). Figure adapted from Wood et al. (2000).

Lee et al., 2006; Smith and Mizumori, 2006b, 2006a; Griffin et al., 2007). Context-dependent firing indicating trajectory encoding has also been shown in the entorhinal cortex (Frank et al., 2000; Lipton et al., 2007; Derdikman et al., 2009). The potential role of this context-dependent firing in memory-guided behavior will be modeled and discussed at greater length in chapter 7.

Theta Phase Precession

In 1993, John O'Keefe made an important finding that strongly supports the existence of phase coding in the brain. O'Keefe noticed that the firing of hippocampal place cells would shift relative to the phase of hippocampal theta rhythm oscillations (O'Keefe and Recce, 1993). He termed this phenomenon theta phase precession. A place cell fires late in the theta cycle when a rat first enters the place field of the cell and then fires at progressively earlier phases as the rat moves through the place field (O'Keefe and Recce, 1993; Skaggs et al., 1996; Huxter et al., 2003) as shown in figure 2.11. These data are tremendously important as they provide concrete evidence for phase coding of continuous dimensions of the environment during behavior, by showing that the phase of spikes relative to network oscillations correlates with the spatial location of the animal during behavior. This phenomenon has been the focus

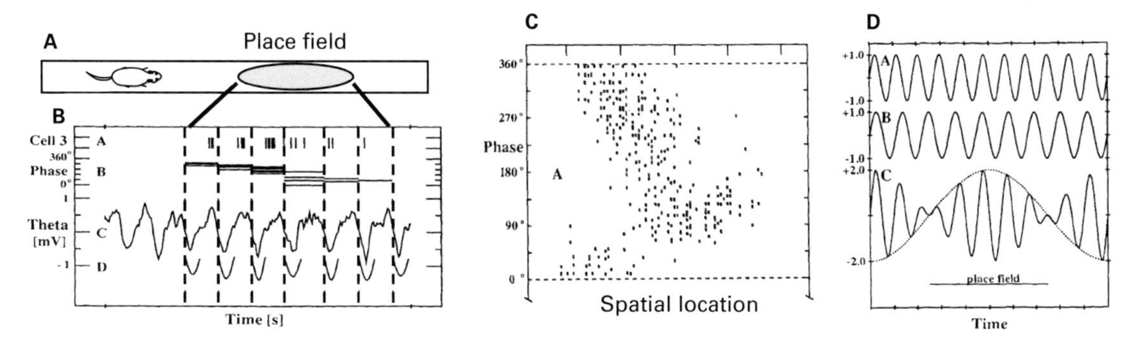

Figure 2.11
(A) Theta phase precession is commonly tested as a rat runs on a linear track. The firing rate of a single hippocampal cell increases in a single place field (gray oval). (B) As the rat crosses the place field, spike times (small A) are plotted relative to one reference phase of theta rhythm (vertical dashed lines). These dashed lines were computed from the EEG (small C) by fitting a template (small D). For each cycle, the phase of spikes are plotted from 0 to 360 (small B). Spikes start at late phases as the rat enters the field and then move to earlier phases. (C) Plotting of the phase of spikes relative to spatial location of the rat during many passes through the field shows a consistent shift from late theta phases early in the field to early theta phases as the rat exits the field. (D) The oscillatory interference model of theta phase precession proposed an interaction between a higher frequency (small A) and a lower baseline frequency (small B). Both the higher frequency and the sum of the two frequencies (small C) shift in phase relative to the baseline frequency. Figure adapted from O'Keefe and Recce (1993).

of extensive computational modeling, resulting in multiple different theories of theta phase precession that are reviewed in chapter 7. The model of episodic memory presented here uses coding of the environment by the phase of spiking relative to network oscillations.

The firing of hippocampal neurons relative to theta rhythm provides other important clues to the dynamics of episodic memory in the hippocampus. Hippocampal pyramidal cells tend to fire most frequently near the trough of the EEG measured in the pyramidal cell layer while showing less firing activity in phases near the peak (Fox et al., 1986; Skaggs and McNaughton, 1996; Csicsvari et al., 1999). Different papers use different reference phases, so it is important to take note of the laminar location of the field potential used as a reference. In anesthetized animals, the firing of interneurons appears to be maximal at the opposite phase (180 degrees) from the pyramidal cell firing (Buzsaki and Eidelberg, 1983; Fox et al., 1986), whereas in awake, behaving animals, firing of a class of interneurons precedes the peak of pyramidal cell firing by about 50 degrees (Fox et al., 1986; Skaggs et al., 1996; Csicsvari et al., 1999). This phase of interneuron firing has been simulated in detailed biophysical simulations of the hippocampal formation (Kunec et al., 2005). In this simulation, the phase of firing of one class of interneurons, the oriens-lacunosum-moleculare (O-LM) cells, might allow selective inhibition of entorhinal input to lacunosum-moleculare at a time when pyramidal cell firing would be dominated by stratum radiatum input. In contrast, during the opposite phase of theta, axo-axonic interneurons may inhibit the output of pyramidal cells (Klausberger et al., 2003; Cutsuridis and Hasselmo, 2011).

The first paper on theta phase precession proposed that it arose from oscillatory interference (O'Keefe and Recce, 1993). The model shown in the O'Keefe and Recce (1993) paper was elaborated in later papers (Lengyel et al., 2003; O'Keefe and Burgess, 2005). As shown in figure 2.11D and described below, this model proposed the interaction of a baseline oscillation (e.g., the network oscillation) and a voltage-controlled oscillator driving spiking that could be pushed to higher frequency by running speed. O'Keefe and Recce showed that a higher oscillation frequency driven by running was consistent with the intervals of cell spiking detected by autocorrelation. Autocorrelation involves testing the correlation of a sequence of spikes with the same sequence of spikes shifted by different time intervals. If the spiking tends to come at specific intervals this appears as peaks in the autocorrelation. In the data, the intervals between the peaks of the autocorrelation were shorter on average than the wavelength of field potential oscillations, that is, the frequency of cell spiking was higher than the frequency of network theta rhythm.

As shown in figure 2.11D, the difference in frequency means that the higher frequency oscillation will shift in phase relative to the baseline. If one sums the two oscillations, the summed oscillation still shifts in phase relative to the baseline. The sum shows a transition in amplitude. When the peaks of the oscillation are out of phase, they do not add together and this results in a small amplitude of the sum (destructive interference). As the peaks move closer in phase, they add together resulting in a progressive increase in amplitude of the sum due to constructive interference, followed by a decrease as they go out of phase again. The change in amplitude was proposed to underlie the change in firing rate within a place field accompanied by the shift in phase of firing. The oscillatory interference model of theta phase precession is discussed further in figures 3.6 and 7.10 and in the appendix section on Theta Phase Precession.

As shown in part D in the figure, the oscillatory interference model of theta phase precession would produce multiple firing fields as the oscillations went in and out of phase. O'Keefe and Recce saw this as a problem and stated that the oscillations should stay out of phase outside the firing field of the place cell. However, this property of the model could instead be seen as an exciting prediction of the model that was experimentally verified when grid cells with a repeating array of firing fields were described in the entorhinal cortex by the Moser laboratory as described below.

Different models of theta phase precession are discussed and compared in detail in chapter 7. An alternate model of theta phase precession uses retrieval of sequences on each cycle of the theta rhythm (Jensen and Lisman, 1996b; Tsodyks et al., 1996; Wallenstein and Hasselmo, 1997; Hasselmo, 2005a). In these models, entry to location 1 causes the readout of locations 2–3–4–5. As shown in chapter 7, if one observes the response of a single cell (coding for location 5), it will initially occur late in theta at the end of the readout sequence, and as the rat moves through the locations 2, 3, and 4, it will move to earlier phases until it is driven by sensory input at the start of the cycle. A variant of this model also performs sequence readout but uses an interaction of forward associative retrieval and context-dependent gating of retrieval (Hasselmo and Eichenbaum, 2005) to address problems with the older model. In particular, this new model can keep the input present during the full theta cycle. In contrast to the older models, this new model can account for the fact that theta phase precession appears on later runs on a linear track but does not appear as strongly on the first run of a day (Mehta et al., 2002) and that place cells show a backward shift which reoccurs on each day of testing (Mehta et al., 1997). In summary, both oscillatory interference models and sequence retrieval models have been used to account for theta phase precession, but they use different coding mechanisms that might not be mutually exclusive. In both cases, theta phase precession indicates possible mechanisms for coding of spatiotemporal trajectories in episodic memory. This topic will be discussed further in chapter 7.

Head Direction Cells

Data indicate that episodic memories often have a perspective, consisting of the point of view of the person retrieving the memory (Nigro and Neisser, 1983; Robinson and Swanson, 1993; Conway, 2009). The firing of cells near the hippocampus may provide a mechanism for coding the direction of view during a memory. While recording unit activity in structures near the hippocampus in rats, Jim Ranck noticed cells with a strong specificity for the head direction of the rat in areas such as the postsubiculum (dorsal presubiculum). He initially presented this work as a poster at the annual Society for Neuroscience meeting (Ranck, 1984) along with videos of the phenomenon. Full quantification and publication of the head direction cell responses required development of new tracking techniques and extensive additional recordings by Jeff Taube (Taube et al., 1990a, 1990b).

Head direction cells respond with strong firing when the rat points its head in a particular direction, and they commonly decrease to no firing for directions more than 45 degrees from the preferred direction. The firing does not depend on the angle of specific sensory cues in the environment, as it occurs for the same head direction angle everywhere in an environment, as if the cell possessed a magnetic compass and

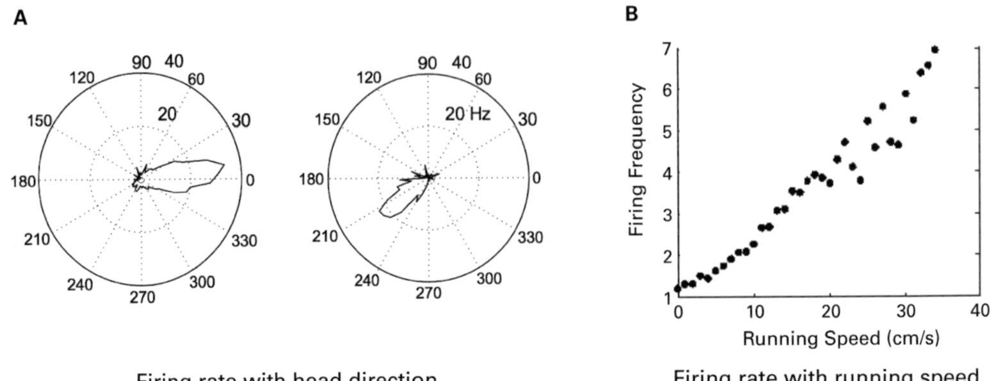

Figure 2.12
(A) Polar plots showing the firing rate for different angles (0 to 360) of two different head direction cells in the postsubiculum recorded by Mark Brandon. These cells increase their firing over a small portion of the full 360-degree range of angles (Brandon et al., 2011). Middle dotted circle is 20 Hz, outer circle is 40 Hz. (B) Plot showing relationship of firing rate to running speed for a single neuron recorded in entorhinal cortex (Brandon et al., 2011b). Some cells show sensitivity to both speed and head direction.

responded when the needle pointed in a specific direction. The preferred direction can differ dramatically between different rooms, so they do not track actual compass direction, but within a single room they are commonly described with compass directions (e.g., a cell may be described as responding when a rat is pointing in the northwest direction). As shown in figure 2.12, the firing drops off relatively rapidly in a linear manner if the rat turns from the preferred direction, which is plotted relative to the 360 degrees of angle within a full circle. The width of tuning is relevant as it appears that most head direction cells in the postsubiculum are not tuned to the cosine of heading (which would be expected from a pure velocity signal) but instead have a sharper, triangular function that only covers about 90 degrees of head directions.

The head direction system appears to be an independent convergent system that brings spatial information to the entorhinal cortex where it can be combined with information about events and items in the environment. Work by Taube and others has traced how the head direction signal propagates through many anatomical stages from the vestibular nuclei to the postsubiculum (Taube and Bassett, 2003). Head direction cells appear in the brain stem, the lateral mammillary nucleus (Blair et al., 1998; Stackman and Taube, 1998), the anterior thalamus (Taube, 1995; Stackman and Taube, 1997), the postsubiculum (dorsal presubiculum; Taube et al., 1990b; Johnson et al., 2005; Brandon et al., 2011), and the medial entorhinal cortex (Sargolini et al., 2006). Models show how input from neurons coding angular velocity can update the activity of head direction cells (Skaggs et al., 1995; Zhang, 1996). Head direction firing in the anterior thalamus is blocked by lesions of vestibular input (Stackman and Taube, 1997) and depends on voluntary movement (Taube, 1995), indicating a direct role of sensory update. In the postsubiculum the vestibular signal appears to be linked to head direction cues from visual stimuli as postsubiculum head direction cells will shift their preferred angle when the angle of visual cue cards is shifted and will persist in their firing specificity even during passive movement (Goodridge and Taube, 1997; Taube and Bassett, 2003).

These data indicate that the head direction signal could provide a self-motion signal based on sensory input to update spatial representations in the hippocampus and entorhinal cortex. Note that one defining feature of the medial entorhinal cortex in rats mentioned above is the input from postsubiculum that targets medial entorhinal cortex but not lateral entorhinal cortex, indicating the selective nature of the head direction input. The important role of both head direction input and theta rhythmic input for episodic memory is highlighted by the phenomenon of diencephalic amnesia, in which anterograde amnesia for episodic memories is caused by damage to structures including the anterior thalamus and the mammillary nuclei (Vann and Aggleton, 2004). As noted in chapter 1, a pair of bizarre accidents resulted in damage to the mammillary nuclei in humans, one involving a miniature fencing foil through the right nostril, and the other involving a pool cue going through the left nostril. Both of these accidents as well as tumor damage to the mammillary nuclei caused robust impairments in the delayed recall of verbal and nonverbal information (Vann and Aggleton, 2004). Further evidence supporting diencephalic amnesia comes from the phenomenon of Korsakoff's amnesia, in which insufficient dietary thiamine causes damage to both the mammillary bodies and the anterior thalamus. These structures contain both nuclei with neurons that code head direction and nuclei that participate in generation of theta rhythm oscillations (Vertes and Kocsis, 1997). Damage to these systems appears to cause impairments in episodic memory (Vann and Aggleton, 2004).

The model of episodic memory described later makes extensive use of head direction cells to code velocity. This requires that the direction signal be combined with a speed signal. The speed signal could arise from cells that fire at different rates with different running speeds as shown in the data from Mark Brandon in figure 2.12. An early model proposed by O'Keefe and Nadel (1978) describes how place cells could be generated by input dependent on speed (as obtained from changes in theta frequency with running speed) combined with a signal for direction. Head direction input was also proposed to provide input for path integration mechanisms for activating place cells in the dark (Samsonovich and McNaughton, 1997). These proposals foresaw the important relationship between head direction cells and place cells that was made clearer by the discovery of grid cells.

Grid Cells

Recently, another influential breakthrough took place in the laboratory of Edvard and May-Britt Moser in Trondheim, Norway. The discovery of grid cells in their laboratory binds together the previous work on place cells and head direction cells. As with many great discoveries it seems surprising that grid cells were not discovered earlier (some researchers came close). Motivated by their earlier work showing a role for direct entorhinal-hippocampal inputs in place cell firing (Brun et al., 2002), the Mosers started recording in the entorhinal cortex to understand the nature of the input driving place cells. They consulted Menno Witter about the region sending input to the dorsal hippocampus, where most recordings of place cells had been performed. He advised them that the dorsal region of the medial entorhinal cortex is the region that sends projections to the dorsal hippocampus and helped them to target electrodes at this region. This area also receives input from the postsubiculum, where head direction cells were initially found. The Mosers used the techniques developed by the O'Keefe and Muller laboratories to record the action potentials of multiple single neurons as a rat runs around actively in an environment. Their initial studies used a 1-meter-square environment surrounded by high walls. They threw little bits of crushed up chocolate cereal into this environment to make the rat run around actively foraging for the cereal. If the rat does not become satiated, then it covers most of the environment, allowing effective mapping of the firing of the cell across a wide range of spatial locations.

The Moser's initial data from this area showed multiple firing fields within an environment (Fyhn et al., 2004). Closer inspection of the pattern of firing fields revealed that they showed a relatively regular pattern of distribution in the environment. This distribution is commonly described as falling on the vertices of tightly packed equilateral triangles (Hafting et al., 2005). An alternate description is that the pattern resembles a series of overlapping hexagons.

An example of grid cell firing recorded by my graduate student Mark Brandon is shown in figure 2.13 (plate 4). The square represents the high walls surrounding an open field in which the rat forages for food. The thin black line shows the trajectory of the rat tracked by a video camera as the rat runs around and forages for food. The dots show the location of the rat each time a single recorded grid cell generated a spike. As can be seen in the plot, the dots cluster together in multiple different firing fields within the environment, showing that the cell fires selectively in an array of locations laid out in a grid pattern.

The regular spatial periodicity is remarkably robust as shown in a series of studies in the Moser laboratory. The first spikes fired by a cell during the first seconds of exploration of a novel environment along a single winding trajectory will usually match with the location of grid cell firing fields later defined on the basis of a 20-minute-long crisscrossing trajectory that fills the space of the environment (Hafting

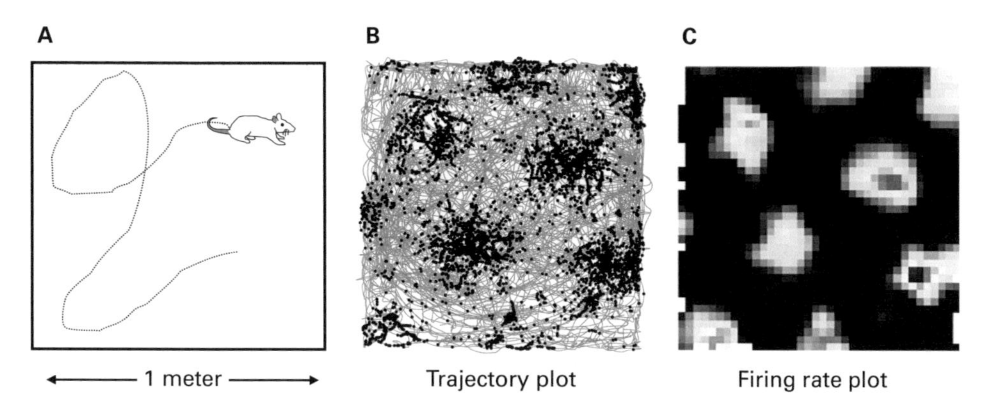

Figure 2.13 (plate 4)
Grid cell firing. (A) Top view schematic of rat foraging in 1 meter square open field with walls. (B) Black line shows trajectory of rat while foraging. Dots show the location of the rat when spikes are generated by a single grid cell. The grid cell fires in multiple firing fields laid out in a hexagonal pattern. (C) Smoothed plot of mean firing rate across all locations for the same grid cell as B, with warm colors showing high firing rate in multiple fields. Data recorded by Mark Brandon in Hasselmo lab (Brandon et al., 2011b) similar to grid cell recordings in Moser laboratory (Hafting et al., 2005; Moser and Moser, 2008).

et al., 2005; Moser and Moser, 2008). Similar to place cells and head direction cells, if the room lights are turned off, the cell will continue to fire in the correct locations for many minutes, indicating that the self-motion cues of the rat (or other nonvisual modalities) are sufficient to maintain a sense of spatial location. Also similar to place cells and head direction cells, the grid cells will rotate with prominent visual cues such as rotation of a white cue card on a black wall. However, the firing does not appear restricted to a familiar space. If the rat explores a large novel open environment, the grid fields appear to tile the environment in a predictable manner anywhere the rat moves, suggesting that grid cell firing corresponds to the rat's own sense of location in the environment.

Grid Cell Spacing

Consistent with their location in the entorhinal cortex, providing input to the hippocampus, the grid cells have properties that would allow them to drive the activity of place cells. Adjacent grid cells tend to have the same overall orientation in the environment, that is, their firing fields line up along lines of particular heading (Hafting et al., 2005; Moser and Moser, 2008). However, adjacent cells show a range of different spatial phases that appear as a spatial offset between the centers of their firing fields. Data show a wide range of spatial offsets that provide coverage of the environment so that any location will be associated with the firing of some grid cells.

However, this signal in itself would be ambiguous. The firing of a single grid cell would only indicate that the rat was in one of the many firing fields of that grid cell.

However, greater spatial specificity can be provided by the fact that grid cells fire with different spatial scales. The progression of spatial scales in medial entorhinal cortex is part of what inspires my previous analogy with the string section of an orchestra progressing from violins to violas to cellos and basses. At the dorsal end of medial entorhinal cortex, near the border with postrhinal cortex, grid cell firing fields are relatively close to each other, with spacing of about 35 to 40 centimeters between firing fields (Hafting et al., 2005; Sargolini et al., 2006) as shown in figure 2.14 (plate 5). As the rat runs through the environment, the firing will transition from firing field to quiescence to firing field in a rapid manner, like the faster melodies played by

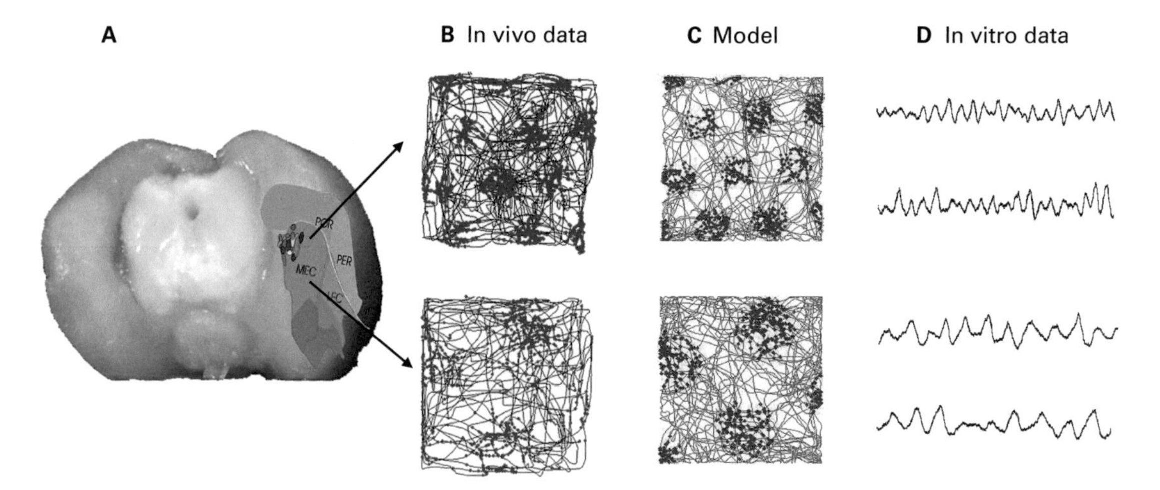

Figure 2.14 (plate 5)
(A) Location of cells recorded in dorsal and ventral medial entorhinal cortex (MEC). Reprinted from Sargolini et al., 2006 with permission from AAAS and E. and M.-B. Moser and M. Witter. (B) In vivo recording shows the trajectory of a foraging rat as gray lines. Dots show the location of the rat when a grid cell fires. The spacing between firing fields is smaller for the grid cell recorded in dorsal medial entorhinal cortex (top) compared to larger spacing for the more ventral (bottom) grid cell (reprinted with permission from Hafting et al., 2005). (C) Simulation (from Hasselmo et al., 2007) of the oscillatory interference model of grid cells (Burgess et al., 2005; Burgess et al., 2007; Burgess, 2008) can replicate the pattern of grid cell firing at different dorsal (top) and ventral (bottom) positions and predicted differences in intrinsic oscillation frequency. (D) In vitro data from intracellular recording in slices supported the prediction by showing higher frequency subthreshold membrane potential oscillations in single stellate cells in dorsal entorhinal cortex (top) versus lower frequency oscillations in more ventral stellate cells (bottom) (Giocomo et al., 2007; Giocomo and Hasselmo, 2008a, 2008b, 2009). Lateral entorhinal cortex, LEC; postrhinal cortex, POR; perirhinal cortex, PER. Reprinted with permission from AAAS.

violins. If the electrode is moved more ventral within medial entorhinal cortex, recording of a more ventral cell reveals a wider spacing between larger firing fields. At 1.5 millimeters more ventral from the postrhinal border, the grid cells have spacing between fields of about 80 centimeters. More recent recordings at more ventral anatomical positions within medial entorhinal cortex show even larger spacing and larger firing fields. A cell in more ventral entorhinal cortex might respond everywhere in a 1-meter environment in one room and respond nowhere in the same size environment in a different room. The Moser lab set up an 18-meter-long linear track through the hallways of their laboratory and found spacings of up to 10 meters between the firing fields of grid cells (Brun et al., 2008). The slow transition between large firing fields is like the slow rhythmic part in an orchestra played by the bass section.

The same 18-meter track was used to show that the size of firing fields for place cells also change their spatial scale, ranging from a size of 40–80 centimeters in the dorsal (septal) end of the hippocampus to firing fields up to 10 meters long in the ventral (temporal) end of the hippocampus (Kjelstrup et al., 2008). This could provide differential scaling for different aspects of memory (Hasselmo, 2008a), as in the analogy of the car keys, where you need to remember both the room where you left the keys (large scale) and the furniture where you left your keys (small scale). The progressive dorsal–ventral difference in spatial periodicity of grid cells, combined with the range of spatial phases at each dorsal–ventral position, means that the population of grid cells along the full dorsal–ventral extent of medial entorhinal cortex could provide accurate information about the location of the rat (Burak and Fiete, 2006; McNaughton et al., 2006; Solstad et al., 2006; Gorchetchnikov and Grossberg, 2007; Fiete et al., 2008; Welinder et al., 2008; Burak and Fiete, 2009; Savelli and Knierim, 2010).

Oscillatory Interference Model of Grid Cells

The spatial scaling of grid cells led to an important prediction of the oscillatory interference model (O'Keefe and Burgess, 2005). As described above, the oscillatory interference model of theta phase precession already predicted multiple firing fields due to constructive interference (O'Keefe and Recce, 1993) and thereby predicted the existence of grid cells. However, the multiple firing fields in the model were seen as a problem until the discovery of grid cells. John O'Keefe tells me that he attended a journal club meeting where Caswell Barry presented the first grid cell paper from the Moser laboratory, and when he heard of the properties of grid cells, he and Neil Burgess both realized that the grid cells could be accounted for with the repeating firing fields of the oscillatory interference model.

Patterns of oscillatory interference provide a simple and effective model of grid cells as described in the next chapter. This simple model coupled with the data on differences in spacing led to an explicit model prediction. In their 2005 paper, O'Keefe and

Burgess stated, "The increasing spatial scale of the grid-like firing as you move from the postrhinal border of the medial entorhinal cortex would result from a gradually decreasing intrinsic frequency..." Explicit quantitative predictions of this sort are the norm in physics but are relatively rare in neuroscience.

Scaling of Intrinsic Frequency and Spacing

I first saw the oscillatory interference model of grid cells in a poster presented by Neil Burgess at the Computational Cognitive Neuroscience meeting in November 2005 (Burgess et al., 2005). I'm glad I saw his poster that day, as it led to the most exciting experimental finding of my career. The model was immediately appealing to me, and along with Lisa Giocomo in my laboratory, I set out to test this prediction of the model. We also tested other physiological parameters that could contribute to the difference in spacing of grid cell firing fields along the dorsal to ventral axis of medial entorhinal cortex.

This required a very simple change in procedures in my laboratory. Surprisingly, most previous slice physiology of structures such as the hippocampus and entorhinal cortex did not keep track of anatomical position within the structure. The structure would be sliced and the slices would be kept together in one holding chamber. To keep track of position, Lisa simply constructed separate holding chambers to keep the slices separated based on their position along the dorsal to ventral axis of medial entorhinal cortex.

Lisa recorded from slices at different distances from the postrhinal border, and she found a decrease (figure 2.14) in intrinsic frequency of neurons at greater distances from the postrhinal border (Giocomo et al., 2007). She showed this difference with measurements of the subthreshold membrane potential oscillation frequency of these neurons measured near the firing threshold of the cells. She also showed the difference in the resonance frequency measured as the peak of amplitude in response to an oscillating input current that increases from zero to 20 Hz (Giocomo et al., 2007). These data form a major motivation for the orchestra analogy that I presented above. The data are analogous to saying that the pitch of the stringed instruments decreases as you move from the left wall of the orchestra stage. In this analogy, Lisa was measuring the pitch of each neuron based on its distance from the dorsal end of medial entorhinal cortex and found a decrease in frequency with distance (figure 2.14D).

As shown in figure 2.14, we used the oscillatory interference model of grid cells (Burgess et al., 2005; Burgess et al., 2007; Burgess, 2008) to immediately show how the in vitro data allowed us to simulate the change in spacing of grid cells along the dorsal to ventral axis (Hasselmo et al., 2007). The model allowed us to show a potential direct link between the intracellular recording data on intrinsic frequency and the unit recording data on the spacing of grid cell firing fields. The model is described in more detail in chapter 3 and the equations of the model are presented in the appendix.

Subsequent studies have addressed the mechanism of the difference in oscillatory frequency, showing a change in the time constant of the h current along the dorsal to ventral axis of medial entorhinal cortex (Giocomo and Hasselmo, 2008b). In biophysical simulations described in the appendix, the difference in oscillation frequency can be simulated with differences in the time constant of the h current, or with changes in magnitude of h current or input resistance. The role of the h current was further supported by recordings in mice with genetic manipulations that remove (knockout) the HCN1 subunit of the H current channel and flatten the gradient of frequency differences along the dorsal to ventral axis of medial entorhinal cortex (Giocomo and Hasselmo, 2009). The dorsal to ventral differences in oscillation and resonance frequency in medial entorhinal cortex have recently been replicated by another group (Boehlen et al., 2010). These differences in oscillation frequency form an important basis for the models of grid cells and episodic memory presented here.

The physiological data also show clear differences in the resonance properties of the entorhinal stellate cells at different positions along the dorsal to ventral axis (Giocomo et al., 2007), as shown in figure 2.15. This is consistent with the difference

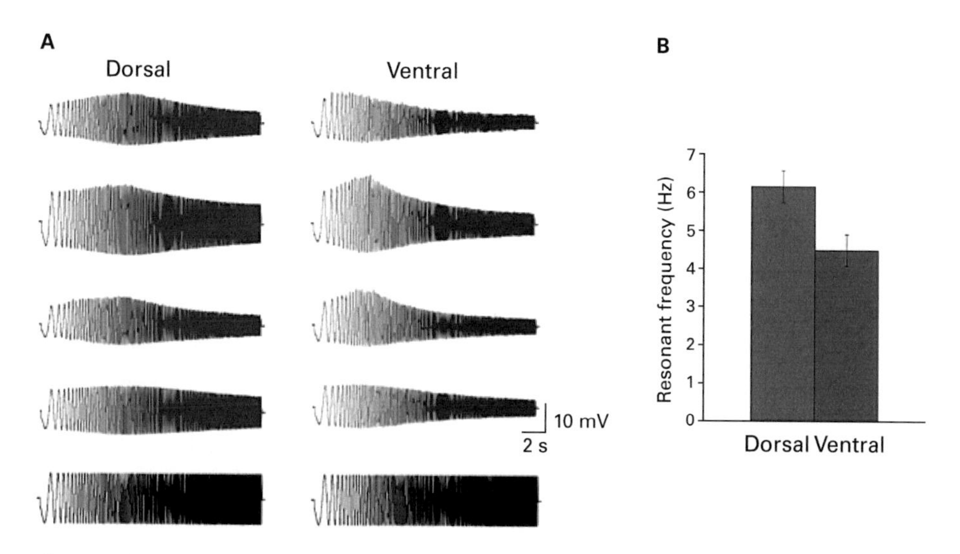

Figure 2.15
(A) Membrane potential responses showing differences in resonance properties of neurons in the dorsal (left) versus ventral (right) medial entorhinal cortex. Neurons are given oscillatory input that starts at low frequencies and increases. Dorsal cells show a peak membrane potential oscillation at higher frequencies (further to the right) compared to ventral cells. (B) Histogram showing the average peak response of dorsal cells is at a higher frequency than the peak response in ventral cells (Giocomo et al., 2007). Reprinted with permission from AAAS.

in membrane potential oscillations arising from a difference in the properties of voltage-gated properties such as the h current. Consistent with this, stellate cells also show differences in the time course of the sag response to hyperpolarization (Giocomo et al., 2007) and in the temporal integration of synaptic input (Garden et al., 2008). As described in the Appendix, biophysical models of the membrane potential oscillations show that they are damped oscillations around a stable focus (Izhikevich, 2007). They require noise to appear in a sustained manner (White et al., 1998a) with intermittent clusters of spiking activity (Alonso and Llinas, 1989). In contrast, the resonance properties of the neuron are easier to measure because they appear away from the spiking threshold and can be measured in a deterministic manner (Giocomo et al., 2007; Heys et al., 2010). The frequency of membrane potential oscillations correlates with the frequency of resonance (Erchova et al., 2004). This resembles the properties of a musical instrument that will resonate at preferred frequencies. Extension of these data suggests the intriguing possibility that anatomical differences in intrinsic frequencies in other structures such as the medial septum, prefrontal cortex, and piriform cortex could underlie differences in the scale of coding for different behaviors (Hasselmo, 2008a).

Theta Phase Precession of Grid Cells

Because the oscillatory interference model of grid cells grew out of the model of theta phase precession, it predicted that theta phase precession should appear in the firing fields of grid cells. Theta phase precession is difficult to observe in two-dimensional environments, so separate studies in the Moser laboratory explored the response of grid cells on a linear track. These data showed the phenomenon of theta phase precession in the firing fields of individual grid cells (Hafting et al., 2008). Consistent with the model, theta phase precession proceeds at different rates in cells with different scales with a slower change in phase relative to position in more ventral cells with larger firing fields. This provides further physiological support for the model of grid cells described further in chapter 3 and in the appendix. Data also indicates that the spacing of grid cell firing fields appears to show discrete quantal increases along the dorsal to ventral axis (Barry et al., 2007; Stensland et al., 2010). This could arise from quantal relationships between integer values of oscillation cycles during interference.

Context Dependence of Grid Cells

Grid cells show consistent location of firing fields within a single unchanged environment, even when they are placed back into the environment at a different time. However, they do respond to changes in sensory cues. When a rat is moved between identical enclosures in two different rooms, hippocampal place cells will remap, and grid cells will show an associated shift in both orientation and spatial phase (Fyhn et al., 2007). When a rat is in the same room with different shaped environments, place cells remap, but grid fields shift only in spatial phase, not orientation. In contrast, when a rat is kept in the same physical location in a single room, but the enclosure walls are changed in color, then the place cells will show a change in firing rate, but grid cells will maintain their same firing pattern, supporting a consistency of spatial coding. However, in a different experiment, insertion of a series of walls into the open field making a hairpin maze causes the grid cells to fire in a different pattern proportional to the distance from turning locations in the maze (Derdikman et al., 2009). This effect is discussed further in chapter 7.

Effect of Environment Size on Grid Cells and Place Cells

Both grid cells and place cells show a systematic change in firing. When the size of the environment is changed by moving the walls, there is a systematic, predictable change in firing of both place cells (O'Keefe and Burgess, 1996) and grid cells (Barry et al., 2007). For example, after recording place cells in a 1-meter-square open field, the east wall could be shifted east to expand this square into a 1- by 2-meter rectangular open field. In this case, a place cell responding near the north wall would often stretch out along the expanded wall (O'Keefe and Burgess, 1996), and grid cells would expand their spacing in the direction of expansion (Barry et al., 2007). This phenomenon was elegantly modeled by Neil Burgess and his colleagues using what they termed boundary vector cells (Barry et al., 2006), which were cells predicted by the model to respond on the basis of distance from one boundary of the environment. In a remarkable experimental confirmation of a modeling prediction, these boundary vector cells were recently confirmed to exist in recordings from the subiculum (Lever et al., 2009) and the medial entorhinal cortex (Solstad et al., 2008). An example of experimental data showing a boundary vector cell in medial entorhinal cortex is shown in figure 2.16 (plate 6).

Unit Recording in Monkeys

Extensive studies have also addressed unit responses during memory tasks in nonhuman primates. Many of these focused on recording during the delayed match and delayed nonmatch to sample tasks used in lesion studies. Unit recording in the temporal lobe and prefrontal lobe show spiking responses to the sample stimulus that persist during the delay period of the task (Fuster, 1995). Spiking activity during the delay period appears in temporal lobe regions including the inferotemporal cortex (Miller et al., 1996) and the entorhinal cortex (Suzuki et al., 1997) but is more sensitive to distractor stimuli than in prefrontal cortex (Miller et al., 1996). More recent studies have analyzed neural activity during learning of new associations between scenes and eye movement responses, showing a correlation of hippocampal unit activity with learning of the association (Wirth et al., 2003) and with trial outcome (Wirth

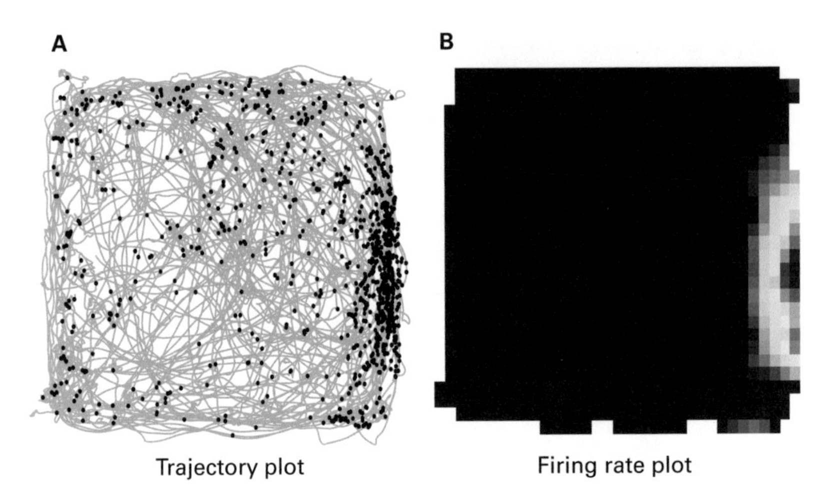

Figure 2.16 (plate 6)
(A) Experimental data supporting the existence of boundary vector cells. Lines show the foraging trajectory of the rat. Dots show the location of the rat when a single border cell fires along the east border of the environment. (B) Firing rate plot showing higher rates near the east border. Entorhinal data from Brandon et al., 2011b in the Hasselmo lab similar to data on border cells (Solstad et al., 2008) and boundary vector cells (Lever et al., 2009). Reprinted with permission from AAAS.

et al., 2009). Recordings in monkeys are usually performed with the monkey stationary and viewing a screen, so place cells are difficult to test, but neuronal responses have been found in monkey hippocampus that appear to depend upon viewing of specific locations in the environment rather than the location of the animal in the environment, called spatial view cells (Rolls and O'Mara, 1995; Rolls et al., 1997) that contrast with head direction cells found in monkey presubiculum (Robertson et al., 1999).

Episodic Replay

Recent recordings of action potentials in the human hippocampal formation have shown dramatic evidence for spiking activity directly associated with memory retrieval. A number of studies have been performed in epileptic patients using depth electrodes implanted for detection of the focus of their seizure activity. These studies have demonstrated spiking activity that appears to directly correlate with retrieval of prior memories. For example, some neurons in the hippocampus show highly specific firing in response to categories of sensory stimuli (Quiroga et al., 2005). A remarkably specific cell appeared to respond selectively to pictures of a famous actress and other cells show selectivity for other stimuli. Further work has shown that individual neurons fire selectively during free recall of stimuli (Gelbard-Sagiv et al., 2008). For example, a neuron that fired while a patient viewed an episode of *The Simpsons* television show also fired in the absence of that stimulus when the patient reported remembering scenes from *The Simpsons* television show.

The analysis of neural activity correlated with previous experience has been performed more extensively in animals, in which the neural firing activity can be selectively linked to specific dimensions in the environment such as spatial location. Despite the early argument by Tulving that animals do not have episodic memory, neurophysiological data show evidence of episodic retrieval in the form of the replay of spiking activity in the hippocampus (Skaggs and McNaughton, 1996; Foster and Wilson, 2006; Diba and Buzsaki, 2007; Johnson and Redish, 2007; Ji and Wilson, 2007; Davidson et al., 2009; Karlsson and Frank, 2009). This research draws on the previous discovery of place cells because the replay or retrieval of a memory can only be detected if the information being coded by the firing of individual neurons can be effectively reconstructed. This work allows direct demonstration of the spatiotemporal trajectory coded by the spiking activity during replay. The firing of place cells is predictable enough that it is possible to predict the location of a rat using the simultaneous firing of a group of place cells. This uses mathematical techniques for predicting location on the basis of the pattern of firing across a large population of place cells (the array of firing rates of different place cells can be described as a place cell vector). This process becomes even more accurate when the rat moves repeatedly along a more restricted, one-dimensional track through the environment. After a period of movement, when the rat is not moving, the sequential reactivation of the same cells can be analyzed to statistically demonstrate the sequential retrieval of a trajectory through the environment.

Episodic replay of neural activity has been observed in region CA3 of the hippocampus during performance of a tone-cued alternation task (Johnson and Redish, 2007). In this task, rats hear a tone that indicates the appropriate direction of response at a later choice point. Hippocampal neurons show selective spatial firing as place cells in different locations in the task, allowing statistical determination of the primary location coded by each neuron. For example, consider neurons A, B, C, and D that code locations A, B, C, and D along the path into one reward arm. At early stages of learning the task, the rat is more hesitant at the choice point and turns between different possible response directions at the choice point (as if exploring possible choices in a phenomenon termed vicarious trial-and-error). During these periods, the spiking activity of a population of neurons shows sequential temporal reactivation of neurons coding spatial locations along individual trajectories to the left or right (Johnson and Redish, 2007). That is, cells coding locations A, B, C, and D would read out in order as the rat sat at the choice point and turned its head toward this segment of the track. This indicates the retrieval of this encoded spatiotemporal trajectory and indicates the precise temporal distinction of sequential places visited on one trajectory, as well as indicating the separation of trajectories encountered at longer temporal intervals (the left vs. right trajectories).

A large percentage of hippocampal neurons show spiking activity during sharp wave/ripple events in the EEG, and several studies have shown replay during these events. The sharp wave events usually occur during periods of time when the animal is stationary. Some studies analyzed the spiking activity of place cells in region CA1 that fire sequentially as a rat runs back and forth between reward locations at each end of a one-dimensional track (Foster and Wilson, 2006; Diba and Buzsaki, 2007; Davidson et al., 2009). During sharp wave/ripple events occurring when the rat is stationary, hippocampal place cells show forward or reversed replay of the sequence of hippocampal place cells that spiked during runs along the one-dimensional track, further indicating the selective spatiotemporal retrieval of encoded trajectories, and the temporal separation of distinct episodes. The replay can even start at locations away from the current location (Davidson et al., 2009; Karlsson and Frank, 2009). In one study where another section of track was visible but not yet visited, preplay was observed for sequences of place cells at locations not yet visited by the rat (Dragoi and Tonegawa, 2011) but later found to be coded by the cells showing preplay.

Further support for episodic memory in rats comes from work on replay of episodes during sleep. Early studies showed reactivation in region CA1 of previously experienced neural ensembles during sharp wave/ripple events in slow wave sleep (Pavlides and Winson, 1989; Wilson and McNaughton, 1994; Skaggs and McNaughton, 1996). Later studies showed that this activity maintains the spatiotemporal structure of experienced episodes. Hippocampal place cells sequentially activated during waking on a linear track appear to fire with the same sequential relationship during the ripple events in slow wave sleep (Skaggs and McNaughton, 1996; Nadasdy et al., 1999). This replay might be episodic or could be based on a representation created over multiple learning experiences. The replay of spiking includes neurons in areas outside of the hippocampus such as primary visual cortex (Ji and Wilson, 2007). In contrast, cells selectively coding head direction in postsubiculum do not show increased spiking activity or replay during sharp wave/ripples (Brandon et al., 2011a).

Usually, replay of sequences is much faster than the sequence of spiking during behavior. However, hippocampal spiking activity during rapid-eye-movement (REM) sleep shows replay with a timescale similar to the spiking during waking behavior (Louie and Wilson, 2001) as simulated in figure 2.17 (Hasselmo, 2008c). This REM sleep replay is temporally structured, with time intervals of spiking activity as well as theta rhythmicity that correspond to the time intervals that the rat spent in particular portions of the behavioral task (Louie and Wilson, 2001). This suggests that neural circuits in the rat may not only encode the order of events but the time interval of events in an episode.

In summary, this chapter described the anatomical and physiological data on the neural circuits of the entorhinal cortex and hippocampus that may support episodic memory function on multiple different levels. The intrinsic cellular properties of

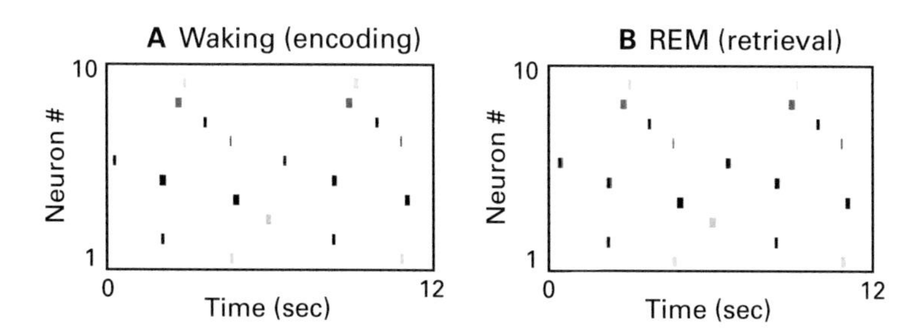

Figure 2.17
Simulation of replay during rapid-eye-movement (REM) sleep. (A) During waking, a set of 10 out of 400 place cells are activated at specific positions during two runs around a circular track. (B) During REM sleep, replay of the same spatiotemporal trajectory activates the same sequence of place cells at the same intervals as during running around the circular track. Figure adapted from Hasselmo (2008c).

neurons may contribute to the response of grid cells in the medial entorhinal cortex, place cells in the hippocampus, and head direction cells in associated structures including the postsubiculum. These neurons may contribute to the encoding and retrieval of episodic memories, as supported by the evidence that place cells show replay of previously experienced sequences. Chapters 3 and 4 will use the physiological data to show different mechanisms of how the intrinsic cellular properties of entorhinal neurons may underlie grid cell firing based on input from cells coding head direction and speed, how grid cells may contribute to place cell firing, and how different circuit interactions of these neurons could contribute to the encoding and replay of episodic memories.

3 Coding of Space and Time for Episodic Memory

How do neurons code space and time along a spatiotemporal trajectory in episodic memory? In the previous chapter, I provided an analogy between neurons and musical instruments. I described how neurons in the entorhinal cortex show persistent spiking at stable frequencies and show systematic anatomical differences in oscillation and resonance frequencies. How could the rhythm and pitch of neurons play a role in episodic memory?

In this chapter, I will describe how the rhythmic activity of populations of neurons could provide a mechanism for encoding changes in continuous dimensions such as time, space, and sensory features and for episodic retrieval of these changes in continuous dimensions. I will present a model using a phase code. In this phase code, sensory experience along continuous dimensions is represented by the firing time of a neuron relative to each cycle of a baseline oscillation. This firing time can change in a continuous manner to represent position in space or time. As an alternative, many features of the episodic memory model could be implemented using a rate code, in which dimensions are coded based on continuous changes in the firing rate of a neuron. However, a phase code has the advantage that a continuous dimension can be coded by single spikes occurring at specific times relative to a baseline cycle. The information is available with each spike rather than requiring multiple spikes over time. The phase code described here has the potential for providing a single mechanism that can code both space and time.

Persistent Spiking Model

To understand this model, it is essential to understand how the entorhinal cortex could provide a phase code for the location and time of events in an episodic memory. There are a number of physiological mechanisms that could underlie such a phase code. I will focus first on my persistent spiking model of grid cells (Hasselmo, 2008b) that was inspired by the oscillatory interference model of grid cells (Burgess et al., 2007). After describing this model, I will then focus on the oscillatory interference model itself. There are a number of different ways that the oscillatory interference model can be implemented. For example, another variant of the oscillatory interference model uses interference between populations of neurons that change their rhythmic spiking activity (Zilli and Hasselmo, 2010).

The persistent spiking model of grid cells uses the cellular capacity of entorhinal neurons to maintain spiking at a stable level for an extended period. As described in the previous chapter, many entorhinal neurons in slice preparations do not shut off after being given a strong but transient depolarizing current (Klink and Alonso, 1997a; Egorov et al., 2002; Tahvildari et al., 2007; Yoshida et al., 2008; Yoshida and Hasselmo, 2009a). After the current injection, they continue to generate persistent spiking at a steady baseline firing rate. Delivery of a small additional depolarizing current injection to such a neuron can temporarily increase the firing rate to a higher baseline frequency. This could shift the relative phase of spiking of one neuron relative to other neurons.

Phase Coding of Spatial Location

Many students initially have difficulty thinking in terms of the relative phase of neuron firing. An example might help. Imagine that a persistent spiking neuron is one of those wind-up toy monkeys that sits and plays the cymbals. As it sits still, the monkey clangs its cymbals at a steady rate, similar to the steady generation of spikes by a persistent spiking cell. Many of these toy monkeys clang their cymbals about once per second (1 Hz). This is slow compared to the persistent spiking cells that fire between 4 and 12 Hz, but imagine this as a neuron in slow motion.

Now imagine a separate toy monkey on a unicycle. (I searched for a toy of this sort but could not find one, so it will have to remain imaginary.) Imagine that this toy monkey clangs its cymbals when stationary at the same rate as the sitting monkey, but when it is moved along on its unicycle, the cymbal clanging increases its rate slightly in exact proportion to the speed that the unicycle wheel turns forward. When the unicycle wheel stops turning, the clanging returns to its baseline rate. This analogy represents a persistent spiking cell that changes its frequency of spiking dependent on the running speed of a rat as it moves. This can be described as a speed-controlled oscillator (the frequency of the rhythmic spiking changes with speed).

Imagine the sitting monkey and unicycle monkey start out next to each other, clanging in perfect unison (with zero phase difference). As shown in figure 3.1 , the time of each clang (each spike) lines up early in time (on the left). Now imagine that you move the unicycle monkey forward by 20 centimeters for 2 seconds with a constant speed of 10 centimeters per second. The figure shows this speed change as an abrupt increase to a higher level and then a decrease. As you see in the figure, the movement causes the clanging to speed up relative to the other monkey (from 1 Hz to 1.25 Hz), and this makes the clanging go out of phase with the sitting monkey.

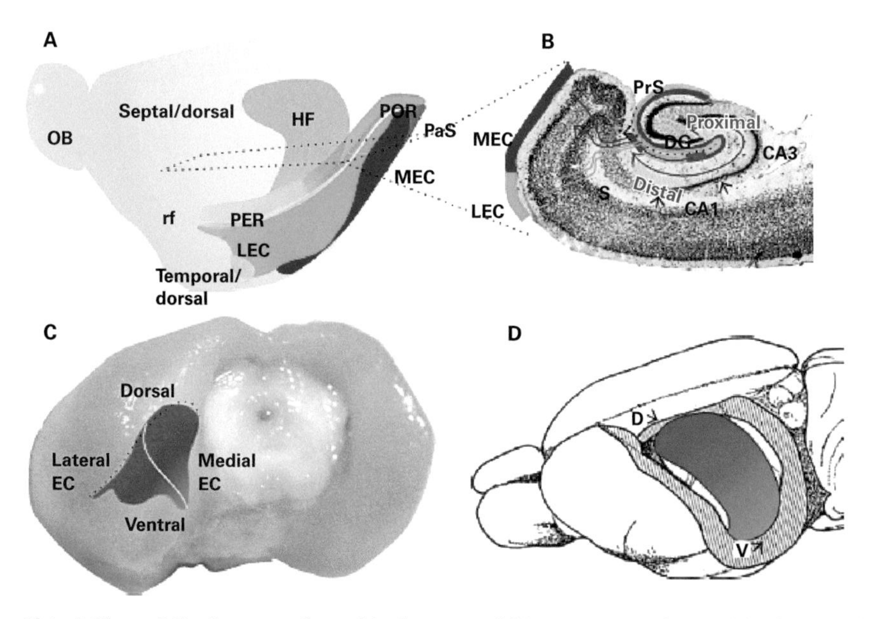

Plate 1 (figure 2.1) Anatomy of entorhinal cortex and hippocampus in the rat. (A) Side view of left hemisphere of rat brain showing location of the rhinal fissure (rf) and the lateral entorhinal cortex (LEC) and medial entorhinal cortex (MEC) receiving input from perirhinal (PER) and postrhinal (POR) cortex and parasubiculum (PaS). Underneath these structures is the hippocampal formation (HF). To the left is the olfactory bulb (OB). Diagram of rat brain reprinted with permission from Menno Witter (Canto et al., 2008). (B) Dotted lines in A show location of a horizontal cross-section showing projections from MEC and LEC into the dentate gyrus (DG) and hippocampal subregions CA1 and CA3 (cornu ammonis = CA). Cross section also shows location of presubiculum (PrS) and subiculum (S). (C) Picture of rat brain from behind. Red to blue shading shows the dorsal to ventral dimension of the lateral and medial entorhinal cortex (EC). (D) The same shading shows the areas of hippocampus receiving input from dorsal (D) and ventral (V) entorhinal cortex and sending back reciprocal connections.

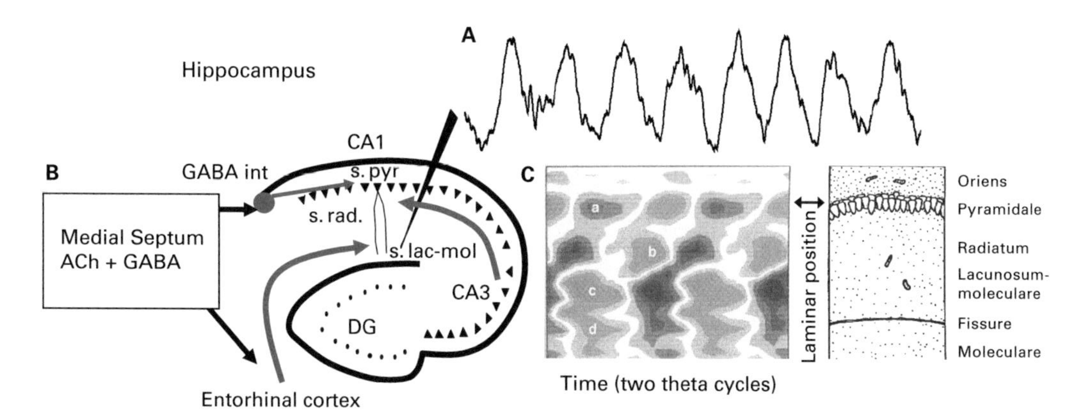

Plate 2 (figure 2.8) Theta rhythm oscillations in the hippocampus. (A) Theta rhythm oscillations recorded in the EEG from stratum lacunosum-moleculare (s. lac-mol) of hippocampal region CA1, where oscillations are largest amplitude. (B) Oscillations are paced by cholinergic and GABAergic input from the medial septum. The GABAergic input causes rhythmic inhibition of GABAergic interneurons in region CA1, contributing to changes in current sinks in stratum pyramidale (s. pyr). (C) The current sources and sinks in different layers of region CA1 are shown over time for two cycles of theta. A current source appears in s. pyr (small a) at the same time that a current sink appears in s. lac-mol (small c). At the opposite phase a current sink (small b) appears in stratum radiatum (s. rad). (C) reprinted with permission from Brancack et al., 1993.

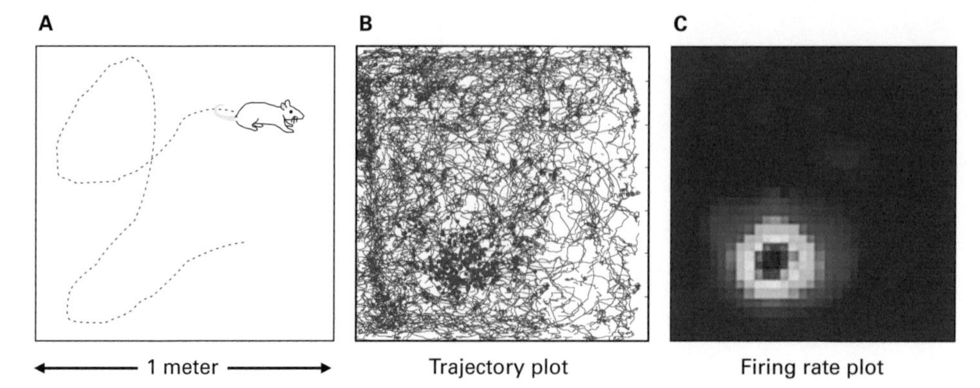

Plate 3 (figure 2.9) Example of a place cell recorded by Mark Brandon in the Hasselmo laboratory. (A) A rat performs a foraging task in a 1-meter-square open field. (B) Lines show the trajectory of the rat during foraging, and dots show the location of the rat when a single place cell fires. Multiple cells were recorded but only a single cell is shown here. (C) Smoothed plot of firing rate in different locations (the number of spikes divided by the time spent in each location) that resembles place cells described in studies by O'Keefe (O'Keefe, 1976) and Muller (Muller et al., 1987).

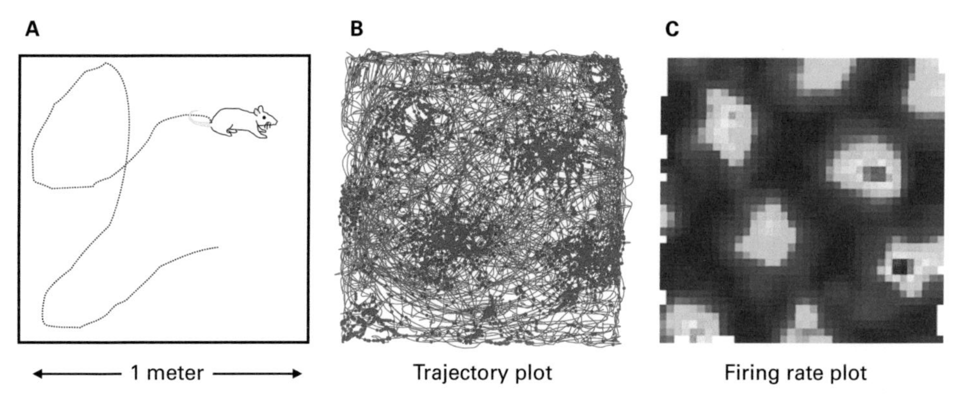

Plate 4 (figure 2.13) Grid cell firing. (A) Top view schematic of rat foraging in 1-meter-square open field with walls. (B) Gray line shows trajectory of rat while foraging. Dots show the location of the rat when spikes are generated by a single grid cell. The grid cell fires in multiple firing fields laid out in a hexagonal pattern. (C) Smoothed plot of mean firing rate across all locations for the same grid cell as B, with warm colors showing high firing rate in multiple fields. Data recorded by Mark Brandon in Hasselmo lab and reprinted from Brandon et al., 2011b with permission from AAAS, showing properties similar to grid cell recordings in the Moser lab (Hafting et al., 2005; Moser and Moser, 2008).

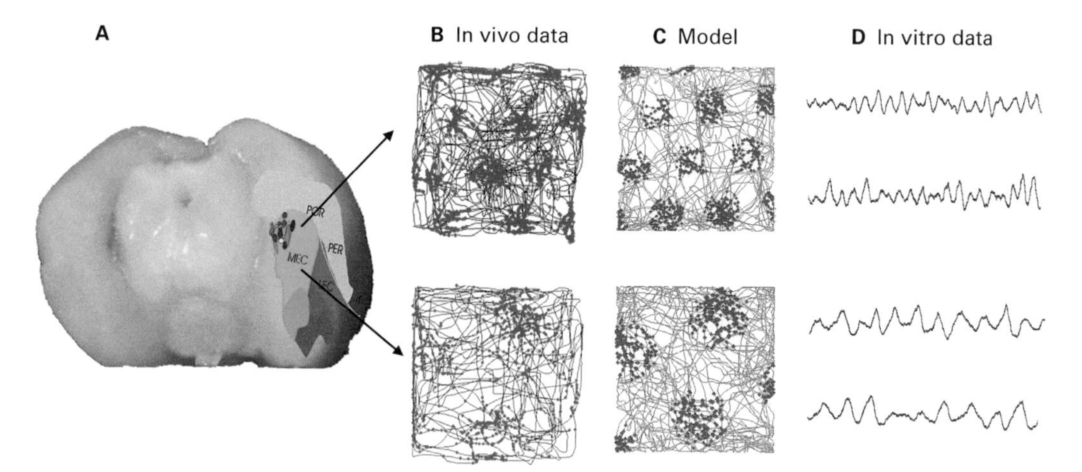

Plate 5 (figure 2.14) (A) Location of cells recorded in dorsal and ventral medial entorhinal cortex (MEC). Reprinted from Sargolini et al., 2006 with permission from AAAS and E. and M.-B. Moser and M. Witter. (B) In vivo recording shows the trajectory of a foraging rat as gray lines. Dots show the location of the rat when a grid cell fires. The spacing between firing fields is smaller for the grid cell recorded in dorsal medial entorhinal cortex (top) compared to larger spacing for the more ventral (bottom) grid cell (reprinted with permission from Hafting et al., 2005). (C) Simulation (from Hasselmo et al., 2007) of the oscillatory interference model of grid cells (Burgess et al., 2005; Burgess et al., 2007; Burgess, 2008) can replicate the pattern of grid cell firing at different dorsal (top) and ventral (bottom) positions and predicted differences in intrinsic oscillation frequency. (D) In vitro data from intracellular recording in slices supported the prediction by showing higher frequency subthreshold membrane potential oscillations in single stellate cells in dorsal entorhinal cortex (top) versus lower frequency oscillations in more ventral stellate cells (bottom) (Giocomo et al., 2007; Giocomo and Hasselmo, 2008a, 2008b, 2009). Lateral entorhinal cortex, LEC; postrhinal cortex, POR; perirhinal cortex, PER. Reprinted with permission from AAAS.

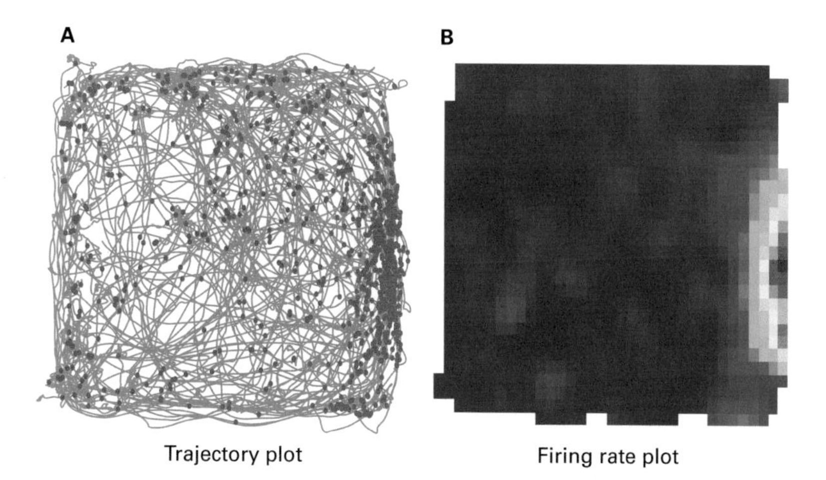

Plate 6 (figure 2.16) (A) Experimental data supporting the existence of boundary vector cells. Lines show the foraging trajectory of the rat. Dots show the location of the rat when a single border cell fires along the east border of the environment. (B) Firing rate plot showing higher rates near the east border. Data recorded in medial entorhinal cortex by Mark Brandon in the Hasselmo lab (Brandon et al., 2011b) similar to data on border cells (Solstad et al., 2008) and boundary vector cells (Lever et al., 2009). Reprinted with permission from AAAS.

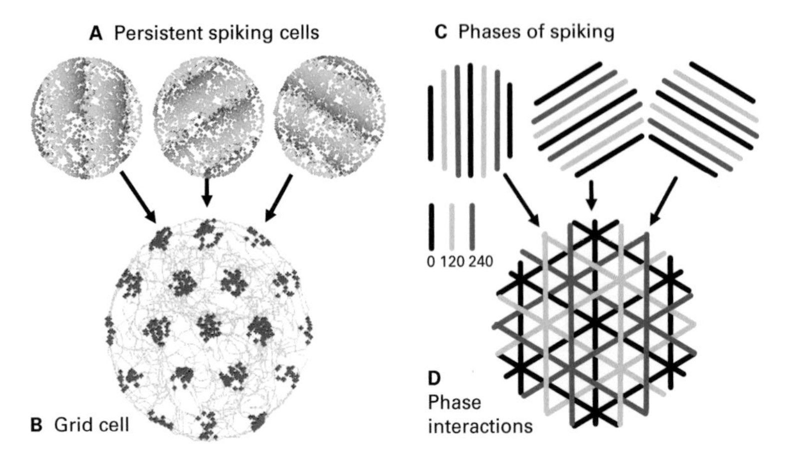

Plate 7 (figure 3.10) (A) Three plots showing rat location during spiking of each of three persistent spiking cells that fire everywhere in the environment. Each dot represents a spike. The shade of the dot represents the phase of persistent spiking relative to baseline (black = 0, light gray = 120, dark gray = 240). (B) Firing of a single grid cell receiving input from all three persistent spiking cells. The persistent spiking cells drive the grid cell to fire when all three persistent spiking cells have a similar phase resulting in a typical grid cell firing pattern (Hasselmo, 2008b). The thin line indicates the rat trajectory from experimental data. (C) Lines showing the distribution of phases at different locations. (D) Overlay of lines shows that grid cell firing occurs when three lines of the same phase (color) cross.

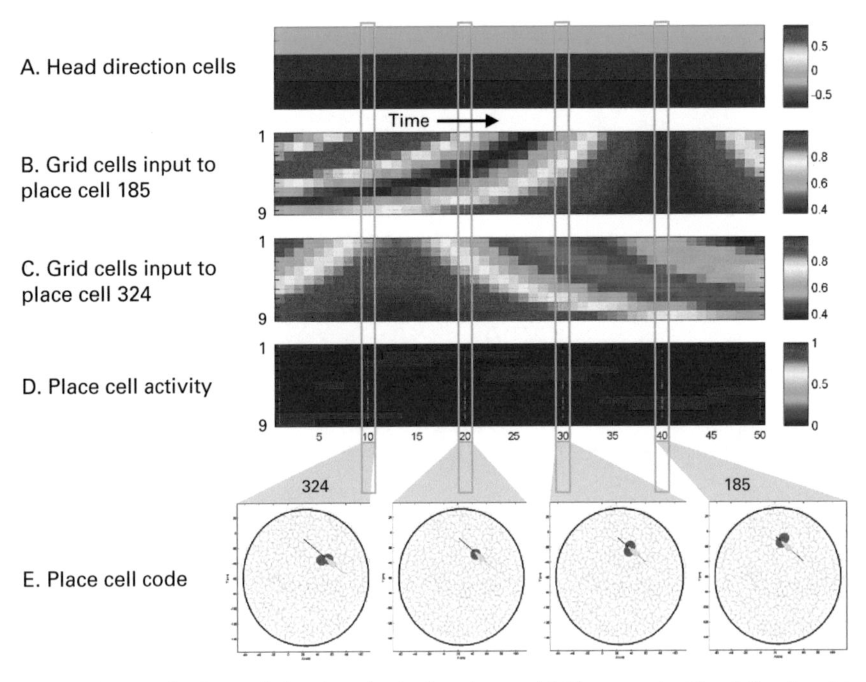

Plate 8 (figure 7.1) Forward planning of a single trajectory. (A) The current rat head direction drives three different head direction cells with static levels of activity. The three head direction cells drive different frequencies in the grid cell model (B and C). (B) The graph shows the activity over time of one group of nine grid cells that drives place cell 185 when all grid cells are simultaneously active (red) late in the trajectory. (C) Another group of nine grid cells (C) drives place cell 324 when all grid cells are active (red) early in the trajectory. Different grid cells transition between constructive and destructive interference at different time points. (D) Input from the grid cells sequentially activates different place cells (red bars) that code sequential locations (filled circles) along a straight trajectory shown in E (figure by Murat Erdem, Erdem and Hasselmo, 2010).

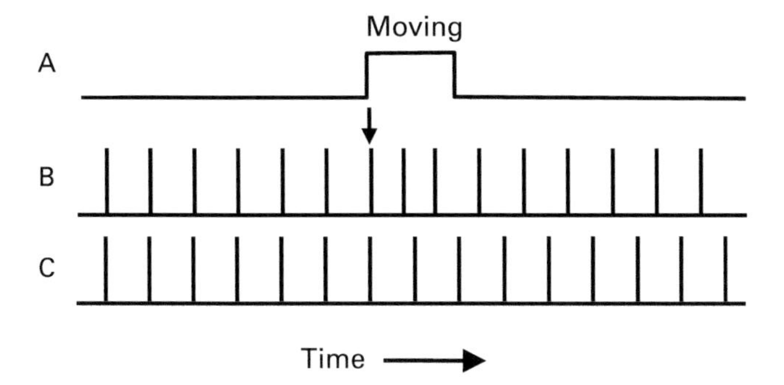

Figure 3.1
Example of shift in phase with a short duration increase in frequency. (A) Speed of movement of the unicycle monkey, which is zero at the start and end and increases to a constant speed of 10 cm per second for 2 seconds during movement. (B) Time of clanging of the unicycle monkey marked by vertical lines. (C) Time of clanging of the stationary monkey at 1 Hz. During movement, the rate of B speeds up to 1.25 Hz and goes out of phase with C, then maintains this phase difference after stopping.

After you stop moving the unicycle monkey, its clanging rate goes back to being the same rate as the sitting monkey (1 Hz). However, in this new position, even though the rate of clanging is the same, the cymbal clashes are out of phase. As shown in figure 3.1 above, the unicycle monkey clangs at an earlier phase relative to the cycle of the sitting monkey. In fact, because the shift in frequency was proportional to the speed of movement, the difference in phase is proportional to the distance you moved the unicycle monkey (i.e., 20 centimeters). If you put a red mark at the top of the unicycle wheel at the start point, then the difference in phase is proportional to the angle that the red mark has turned with the wheel. This illustrates how the difference in timing and the difference in location can both be represented by a phase angle. If the clanging comes at exactly the middle of the time between clangs of the sitting monkey, this would be a 180-degree phase difference (shown in the figure). Imagine that the radius r of the wheel is very large (about 6.37 centimeters), then after moving 20 centimeters (π multiplied by r), the wheel has moved half a turn (180 degrees).

Move the unicycle monkey forward again at a constant speed of 10 centimeters per second, and its clanging rate speeds up again (to 1.25 Hz), and the phase of the clanging shifts earlier and earlier relative to the stationary monkey. If you stop at exactly 40 centimeters ($2\pi r$) from the sitting monkey, the clanging is at exactly the same time as the sitting monkey, and they clang happily in unison, as shown in figure 3.2. This could be considered a 360-degree phase difference, which is equivalent to the zero-degree phase difference they showed at the start.

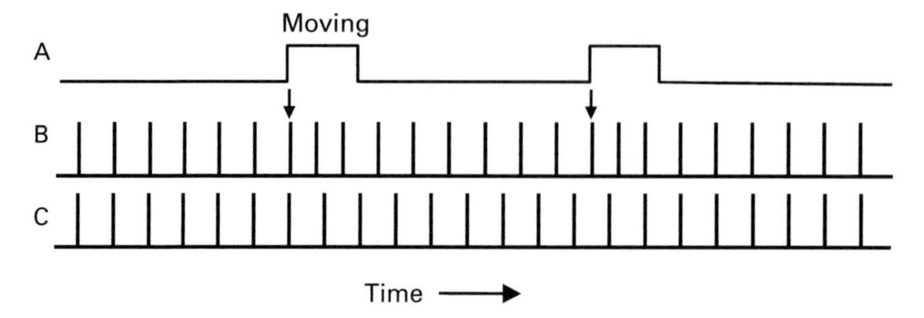

Figure 3.2
Example showing the initial movement of 20 cm causing a 180-degree phase difference, followed by a second movement of 20 cm adding another 180-degree phase shift so the phase difference is 360 degrees, which is equivalent to zero degrees.

If you could measure the time interval between the clangs of the unicycle monkey and the sitting monkey, you would find that the time interval between the cymbal clangs of the sitting monkey and the unicycle monkey is proportional to the distance between the two monkeys. Thus, the relative phase of the clangs provides a phase code for spatial location. The term relative refers to the fact that comparison with the phase of a second rhythm is essential. Without the sitting monkey or some other comparison rhythm you cannot code location. This example shows how the persistent spiking model could provide phase coding of spatial location in the form of shifts in relative phase as a rat moves around an environment.

This example shows how relative phase can code a continuous dimension. But how can one read out the relative phase? Figure 3.3A shows how the relative phase can be read out in a binary manner by listening for when the clanging occurs at the same time. This is analogous to the spikes of two neurons causing synaptic potentials in a third cell at the same time that sum together and cross threshold to cause a spike in the third cell. When the spikes occur at the same time, then they sum together to cause spiking. When the spikes are out of phase with each other, then they do not sum together in the third cell and do not cause spiking. The output from a single pair of persistent spiking cells will cause spiking in a third cell that repeats at regular intervals, simulating one dimension of grid cell firing as seen in figure 3.4 below. The persistent spiking cells that change frequency with speed are not grid cells, but they can drive the activity of a cell that spikes as a grid cell. This does not give an exact measure of location, but as described below when combined with output from multiple different grid cells, this can be used to give an accurate readout of position.

What if the unicycle monkey moves at a different speed? Remember that the clanging changes in proportion to the speed of the monkey, so slower movement will cause a smaller increase in clanging rate (frequency) and a slower shift in phase. This will ensure that the relative phase still accurately indicates position, as shown in figure

3.3. The regulation of spiking frequency based on translational speed is neurophysiologically realistic, as neurons responding to translational speed have been shown in the postsubiculum (Sharp, 1996) in the mammillary bodies (Sharp and Turner-Williams, 2005; Sharp et al., 2006) in the entorhinal cortex (Brandon et al., 2011c) and in axons traversing the hippocampus (O'Keefe et al., 1998). An example of a speed neuron recorded by Mark Brandon was shown in the previous chapter and is included in figure 3.3.

This theoretical framework could use rhythmic activity of many sorts. The two monkeys could be two different persistent spiking cells (Hasselmo, 2008b), or each monkey could be a population of neurons spiking in unison (Zilli and Hasselmo, 2010), or the unicycle monkey could be a cellular oscillation and the sitting monkey could be theta rhythm oscillations in the local field potential (Burgess et al., 2007). It is common to mathematically describe rhythmic activity using a cosine function of time. The Appendix shows how the rhythmic activity in this type of model can be described by a cosine function of time.

The examples shown here illustrate a situation where the rat changes from being stationary to moving at a constant velocity to being stationary again. If the rat continues moving at a steady speed, then the cells will go in and out of phase repeatedly (figure 3.4A). As shown in the figure, interactions of single persistent spiking cells will often yield only a single spike by the grid cell in each firing field. However, if the persistent spiking cells cause postsynaptic potentials that have a slower time course, or if there are multiple persistent spiking cells generating spikes at slightly different times (phases), this can generate spiking in a larger grid cell firing field (figure 3.4B). In figure 3.3, firing depends on the quantal ratio of frequency.

Phase Coding and Theta Phase Precession

The strongest evidence for the phase coding of spatial location comes from the neural data on theta phase precession, which shows a direct relationship between spatial location and the relative phase of spiking of entorhinal neurons.

Imagine lifting the unicycle monkey (so that the unicycle wheel does not turn) and placing it –20 centimeters on the other side of the sitting monkey. The clanging is still in unison. Now move the unicycle forward toward the sitting monkey at a speed of 4 cm per sec, and then another 20 centimeters past the sitting monkey. As you moved the unicycle monkey, you would hear the unicycle monkey clangs gradually shift away from the clangs of the sitting monkey, shifting earlier and earlier until they are again in unison at +20 centimeters. As the phases start to shift, you can think of the unicycle monkey clangs as earlier in phase (e.g., minus 36 degrees relative to the clangs of the sitting monkey) or as much later in phase (e.g., 324 degrees later than the preceding clang). If you only plot the positive values of phase between 0 and 360 degrees, then as you move the unicycle monkey, the firing shifts from late phases (350 degrees) to earlier phases as shown in figure 3.5.

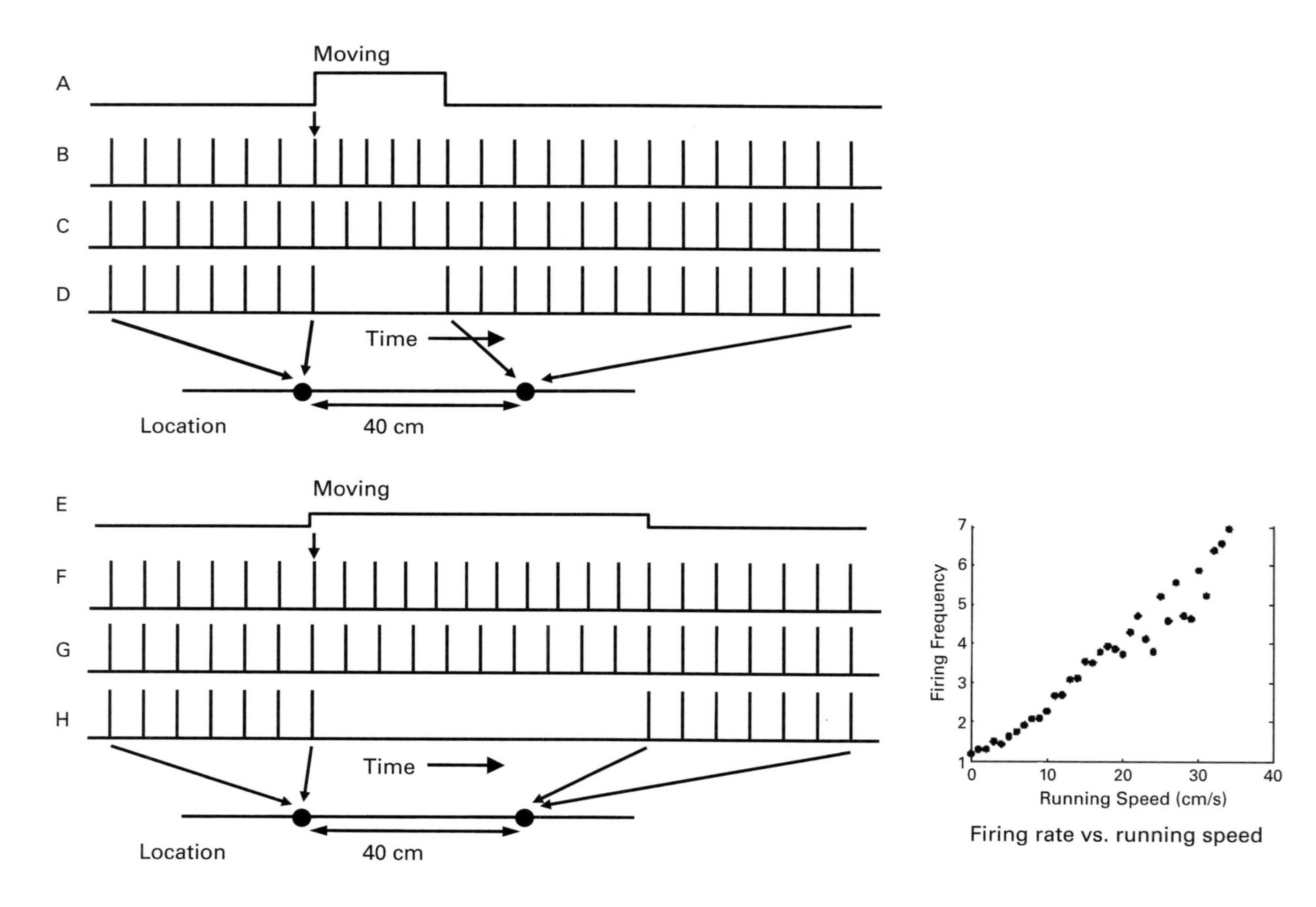
A
B
C
D
Moving
Time
Location
40 cm
E
F
G
H
Moving
Time
Location
40 cm
Firing Frequency
Running Speed (cm/s)
Firing rate vs. running speed

Figure 3.3
Top: (A) Rapid movement (10 cm/sec) causes a frequency change in B and rapidly shifts the phase relative to C. B fires 5 spikes at 1.25 Hz in the 4 sec period that C fires 4 spikes at 1 Hz. D shows spiking when phases are in synchrony when the monkeys are stationary at the first location (dot). Spiking in D stops when the neurons go out of phase during movement but then starts again when they come back into phase during movement. They are in phase when the monkey stops at the second location. Bottom: (E) Slower movement (4 cm/sec) causes a smaller frequency change in F (from 1 Hz to 1.1 Hz) and a slower change in phase relative to G. Because the frequency change is smaller, H turns off for a longer time because the phase shift is slower. The phases come back into synchrony at the same spatial location because the frequency change was proportional to speed. (Right) Experimental data showing change in firing rate of a cell that depends on running speed of a rat (Brandon et al., 2011c).

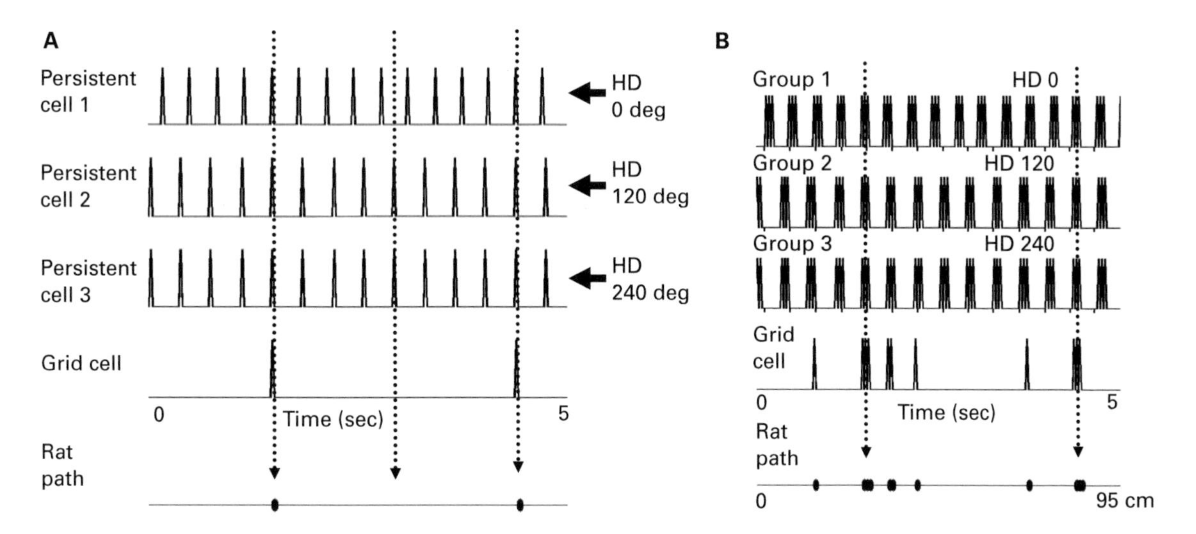

Figure 3.4
Persistent spiking model of grid cells. (A) This example shows the phase shift of three persistent spiking cells getting input from three different cosine tuned head direction cells with three different preferred directions (0, 120 degrees, and 240 degrees). As the rat moves in an eastward direction (0 degrees), the cell driven by head direction zero (HD 0 deg) shifts from being in phase with the others to being out of phase and then back in phase with the others (Hasselmo, 2008b). When they are all in phase together, they drive the spiking of a Grid cell. (B) Three different groups of persistent spiking neurons with the same baseline frequency get driven to different phases by input from head direction (HD) cells with 0-degree preferred angle for group 1, 120-degree angle for group 2, and 240-degree angle for group 3. Grid cell firing arises from the convergent spiking of the three groups of persistent firing neurons. When all three persistent firing groups fire in synchrony, the grid cell will fire (dots on rat path).

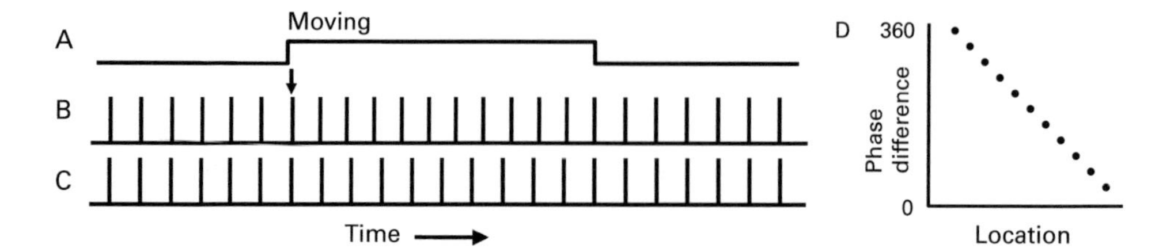

Figure 3.5
Simple model of theta phase precession. (A) The unicycle monkey starts at a position 20 cm behind the stationary monkey. Then it moves at a constant slow speed until it is 20 cm past the stationary monkey. (B) During movement, the time of clanging by the unicycle monkey steadily shifts in phase relative to the time of clanging by the stationary monkey shown in C. (D) The difference in phase of the clangs by the unicycle monkey and the sitting monkey plotted relative to the location of the moving monkey, showing a systematic shift to earlier phases resembling theta phase precession.

This phase shift gives a simple sense of what occurs during theta phase precession. As the unicycle monkey approaches the sitting monkey, the clanging shifts from late phases to earlier phases in proportion to the spatial location of the unicycle monkey relative to the sitting monkey. If one imagines that the unicycle monkey only makes a sound when it is within 20 centimeters of the sitting monkey, then this is analogous to the theta phase precession of a place cell in its single firing field.

This example is simplified to make the phase difference clear between a single cell and a stable baseline phase. In the original model of theta phase precession shown in chapter 2, the center of the firing field would be where the two phases were closest to each other (O'Keefe and Recce, 1993). The original model of theta phase precession is described in more detail in the appendix. As mentioned above, one apparent problem with this early model was that the spiking phase would continue to shift in and out of phase multiple times. Thus, over an extended distance, this model predicted that there should be multiple firing fields at regular intervals. This prediction was inconsistent with the single firing fields of many place cells, and the model was modified to prevent out-of-field firing. However, when the data on grid cells were first presented, Burgess and O'Keefe immediately realized that their model of oscillatory interference could address grid cell firing. In fact, their model essentially predicted the existence of grid cells at least in one dimension. Figure 3.6 shows how a simulation of the persistent spiking model can generate phase precession similar to the precession observed in entorhinal grid cells.

Two Dimensions Coded by Relative Phase

The example above describes how location could be coded in a single dimension. But how would it code for the two-dimensional spaces that rats usually explore?

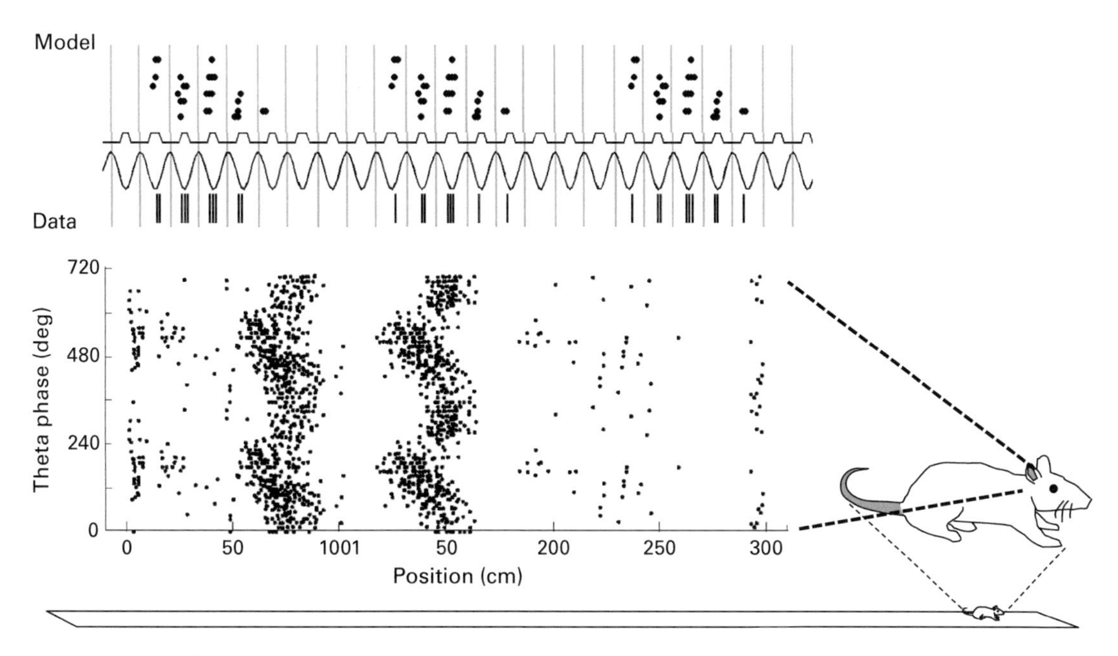

Figure 3.6
Model: Simulation of theta phase precession in the persistent spiking model (Hasselmo, 2008b) during movement of a simulated rat along a linear track. Data: Figure of theta phase precession in an entorhinal grid cell recorded on a linear track in the Moser laboratory (Hafting et al., 2008).

(Two-dimensional representations appear to be sufficient as studies have shown that cell firing in rats seems to map three-dimensional space onto a two-dimensional representation. Thus, grid cells seem to also respond in locations above a given firing field.) A code for two dimensions can be obtained by adding additional neurons to code the additional spatial dimension based on relative phase. This requires considering the pattern of firing activity of individual neurons in a group of neurons that change their frequency dependent on input.

Selective coding of one spatial dimension within a two-dimensional space requires that the effect of speed on frequency be combined with an effect of direction, so that frequency changes in proportion to movement in one particular direction. In physics, combining speed with direction gives velocity. For example, imagine that the unicycle monkey always faces in the eastward direction, but its unicycle wheel can swivel in different directions. When the unicycle wheel faces the same direction as the monkey, then the clanging rate changes directly in proportion to the movement speed (as described above). However, in this analogy when the unicycle wheel is pointed in a different direction, the clanging rate changes in proportion to the cosine of the angle relative to east. Figure 3.7 shows an example of this change. When the monkey is

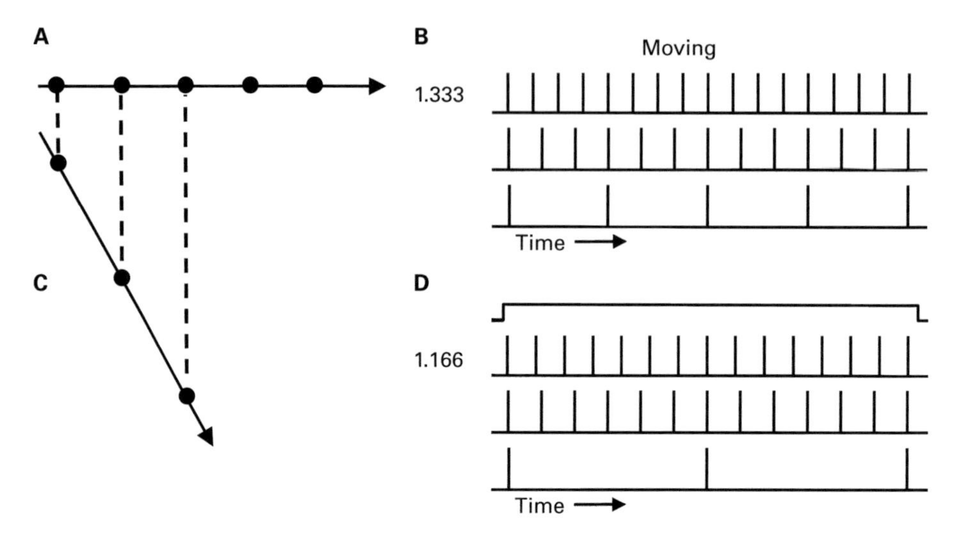

Figure 3.7
Effect of direction of movement on persistent spiking frequency in the model. (A) Moving in the preferred direction results in a steady shift in frequency from 1 Hz to 1.333 Hz that causes spiking at regular locations. Frequency in time shown in B. (C) For a different direction of movement, the change in frequency is scaled by the cosine of the difference in angle (60 degrees) between the direction of movement and the preferred direction of the head direction cell. Cosine (60 degrees) = 0.5, resulting in a smaller change in frequency from 1 Hz to 1.166 shown in D. This causes the spikes occur at the same location in one dimension (west to east).

going east, the clanging rate is 1.33 Hz. When the monkey faces east but the wheel is pointed southeast at an angle of 60 degrees from east, then movement in this direction causes the clanging rate to be 1.166 Hz, as shown in figure 3.7.

When the monkey is turned to the south and the monkey moves sideways to the south (90 degrees from the eastward direction it faces), the clanging rate does not change at all because the cosine of 90 is zero. Imagine that any angle between east and south changes the clanging rate at an intermediate level (proportional to the cosine of the angle relative to east). Similarly, when the wheel is turned for sideways movement toward the north, there is again no change in clanging rate. Angles between north and east again have intermediate values. When the unicycle wheel turns backward to go in the westward direction, then the clanging rate decreases in proportion to the movement speed. For different directions, the rate decreases by intermediate values at intermediate angles. A unicycle monkey with these properties can effectively code one dimension of space (the distance from west to east), as long as the intermediate angles cause a rate change proportional to the cosine of the angle of movement measured relative to moving east. This causes the clanging rate to change in

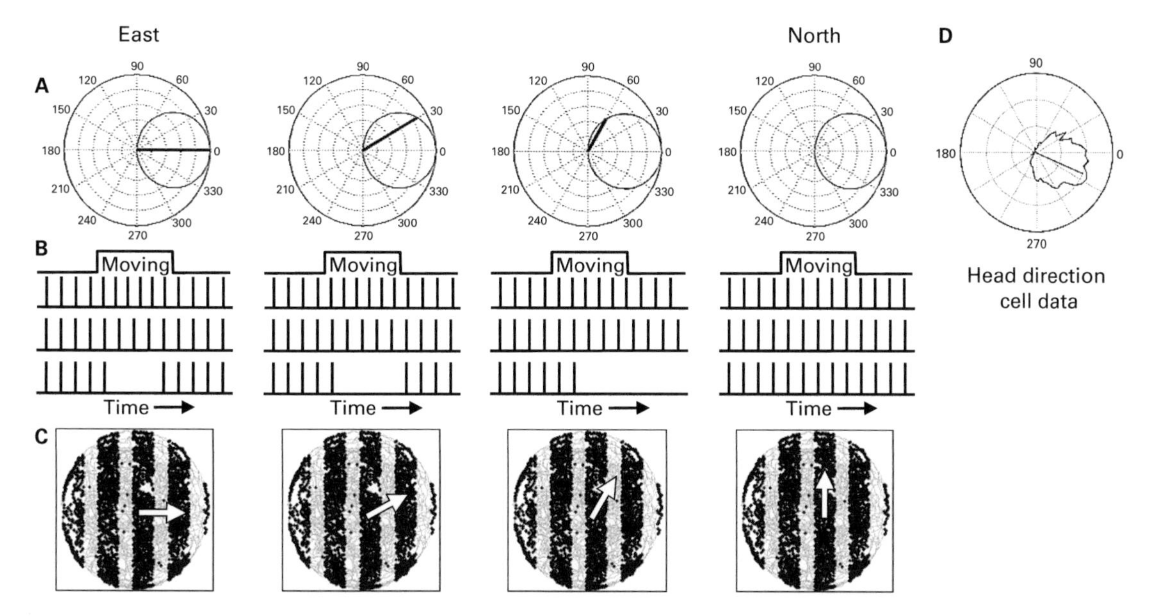

Figure 3.8
(A) Different directions result in different firing rates for a single speed-modulated head direction cell. The small circle is a cosine tuned head direction cell plotted in polar coordinates. The length of the line represents firing rate of the head direction cell for different angles of movement. (B) Persistent spiking model showing large frequency change during movement in the preferred direction of east (far left) and smaller frequency changes for three directions further from the preferred direction (ending with no change for north). (C) Black dots show areas in the environment where the clanging occurs close together. Arrow shows movement corresponding to examples in B. Left: For movement in the preferred direction (east), the frequency change is large and causes rapid transitions between synchronous firing. Right: For movement to the north, the head direction cell does not fire at all and firing stays synchronous. (D) Experimental data showing the firing rate of a head direction cell for different directions of movement. Line shows preferred direction (center of firing rate distribution).

proportion to how much the velocity overlaps with the one dimension of movement (to the east). The clanging rate is now a velocity controlled oscillator. The frequency of the clanging depends on velocity. With this feature, the unicycle monkey can be moved anywhere in the environment and will only clang in unison with the sitting monkey at certain spatial bands within the environment as shown in figure 3.8.

This coding of clanging rate by direction might seem arbitrary, but it actually resembles the coding of head direction cells. Head direction cells respond most strongly when the rat faces in one allocentric direction (e.g., east) and fire at lower rates at larger angles from this preferred direction. At 90 degrees from the preferred direction (i.e., north or south), the head direction cells will usually not fire at all (many of them drop off even faster and stop firing at 45 degrees from the preferred direction). As described in the section on anatomy in chapter 2, head direction cells appear in the postsubiculum (Taube et al., 1990b, 1990a; Sharp, 1996; Taube, 1998). The postsubiculum provides an excitatory input that is used to define the extent of the medial entorhinal cortex (Caballero-Bleda and Witter, 1993). Cells sensitive to head direction also appear in deep layers of the entorhinal cortex (Sargolini et al., 2006) that provide excitatory input to layers containing grid cells. As described in figure 2.12, some head direction cells are sensitive to both speed and head direction.

The specific quantitative properties of head direction cells in relation to this model will be described further below. In particular, some head direction cells in entorhinal cortex appear to respond in a manner close to the cosine of direction angle, but most of them respond with a narrower range of angles. In addition, head direction cells primarily show excitatory increases in response to head direction, so the movement in the opposite direction from preferred direction needs to be coded with excitatory input.

In figure 3.8 I have described how the sensitivity of the unicycle monkey to a combination of speed and direction allows its relative phase to code one dimension in a two-dimensional space. But how would two dimensions be coded? Coding two dimensions would require a second monkey with another pair of cymbals, perhaps perched on top of the first unicycle monkey. The two unicycle monkeys would move together, but each of the two unicycle monkeys would change their clanging frequency dependent only on movement in one dimension, with a difference in the dimension that they encode.

The top unicycle monkey could face south and selectively change its clanging frequency when the monkeys on the unicycle wheel move in the south direction (in that case the east facing monkey does not change phase). This would cause the clanging of the south facing unicycle monkey to shift in phase relative to the separate sitting monkey in proportion to how far the unicycle wheel had turned along the south to north dimension. The clanging of the two unicycle monkeys relative to the sitting monkey would code the separate two-dimensional Cartesian coordinates x and y (analogous to positions on a chessboard or the longitude and latitude on a map).

Thus, as shown in figure 3.9, a two-dimensional environment can be coded by progressive shifts in relative phase between three persistent spiking neurons. The phase of the baseline neuron is illustrated by the spikes next to the letter "b." The phase of the neuron coding the spatial dimension x is illustrated by the spikes next to the letter "x," and the phase of the neuron coding the spatial dimension y is illustrated by the spikes next to the letter "y." Imagine that movement in the x dimension shifts the firing frequency of the x neuron. As shown in the figure, this results in the x neuron's coding a shift in location along the x dimension by a shift in phase relative to the

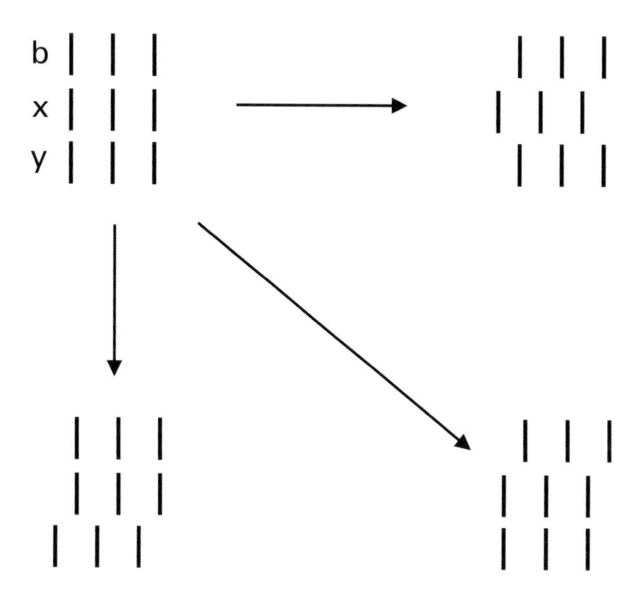

Figure 3.9
Two-dimensional phase code of an environment. Upper left: The northwest position is coded by units x and y showing persistent spiking at the same phase as the baseline b. Upper right: When the animal moves to the east, this shifts just unit x to an earlier phase relative to b that is proportional to the distance traveled. Lower left: When the animal moves to the south, this proportionally shifts just unit y to an earlier phase relative to b. Lower right: When the animal moves both east and south, this proportionally shifts both unit x and unit y to earlier phases relative to b.

baseline b. Similarly, movement in the y dimension shifts the frequency of the y neuron, resulting in a shift in the spiking phase of y relative to baseline. Diagonal movement shifts the relative phase of both neurons. Thus, the firing phase of these three neurons can code two dimensions. The pattern of phases in the array of different neurons can be referred to as a vector of phases.

Simulation of Grid Cell Firing

The phase code described here provides a framework for simulating the firing patterns of entorhinal grid cells. As shown in figures 3.8, 3.9, and 3.10 (plate 7), different populations of persistent spiking neurons with the same baseline firing frequency can drive the activity of a simulated grid cell (Hasselmo, 2008b).

To understand this model, let's return to the example of the two unicycle monkeys again. Recall that the first east-facing monkey would start to clang in unison with the sitting monkey at different intervals in bands alternating along the west-east dimension in the environment. Similarly, the second, south-facing monkey will clang in

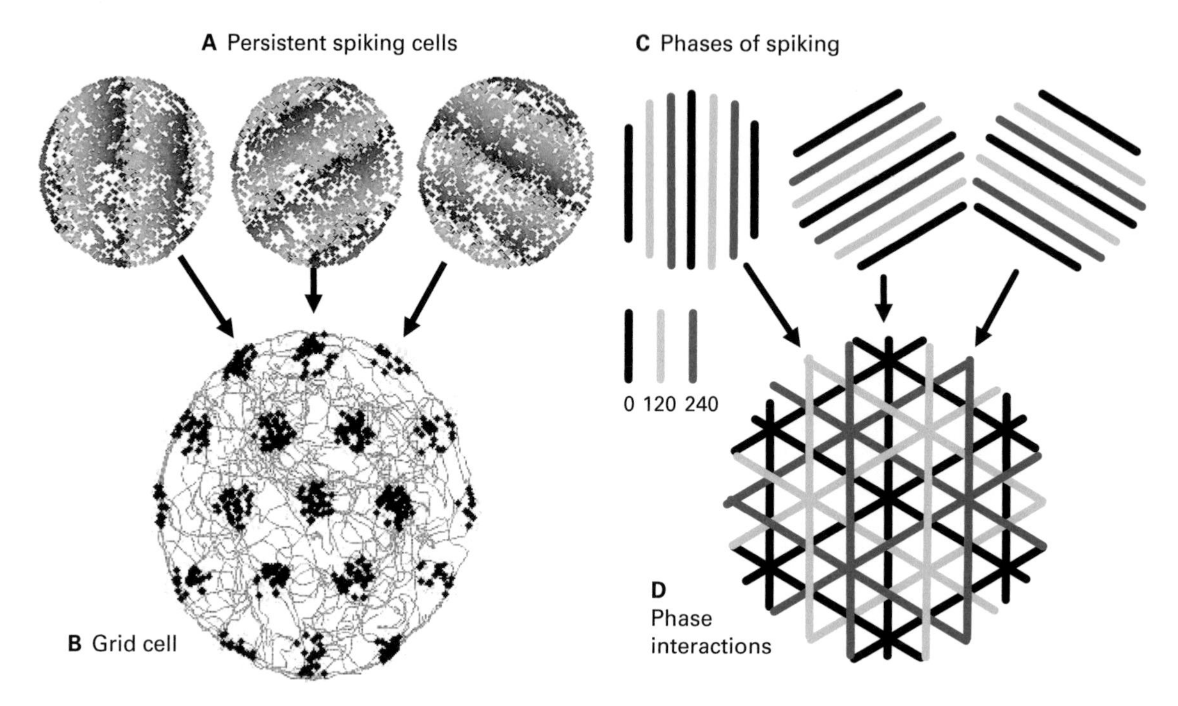

Figure 3.10 (plate 7)
(A) Three plots showing rat location during spiking of each of three persistent spiking cells that fire everywhere in the environment. Each dot represents a spike. The shade of the dot represents the phase of persistent spiking relative to baseline (black = 0, light gray = 120, dark gray = 240). (B) Firing of a single grid cell receiving input from all three persistent spiking cells. The persistent spiking cells drive the grid cell to fire when all three persistent spiking cells have a similar phase resulting in a typical grid cell firing pattern (Hasselmo, 2008b). The thin line indicates the rat trajectory from experimental data. (C) Lines showing the distribution of phases at different locations. (D) Overlay of lines shows that grid cell firing occurs when three lines of the same phase (color) cross.

unison with the sitting monkey at regular intervals in bands alternating along the south-north dimension in the environment. This means that the two unicycle monkeys will also clang in unison with each other at locations where the bands intersect in the environment. The locations where the two unicycle monkeys clang in unison will be laid out in a regular grid of squares in the environment. If we imagine that clanging in unison causes the toy monkeys to grin in their menacing way, then the monkeys will grin in a square grid of locations in the environment.

This is the essence of the function of the persistent spiking model of grid cells. The only remaining feature is that grid cells do not fire in a square grid of locations, but instead fire in a triangular grid of locations. This can be obtained if we combine the east-facing monkey with a monkey facing to the northwest at 120 degrees or to the southwest at 240 degrees. The monkey facing to the northwest will clang in unison with the sitting monkey at regular interval bands that are laid out at a 120 degree angle in the environment, and the crossing points where the unicycle monkeys clang in unison and grin will be laid out in a hexagonal pattern that matches grid cell firing (it can also be described as an array of tightly packed equilateral triangles). As shown in figures 3.8, 3.9, and 3.10, modeling of broader grid firing fields requires the use of groups of persistent spiking cells with slightly different phases.

Phase Coding of Time

The mechanisms described above can code two-dimensional spatial location. However, the coding of episodes also requires coding of time intervals. For example, I can remember the time interval between different people coming to visit my office, even when I sit in the same location facing the same direction. This would require coding of time that does not depend upon my own movement.

The same mechanisms described above can be used for coding of time. Imagine persistent spiking cells with slightly different firing frequencies that stay fixed in frequency over time. As shown in figure 3.11, these could correspond to the clanging of cymbals by wind-up monkeys (A and B) with cymbals clanging at slightly different frequencies as they sit stationary. Some triggering event could start both cymbals clanging at the same time. After this, the clanging of the higher frequency cymbal (A) will gradually shift in phase relative to the lower frequency cymbals (B), resulting in

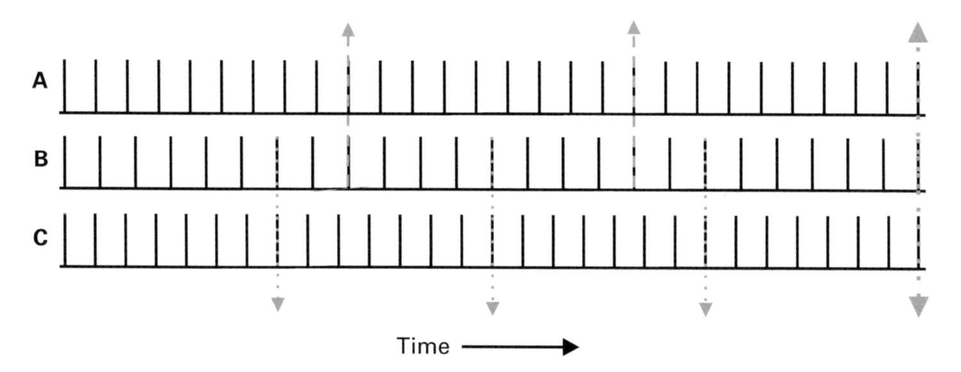

Figure 3.11
Example of shift in phase over time for persistent spiking at different stable frequencies. Row A shows persistent spiking at a slightly higher frequency than row B. The spikes come into phase at the upward arrows (every 8 clangs of B). Row C shows persistent spiking at a frequency higher than both B and A. This comes into phase with B at shorter time intervals shown by downward arrows (every 6 clangs of B). All three are simultaneous at the longer interval of 24 spikes by B (large arrows on right end). This could drive a time cell coding that temporal interval.

simultaneous clanging at regular time intervals (every 8 clangs by B). The figure shows the timing of a third monkey (C) with cymbals clanging slightly faster than either of the other two. These clang simultaneously with B at every 6 clangs by B. All three monkeys clang together on the 24th clang by monkey B.

If these are neurons, they could cause spiking in other neurons at specific time intervals. Neurons A and B could cause spiking in another neuron (not shown) each time they spike simultaneously (upward arrows). Neurons B and C could cause spiking in a different neuron (not shown) each time they spike simultaneously (downward arrows). Spiking in yet another neuron could depend on all three inputs (A, B, and C), resulting in spiking only at the longest interval (large arrows at right). In this manner, the mechanism can provide coding for multiple different time intervals. Multiple neurons with different frequencies could potentially code other time intervals at higher resolution between the intervals shown in this example.

Oscillatory Interference Model of Grid Cells

The mechanism of coding spatial location by relative phase was initially proposed by Neil Burgess, Caswell Barry, and John O'Keefe in a model of grid cell firing properties (Burgess et al., 2005; Burgess et al., 2007; Burgess, 2008). A closely related model was also proposed by Hugh T. "Tad" Blair that utilized elements of interference (Blair et al., 2007; Blair et al., 2008). The essential feature of the original oscillatory interference model is that velocity shifts the oscillation frequency of neurons and determines their oscillatory interference, resulting in spiking when neurons are in phase and the absence of spiking when cells are out of phase (Burgess et al., 2007; Hasselmo et al., 2007; Burgess, 2008). The oscillatory interference model was then modified to use persistent spiking cells (Hasselmo, 2008b). Because these models are based on patterns of interference, I will refer to them collectively as interference models.

The original description of the oscillatory interference model was neutral as to whether it was implemented at the cellular level or the network level (Burgess et al., 2007). There are a number of ways a phase code could be used for an oscillatory interference model. One possible implementation of the oscillatory interference model could utilize interactions of membrane potential oscillations at different frequencies in a single neuron that could result in a grid cell firing pattern (Burgess et al., 2007; Giocomo et al., 2007; Hasselmo et al., 2007) based on properties of subthreshold membrane potential oscillations in entorhinal cortex (Giocomo et al., 2007; Giocomo and Hasselmo, 2008a). However, difficulties with this mechanism suggest the need for implementations using spiking at a network level. Differences in spike timing of different excitatory neurons was used in the persistent spiking model described at the start of this chapter (Hasselmo, 2008a). A more recent model used differences in the spike timing of different populations of excitatory or inhibitory neurons undergoing network oscillations (Zilli and Hasselmo, 2010). Models can also use differences in spike timing of internal entorhinal populations versus the network oscillation induced by medial septal input, as supported by recent data (Brandon et al., 2011b).

The oscillatory interference model of grid cell firing can effectively simulate the difference in spacing of firing fields for grid cells recorded at different positions along the dorsal to ventral axis of medial entorhinal cortex (Hafting et al., 2005; Sargolini et al., 2006). In the oscillatory interference model, the spacing of grid cell firing fields depends upon the change in frequency caused by a given velocity. Similar to the examples of the persistent spiking model shown above, the oscillatory interference model involves a baseline oscillation interacting with other oscillations that shift in frequency with the running velocity of a rat. In the examples shown here, the oscillations are summed together, and when the sum goes over a threshold, a spike in the grid cell is produced.

If the frequency of a single cell increases linearly with velocity, then the firing fields of this cell will have the same spacing for different velocities as shown in figure 3.3, with spacing dependent on the slope of frequency versus velocity. However, if different cells have different slopes of the frequency to velocity, the spacing of their firing fields will differ. As shown in figures 3.12 and 3.13, a larger frequency change with velocity causes a faster phase shift, resulting in a shorter distance between firing fields. In contrast, a smaller frequency change with velocity results in a slower phase shift that causes a larger distance between firing fields (Burgess et al., 2007; Giocomo et al., 2007; Hasselmo et al., 2007).

Based on the ability of the oscillatory interference model to account for the change in spacing of grid cell firing fields, Burgess and O'Keefe generated the explicit prediction (O'Keefe and Burgess, 2005) that the increased spacing between firing fields of grid cells as you move from the postrhinal border of the medial entorhinal cortex would result from a gradually decreasing intrinsic frequency of entorhinal neurons. Lisa Giocomo tested this prediction in my laboratory by recording from layer II stellate cells in slice preparations of medial entorhinal cortex, and she found the predicted dorsal to ventral difference in frequency of membrane potential oscillations induced by depolarization (Giocomo et al., 2007) as shown in figure 3.14. As described in the analogy with an orchestra, it is as if neurons at different dorsal to ventral positions played at different preferred pitches.

As shown in figures 3.14 and 3.15 and earlier in figure 2.14, the differences in the membrane potential oscillation frequency generated by depolarization were implemented in the oscillatory interference model (Giocomo et al., 2007; Hasselmo et al., 2007). This effectively replicated the differences in the size and spacing of grid cell firing fields at different dorsal to ventral anatomical positions as shown in the neurophysiological data (Hafting et al., 2005; Sargolini et al., 2006; Giocomo et al., 2007; Brun et al., 2008).

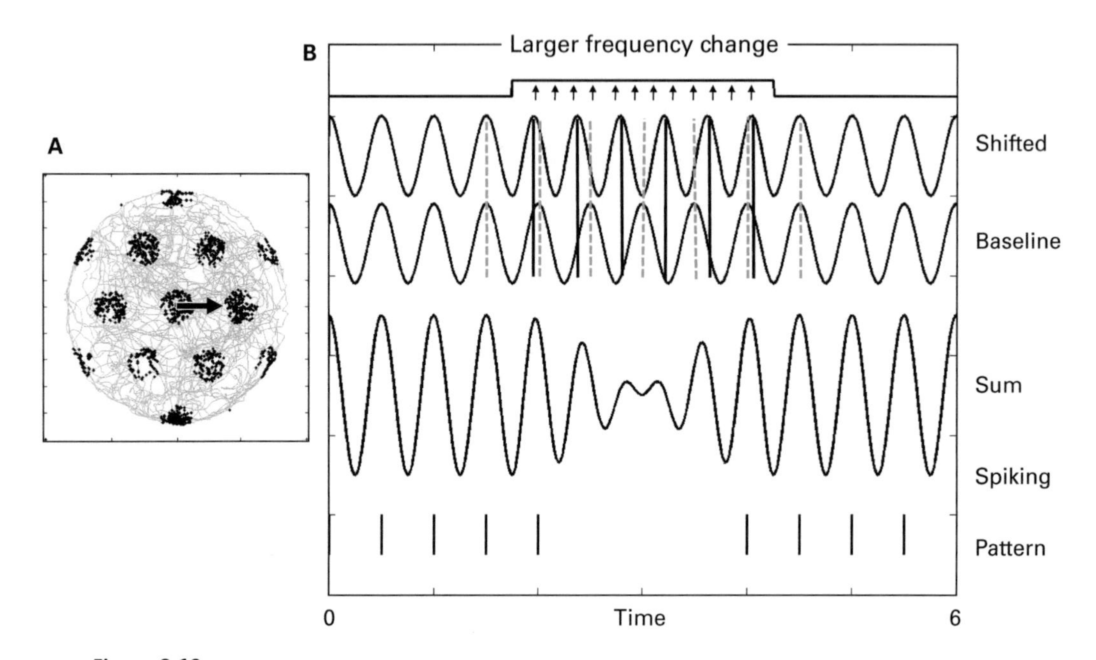

Figure 3.12
(A) Narrow spacing of firing fields occurs in the oscillatory interference model when there is a large change in frequency with velocity (B). The large change in frequency of the shifted oscillation causes a rapid change in phase relative to baseline (baseline phase shown with gray dashed lines, shifted phase shown with solid black lines). When the oscillations are summed (Sum), they rapidly transition between constructive interference (large oscillations that cause spikes) to destructive interference (smaller oscillations) to constructive interference (large oscillations causing spikes). This causes a short spacing between firing fields (Spiking pattern and A).

Strengths of the Oscillatory Interference Model

I have worked with many models over the years, but the oscillatory interference model of grid cells has provided the most impressive link between different sets of neurophysiological data. As described above, the model generated predictions about the intrinsic frequency of entorhinal neurons that were supported by the intracellular recording data from stellate cells (Giocomo et al., 2007; Giocomo and Hasselmo, 2008a). In addition, when the intracellular recording data are implemented in the model, the model can generate a remarkably realistic match to the data on unit recording from foraging rats (Hasselmo et al., 2007). This is the most dramatic example that I have experienced of a model directly linking cellular physiological properties to the network spiking activity. The model is not yet a complete, correct model of the mechanism of grid cell firing, but before addressing the challenges for the model, I want to enumerate some of the strengths of the model that make it so appealing. Another

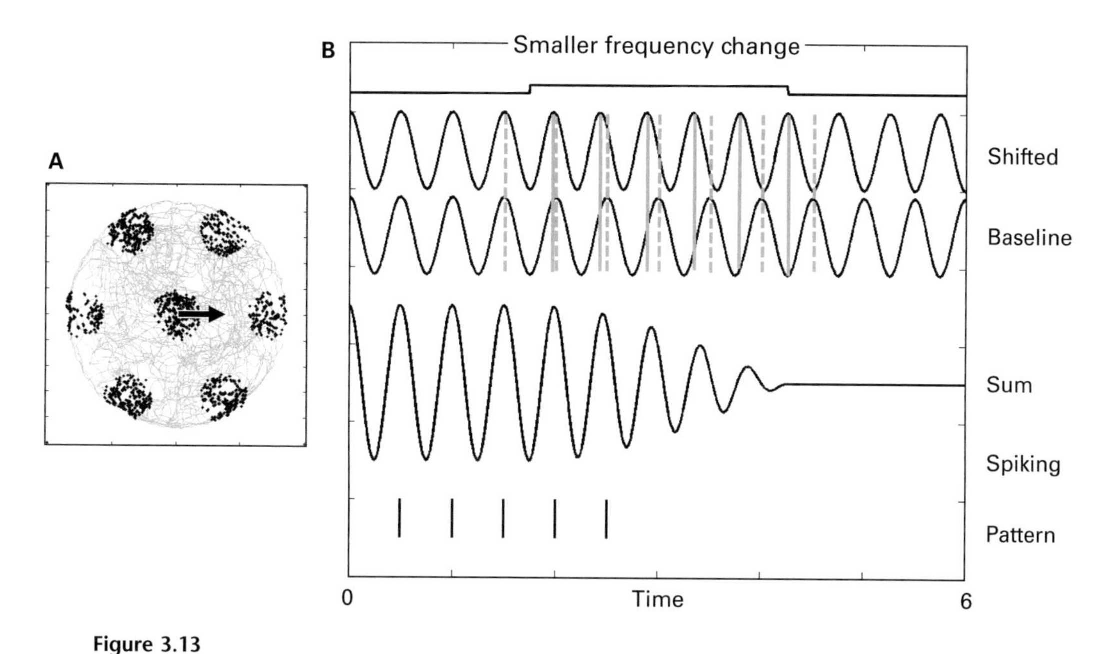

Figure 3.13
(A) Wider spacing of firing fields occurs in the oscillatory interference model when there is a smaller change in frequency with the same velocity (B). The smaller change in frequency causes a slower change in the phase of the shifted oscillation relative to baseline (baseline phase shown with gray dashed lines, shifted phase shown with solid gray lines). The same movement causes a smaller phase shift of only 180 degrees, so the sum goes from constructive interference to destructive interference (lack of spiking between firing fields). Overall, the smaller change in frequency with velocity causes a larger spacing between firing fields (Spiking pattern and A).

class of models that have proven effective at addressing grid cell firing are attractor dynamic models described in a later section (Fuhs and Touretzky, 2006; McNaughton et al., 2006; Burak and Fiete, 2009). These models also have a number of strengths, so I will contrast the oscillatory interference models with the attractor dynamic models:

1. *Simplicity* A strength of the oscillatory interference model is simplicity. It is described with a single equation presented in the appendix in the section Equation for Grid Cell Model (Burgess et al., 2005; Burgess et al., 2007; Burgess, 2008; Giocomo and Hasselmo, 2008a). This eased the initial communication about this model for researchers to implement the same model, in contrast to more complex biophysical models that can be difficult to quickly replicate. The model has only three parameters for fitting the unit recording data on grid cell firing fields as described in the Appendix. The first is spatial frequency, which can be directly scaled to the data on intracellular oscillation frequency during depolarization (Hasselmo et al., 2007), the second is

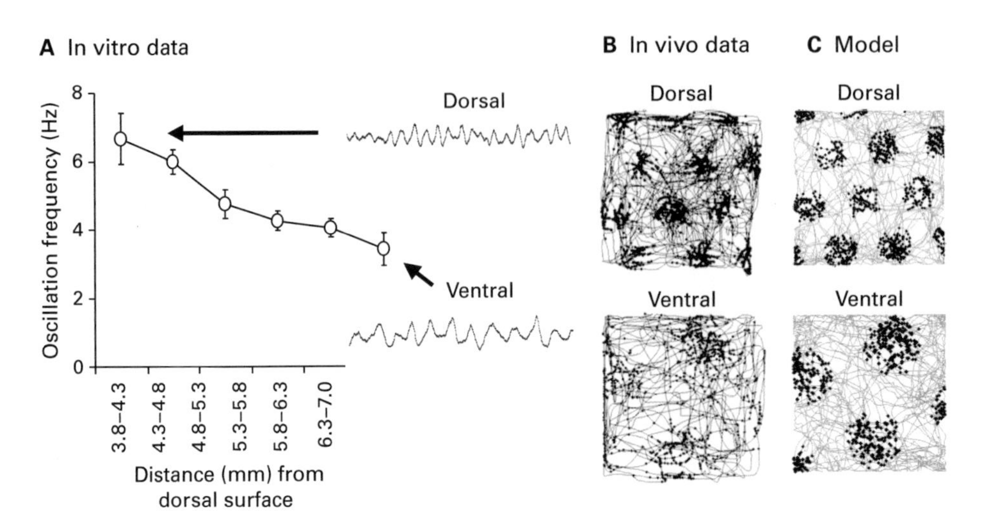

Figure 3.14
(A) Experimental data showing a progressive decrease in the mean frequency of subthreshold membrane potential oscillations in layer II stellate cells at greater distances from the dorsal surface (Giocomo et al., 2007; Giocomo and Hasselmo, 2008b), with examples of individual oscillations in dorsal and ventral neurons on right. (B) This decrease in frequency scales with the increase in spacing between grid cell firing fields in data from awake behaving rats (Hafting et al., 2005). (C) The oscillatory interference model uses the in vivo data in A to simulate the data on grid cell firing fields in B (Hasselmo et al., 2007).

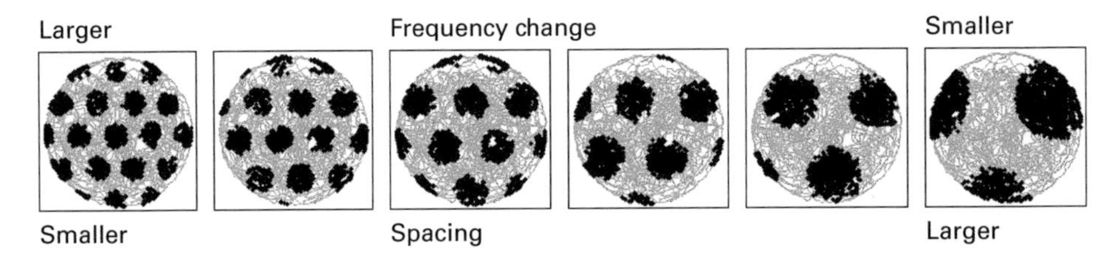

Figure 3.15
Differences in grid cell spacing obtained with the oscillatory interference model using oscillations that shift in frequency by different amounts for a given input velocity (Hasselmo et al., 2007).

spatial phase, which is determined by the initial phase of oscillations or spiking, and the third is the orientation, which can be linked to the orientation of the head direction cells providing input to the model. The simplicity of the model makes it much easier to implement in a model of episodic memory function.

2. *Theta phase precession* The oscillatory interference model of grid cells (Burgess et al., 2007) arose naturally out of models simulating the theta phase precession of hippocampal place cells (O'Keefe and Recce, 1993; Lengyel et al., 2003; Burgess et al., 2005; O'Keefe and Burgess, 2005; Burgess et al., 2007) and effectively simulates theta phase precession in entorhinal cortical neurons (Burgess et al., 2007; Burgess, 2008; Hasselmo, 2008a; Hasselmo and Brandon, 2008). In fact, as described in chapter 2, the models of theta phase precession in place cells automatically generated multiple firing fields and had to be modified to prevent multiple fields. Thus, the oscillatory interference model of theta phase precession effectively predicted the existence of grid cells. In contrast, recurrent attractor models of grid cells effectively account for grid cells, but in their initial form they do not require theta rhythm. Only recently have attractor models started to address theta phase precession (Navratilova et al., 2011).

3. *Predicted spatial scaling mechanism* As described above, the oscillatory interference model already generated a prediction that differences in spatial scaling of grid cells along the dorsal to ventral axis of entorhinal cortex arises from differences in intrinsic oscillation frequency (O'Keefe and Burgess, 2005; Burgess et al., 2007). This explicit prediction has been supported by experimental data showing differences in membrane potential oscillation frequency along the dorsal to ventral axis of entorhinal cortex (Giocomo et al., 2007; Hasselmo et al., 2007; Giocomo and Hasselmo, 2008a).

The oscillatory interference model comes in two variants, an additive version, in which the change in frequency is added to a baseline frequency, and a multiplicative version, in which the change in frequency is obtained by multiplying the baseline frequency by a scaling factor (Burgess et al., 2007; Giocomo et al., 2007; Giocomo and Hasselmo, 2008a). We initially used the multiplicative model to link our membrane potential oscillation data to grid cell firing (Giocomo et al., 2007). However, the experimental data on oscillation frequencies at different membrane potentials in different neurons supported the additive model (Giocomo and Hasselmo, 2008a). The additive model also allows us to more effectively simulate (Giocomo and Hasselmo, 2008b) the very large spacing of up to 10 meters seen between the firing fields of grid cells recorded in the most ventral portion of the medial entorhinal cortex (Brun et al., 2008). Large firing fields can be obtained with a small frequency addition in the additive model. In contrast, large firing fields would require very low baseline frequencies in the multiplicative model.

Experimental data provide a parameter B in the oscillatory interference model that describes the relationship of frequency to velocity and accounts for both the scaling of field size and spacing along the dorsal to ventral axis in (Giocomo and Hasselmo,

2008a). As noted above, the parameter B could change in a gradient that depends on the intrinsic resonant and membrane potential oscillation frequency of neurons (Giocomo et al., 2007; Giocomo and Hasselmo, 2008a), and might arise from the cellular properties of the h current that appears to change in time constant and magnitude along the dorsal to ventral axis (Giocomo and Hasselmo, 2008b). Knockout of the HCN1 subunit which contributes to a faster time constant of the h current flattens this gradient (Giocomo and Hasselmo, 2009) and causes an increase in the spatial scale of grid cells during behavior (Giocomo et al., 2010).

The simulation of both field size and spacing with the same parameter differs from recurrent attractor models of grid cells that model grid field size with the pattern of synaptic connectivity but model the spacing between firing fields based on the magnitude of velocity input. As another alternate mechanism, the biophysical timescale of spike frequency accommodation has been used to account for differences in grid cell spatial scaling (Kropff and Treves, 2008).

Recent data show theta cycle skipping in some grid cells and head direction cells (Deshmukh et al., 2010; Brandon et al., 2011b). That is, instead of firing on each cycle of theta rhythm, these cells fire on alternate cycles, suggesting possible mechanisms using phase coding by spiking on discrete, noncontiguous cycles of theta. In simulations, the cycle skipping is easily modeled in cells with h current resonance receiving an oscillatory inhibitory input. The number of available cycles, or the integer values for interactions of one spiking frequency and another might result in quantal properties of grid cell interactions that could underlie the apparent quantal steps in grid cell size along the dorsal to ventral axis (Barry et al., 2007; Stensland et al., 2010).

4. *Predicted grid cell firing frequency* In addition to the experimental prediction about the intrinsic frequency of neurons measured intracellularly, the oscillatory interference model also correctly predicted differences in intrinsic spiking frequency measured with autocorrelations of unit recording data in behaving rats (Jeewajee et al., 2008a). The oscillatory interference model (Burgess, 2008) predicted that the unit recording data on grid cells should show a difference of the within-field firing frequency consistent with the dorsal to ventral differences in spatial scaling of grid cell firing fields. The model also predicted that grid cells should show changes in within-field intrinsic firing frequency that are proportional to running speed. Both of these predictions were supported by analysis of experimental data (Jeewajee et al., 2008a). The model also accounts for the change in network theta rhythm frequency that has been shown to be correlated with running speed (Maurer et al., 2005; Hasselmo et al., 2007; Jeewajee et al., 2008a; Jeewajee et al., 2008b).

5. *Expansion in novel environments* The model (Burgess et al., 2007) predicted that the expansion of grid fields in novel environments (Barry et al., 2008) should be accompanied by a reduction in network theta rhythm frequency as supported by experimental data (Jeewajee et al., 2008b). The reduction in theta rhythm frequency could arise

from the effects of acetylcholine. Microdialysis shows increases in levels of the neuromodulator acetylcholine in the cortex when an animal is in a novel environment (Acquas et al., 1996). Acetylcholine has been shown to regulate theta rhythm oscillations in the hippocampus (Bland, 1986; Konopacki et al., 1987), and the slower, type II theta rhythm oscillations have been shown to be sensitive to the muscarinic acetylcholine receptor blocker atropine (Kramis et al., 1975). On a single-cell level, cholinergic modulation lowers the resonance frequency of entorhinal stellate cells (Heys et al., 2010). By reducing neuronal intrinsic frequencies, acetylcholine could cause an increase in the size and spacing of grid cell firing fields observed in novel environments (Barry et al., 2008).

6. *Path integration* The same mechanism that produces the pattern of grid cell firing in the model also performs path integration (Burgess et al., 2007; Burgess, 2008; Hasselmo and Brandon, 2008). This contrasts with recurrent attractor models in which the pattern of firing is created by the pattern of excitatory recurrent connectivity (Fuhs and Touretzky, 2006; McNaughton et al., 2006; Burak and Fiete, 2009), and the network mechanism of path integration is a separate velocity-dependent shift in representation that works in the same manner as in models of place cell representation (Samsonovich and McNaughton, 1997).

7. *Blockade of theta rhythm blocks grid cell firing* Recent experiments performed by Mark Brandon in my laboratory have shown that grid cell firing patterns appear to depend upon the presence of network theta rhythm oscillations (Brandon et al., 2011b). Similar results were independently obtained in the Leutgeb lab (Koenig et al., 2011) using infusions of lidocaine. While recording from grid cells in rats, Mark performed infusions of the GABA agonist muscimol into the medial septum to shut off network theta rhythm oscillations in the entorhinal cortex. Grid cells recorded before the infusion show a loss of grid cell firing pattern during the reduction in theta rhythm oscillations (Brandon et al., 2011b), suggesting the network theta rhythm oscillations are necessary to generate grid cell responses. In contrast, entorhinal head direction cells maintain their firing selectivity after the infusion, and conjunctive grid by head direction cells also maintain their head direction selectivity while losing their grid cell spatial periodicity (Brandon et al., 2011b).

Challenges for the Oscillatory Interference Model

After we published our paper in *Science* supporting the prediction from the oscillatory interference model about intrinsic properties (Giocomo et al., 2007), I thought that we could immediately progress from using an abstract version of the model to using a detailed biophysical simulation to link a full range of cellular mechanisms to the range of unit recording data on grid cells. However, there are a number of challenges for the model that became clear to me as we have worked with the model and performed further experiments. While there are many strengths of the oscillatory interference model and related models, we have clearly not yet found the complete model of the mechanism of grid cell firing. There are a number of challenges that concern the oscillatory interference model that are currently being addressed in both modeling and experimental work.

1. *Synchronization of oscillations in single neurons* The initial simulations of the oscillatory interference model used simple mathematical models of oscillations as cosine functions. Later, we used compartmental biophysical simulations of entorhinal stellate cells to simulate the oscillatory interference model based on the interactions of membrane potential oscillations on different dendrites within a single neuron. However, we were not able to obtain oscillations of different phase and frequency within a single simulated cell. Simulations demonstrate that membrane potential oscillations within a single neuron have a strong tendency toward synchronization of phase and frequency (Remme et al., 2009, 2010). This prevents implementation of the model on a single-cell level and suggests that the influence of the intrinsic frequency of neurons requires network interactions between cells, such as those in the persistent spiking model of grid cells (Hasselmo, 2008b) or in simulations of oscillations arising within networks of neurons (Zilli and Hasselmo, 2010).
2. *Sensitivity to noise* Both the oscillatory interference model and the persistent spiking model are sensitive to noise, including both noise in the phase and noise in the speed modulated head direction input. Experimental data on membrane potential oscillations show high variance in the wavelength of membrane potential oscillations that we showed can disrupt the grid pattern in the model (Giocomo and Hasselmo, 2008a). In addition, spiking will disrupt phase in the membrane potential oscillation model but not in the persistent spiking model. Whole cell patch recordings of persistent spiking show smaller variance than membrane potential oscillations (Yoshida and Hasselmo, 2009a), and sharp electrode recordings of persistent spiking show even lower variance of interspike intervals (Fransén et al., 2006; Tahvildari et al., 2007). However, a detailed analysis of the variance in experimental data on membrane potential oscillations and persistent spiking shows that even the lower level of variance in persistent spiking neurons would degrade accurate grid cell firing in a matter of seconds (Zilli et al., 2009).

This problem of intrinsic noise can be solved in network simulations, which demonstrated that firing frequencies with sufficiently low variance could be obtained with networks of neurons even if the network was made up of individual neurons showing noisy and variable spiking intervals (Zilli and Hasselmo, 2010). The network implementation worked either with excitatory recurrent connections or with inhibitory interneurons driven at different frequencies by the velocity input (Zilli and Hasselmo, 2010). The network implementation is able to overcome the problem of intrinsic noise as shown in figure 3.16.

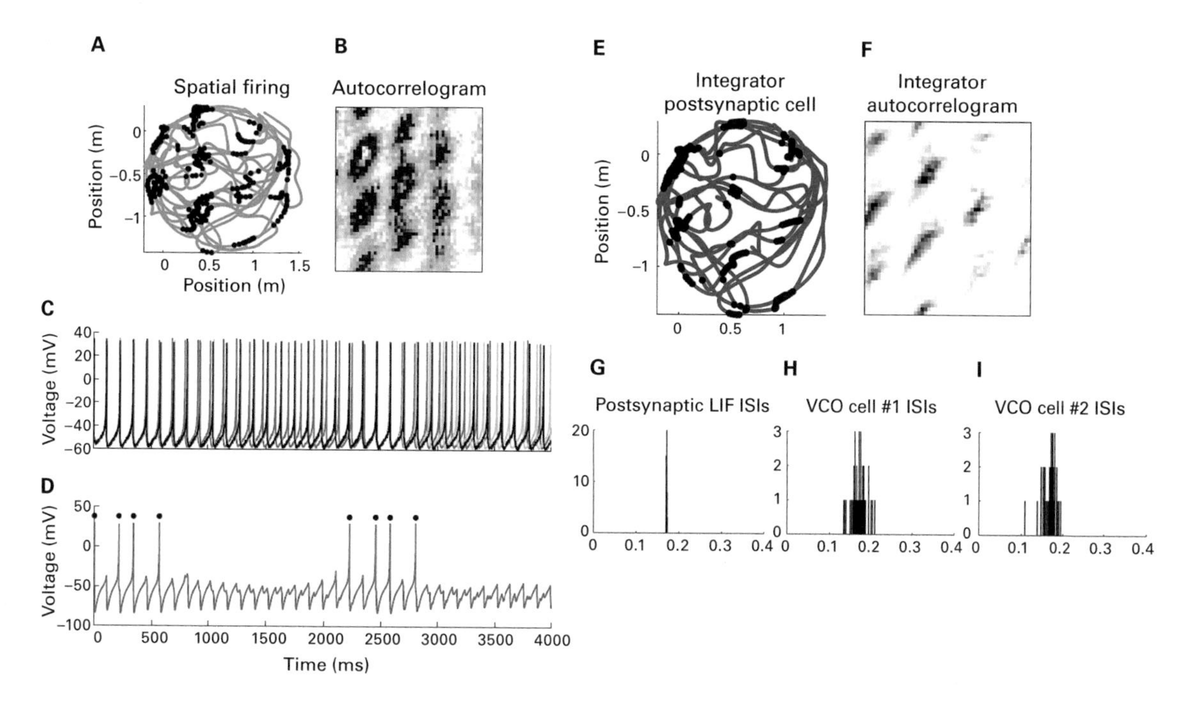

Figure 3.16
Simulations of oscillatory interference overcoming noise with populations of biologically detailed neurons. Effective simulations of spiking activity on a rat trajectory A and in the two-dimensional spatial autocorrelation of this activity B can be obtained with velocity input regulating the frequency of noisy inhibitory interneurons (C) that allow a grid cell (D) to spike when they are in phase. A separate simulation shows that large populations of sparsely connected noisy excitatory neurons can generate a grid cell firing pattern (E) and spatial autocorrelation (F). In this model, the interspike intervals (ISIs) between postsynaptic spikes computed by a leaky integrate-and-fire (LIF) grid cell (G) are very consistent even though the interspike intervals (ISIs) of individual neurons (H, I) in the velocity-controlled oscillator populations (VCOs) are highly variable.

As shown in the figure, the problem of intrinsic noise can be overcome by network dynamics. A separate problem concerns the noise in the velocity signal to the network. If this noise is correlated on all inputs, it will disrupt both interference models and attractor models. A possible solution to the problem of noise in the velocity signal could be resetting of phase based on sensory input from boundary vector cells (Burgess et al., 2007).

3. *Requirements for initial phase and phase reset* In his first descriptions of the model, Neil Burgess showed how sensitivity to noise in the oscillatory interference model can be counteracted by resetting of phase based on current place cell activity that is driven by sensory input (Burgess et al., 2005; Burgess et al., 2007; Burgess, 2008). However,

the oscillatory interference model requires specific phase relationships in the initial phase. A grid cell model starting with all phases at zero will not show the same grid cell pattern if all phases are shifted by 180 degrees (Burgess et al., 2005; Hasselmo et al., 2007). This problem is avoided if the soma reference phase is not used, as in the persistent spiking model presented here, or if phase is only reset at specific locations where all phases are zero. It may be simpler to reset phase in the persistent spiking model with a single strong input.

Phase reset may allow the interference models to address an additional set of data on context-dependent changes in firing properties. The possible role of phase reset for context-dependent responses will be addressed in more detail in chapter 7. Phase reset in the oscillatory interference or persistent spiking model allows simulation of the context-dependent firing properties of grid cells (Frank et al., 2000; Lipton et al., 2007; Derdikman et al., 2009) and place cells (Markus et al., 1995; Wood et al., 2000; Lee et al., 2006; Griffin et al., 2007). Phase reset could also provide a mechanism for changes in grid fields associated with changes in the size of the environment including expansion (Barry et al., 2007) or remapping (Savelli et al., 2008).

4. *Head direction preference intervals* To obtain the hexagonal pattern of grid fields, the oscillatory interference models require that head direction inputs have preference angles at multiples of 60 degrees (Burgess et al., 2007; Hasselmo et al., 2007) or at 120-degree intervals in the persistent spiking model. This pattern was proposed to arise from self-organization of the head direction input to grid cells (Burgess et al., 2007). Unit recording has not shown such a periodic distribution, but the head direction selectivity of grid cells aligns with their orientation, and human imaging data from entorhinal cortex shows a rotational symmetry of periodic increases in activity associated with running in different directions at 6 intervals at 60 degrees (Doeller et al., 2010). The assumption of head direction input angle applies to both the persistent spiking model presented here and the oscillatory interference model. This problem specifically does not apply to the recurrent attractor models of grid cells, which obtain hexagonal firing patterns from circularly symmetrical excitatory connectivity and do not require a specific distribution of head direction input. Variants of the oscillatory interference model show that self-organization can selectively strengthen synaptic input from head direction cells with the appropriate difference in preference angles.

5. *Requirement of cosine function of head direction tuning* Both the oscillatory interference models and the attractor dynamic models require input from head direction cells that fire at a rate determined by a cosine (or sine) function of actual head direction (covering 180 degrees). However, most actual head direction cells have much narrower tuning with a Gaussian shape covering about 90 degrees (Taube et al., 1990a; Taube and Bassett, 2003). In addition, head direction cells lack the negative component of the cosine function. Their firing goes down to a very low baseline outside of 90 degrees but does not show inhibition below baseline. This property of all the models has not attracted much attention, but I feel it is important to understand why there is such

narrow tuning of so many head direction cells. Possibly, the head direction cells are updating an internal representation in non-Euclidean space (possibly spherical space). Alternately, the system might utilize head direction and angular head velocity to update internal states based on the viewing angle of visual stimuli as a state instead of the allocentric location.

The lack of a negative component of firing in head direction cells has partly been solved by using a rectified cosine function that removes the negative component of the cosine. Initially it appeared that this requires coupling of pairs of head direction inputs that are tuned at 180-degree differences, but my recent simulations without a single baseline frequency show that this is not necessary. Another reason for not including the negative component of the cosine function is that it causes the model to show incorrect theta phase precession with a phase shift from early to late. The removal of the negative component of the input cosine functions allows simulation of theta phase precession that shows the correct phase shift (from late to early) regardless of the direction that the rat runs through the place field (Burgess et al., 2007; Burgess, 2008; Hasselmo and Brandon, 2008). However, the theta phase precession in the model still differs in certain details from the experimental data. Another solution to the problem of the negative component is to increase all frequencies including baseline proportional to speed (Burgess, 2008), so that head direction input does not cause a net negative shift.

6. *Requirement of linear frequency change with depolarization* The interference models of grid cells require a velocity input that drives the shift in frequency of neurons. Neurophysiological data on membrane potential oscillations indicated that different neurons at different positions along the dorsal to ventral axis of medial entorhinal cortex responded with different magnitudes of frequency change at different membrane potentials (Giocomo et al., 2007; Giocomo and Hasselmo, 2008a, 2009). However, the initial data were shown for different neurons at different membrane potentials. More recent data from my laboratory indicate that single neurons show a change from broadband frequencies of fluctuation to a single narrow band that does not change in a linear manner with depolarization as required by the model (Yoshida et al., 2010). Instead of changing in a linear manner, dorsal neurons appear to go to a single narrow band at higher frequencies, and ventral cells go to a narrow band at lower frequencies. This raises interesting implementation problems for the oscillatory interference model, suggesting that individual stellate cells might go to the same frequency each time a depolarizing input arrives from a head direction cell. In this case, the phase shift might arise from the difference in the characteristic frequency from the network frequency or from an interaction with a change in network frequency caused by speed input. It is notable that many of the cells showing strong speed modulation in our data (Brandon et al., 2011c) appear to be inhibitory interneurons that could mediate a shift in local circuit frequency relative to fixed frequencies of stellate cells. Interneurons could also fire in anti-grid patterns.

7. *Requirement of combination of speed and head direction* Both the interference models and the attractor models require a velocity input that combines speed and head direction. Some head direction cells show modulation with speed in our data (Brandon et al., 2011c) and in previous work (Sharp, 1996; Sargolini et al., 2006), but many head direction cells show little change in firing rate with running speed and remain highly active when the rat is stationary facing in the preferred head direction (Taube et al., 1990a). On the other hand, many cells show sensitivity to speed in the hippocampus (O'Keefe et al., 1998), entorhinal cortex (Sargolini et al., 2006; Brandon et al., 2010), and subcortical areas (Sharp et al., 2005) but do not alter firing with head direction. Input from separate groups of speed and head direction cells might undergo spatial summation within individual neurons, but the separation suggests the two signals might be used independently. Speed modulation of interneurons may influence network baseline frequency whereas head direction selectively influences individual velocity-controlled oscillators.

Overall, considerable data supports the general features of the oscillatory interference model, including the dorsal to ventral gradient of cellular resonance (Giocomo et al., 2007) and intrinsic frequencies (Jeewajee et al., 2008) and the theta phase precession of grid cells (Hafting et al., 2008), but the challenges described above suggest we have not yet found the variant of the model that accounts for the full range of data with phase coding. Variants of phase coding already analyzed include interactions of intrinsic membrane potential oscillations (Burgess et al., 2007; Hasselmo et al., 2007), interactions of rhythmically spiking cells (Hasselmo, 2008; Burgess, 2008), or interactions of populations of spiking cells (Zilli and Hasselmo, 2010). Other phase coding options include the phase relationship between different permutations of excitatory cells, inhibitory cells, local field potentials and medial septal cells. Medial septal neurons fire at different phases relative to theta rhythm (Bland, 1986; Brazhnik and Fox, 1999) and could regulate entorhinal populations with a range of different phases. Amplitude of oscillations could contribute to maintenance of phase information. Tonic speed and head direction input could shift activity between different populations of conjunctive grid by head direction cells with different temporal and spatial phases, with slower inhibition from the overall population inhibiting a subset of phase populations, and interactions of different conjunctive cells driving layer II grid cells with precession. This would combine elements of oscillatory interference with network dynamics of feedback inhibition and slow integration.

Attractor Models of Grid Cells

The phase coding models of grid cells described in this chapter are not the only existing models of grid cells; another class of models simulate grid cell firing through attractor dynamics based on recurrent connectivity (Burak and Fiete, 2006; Fuhs and Touretzky, 2006; McNaughton et al., 2006; Guanella et al., 2007; Fiete et al., 2008; Welinder et al., 2008; Burak and Fiete, 2009). In fact, the episodic memory model described here could be implemented with the attractor dynamic model of grid cells in place of the interference models. The attractor dynamic models differ from the interference models in that they focus on the firing rate of neurons based on a structured pattern of synaptic connectivity between neurons within the entorhinal cortex. Even though grid cells show systematic dorsal to ventral differences in the size and spacing of firing fields, they do not show anatomical topography for location in the environment. Cells that are anatomically adjacent can have grid cell firing fields centered far from each other. Therefore, the synaptic interactions in these models do not depend upon anatomical position within the cortex but depend upon the location coded by the grid cells in the rat's spatial environment. In these models, an entorhinal neuron coding one spatial phase sends excitatory connections to neurons coding very nearby spatial phases in the environment within a given radius of that spatial phase and sends inhibitory connections to neurons coding slightly more distant spatial phases in a ring of a larger radius. The connections are circularly symmetrical within the spatial environment being encoded by the neurons. The network then builds up to a steady state level of feedback inhibition and excitation between neurons that generates the characteristic hexagonal pattern of grid cell firing fields as shown in figure 3.17.

The feedback synapses will generate a grid of cells for one location. The movements of a simulated rat can move this attractor state around due to input from speed-modulated head direction cells. As shown in figure 3.17C, speed-modulated head direction cells modulate the activity of neurons that connect grid cells in the network to adjacent cells in the appropriate direction. For example, a speed-modulated head direction cell coding a movement east will activate a cell that receives input from one set of grid cells and sends excitatory output to another set of grid cells to the east of the input cells (shown in upper left corner of figure 3.17C). When the rat moves east, the east head direction cell activates this neuron and causes the current grid cell attractor to activate an attractor state slightly to the east. This shifts the attractor state in that direction. Each step of the trajectory involves speed-modulated head direction cells shifting the attractor in the appropriate direction. Recording of one neuron during a trajectory then yields a characteristic grid cell pattern of activation.

These attractor network models have certain strengths in comparison to the phase coding models. In particular, the network models are less sensitive to intrinsic noise as the noise in any individual neuron will be dominated by the network synaptic interactions. As described above, population interactions have been used to overcome noise problems in the oscillatory interference model as well by enhancing the stability of phase differences (Zilli and Hasselmo, 2010). There are other ways in which features of these types of models could be combined.

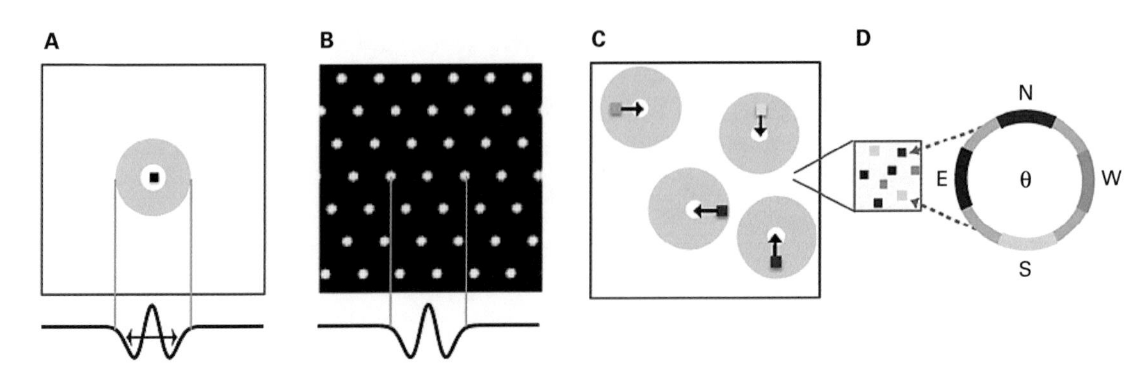

Figure 3.17
Overview of attractor models of grid cells. (A) Connectivity is based on the location in the environment coded by individual cells. The cell coding the location marked by the black square will send excitatory connections to neurons coding nearby locations and inhibitory connections to neurons coding a ring of more distant locations (shown in gray in two-dimensional plot). The line shows a cross section of synaptic strength. (B) When all neurons are connected with this same ring pattern of connectivity, the interactions of active neurons will cause a stable attractor state of activity in a hexagonal pattern. Active neurons in the pattern are coding locations in the environment that are just outside of the radius of inhibition of other firing neurons. (C) The attractor state can be moved around by excitatory input to individual neurons in the attractor states from head direction cells coding directions shown in D. Figure reproduced with permission from Burak and Fiete (2009).

Attractor dynamics models require interactions of a population of neurons with the same spatial scale and spatial orientation. This is a strength of the model as data indicates that grid cells within a local region at one dorsal to ventral position have the same scale and orientation, and there appear to be sharp transitions between the spatial scale at different anatomical positions (Barry et al., 2007; Stensland et al., 2010), rather than a smooth gradient of spatial scales. Alternately, the quantal transitions could arise from interactions of integer values of frequency in oscillatory interference.

Another strength of the attractor network models is that they provide a mechanism to ensure an even distribution of different spatial phases within a single region as observed in the experimental data (Hafting et al., 2005), the network interactions in attractors prevent all neurons from coding the same spatial phase, and ensure that neurons maintain the same relative spatial phase. Single-neuron oscillatory interference models cannot directly account for the distribution of spatial phases within a single region, but they again could be supplemented by network interactions to account for this property of the data. Because network models do not use oscillations, they do not have the requirements for specific phase relationships among oscillations that the interference models have.

The hexagonal pattern of firing in attractor network models arises from the circularly symmetrical pattern of recurrent excitation and inhibition (Fuhs and Touretzky, 2006; McNaughton et al., 2006). This avoids the problem of head direction input with specific preference angle relationships but replaces it with the requirement of specific synaptic connectivity within entorhinal cortex. Some studies have not shown excitatory connections between excitatory cells in layer II of entorhinal cortex (Dhillon and Jones, 2000) whereas other studies have indicated such connections (Kumar et al., 2007). However, excitatory recurrent connections within layer II are not essential for attractor models. Attractor dynamic models could function with purely inhibitory connectivity (Burak and Fiete, 2009) or could depend upon connectivity between layer II and deeper layers such as layer III of medial entorhinal cortex.

The attractor dynamics models are not the only alternatives to oscillatory interference models. Another alternate network model avoids the need for specific recurrent connectivity by obtaining hexagonal patterns of synaptic input by self-organization of afferent input regulated by the time course of spike frequency accommodation (Kropff and Treves, 2008). In that model, the biophysical timescale of spike frequency accommodation can account for differences in grid cell spatial scaling (Kropff and Treves, 2008). Differences in spike frequency accommodation along the dorsal to ventral axis have been shown experimentally (Yoshida and Hasselmo, 2009b).

The attractor dynamic models suffer some disadvantages relative to the oscillatory interference and persistent spiking models. In particular, the initial formulation of the attractor dynamic models does not require theta rhythm oscillations. This means that the initial attractor network models of grid cells (Fuhs and Touretzky, 2006; McNaughton et al., 2006; Welinder and Fiete, 2008) do not yet simulate theta phase precession of grid cells (Hafting et al., 2008), though recently developed modifications of these models have addressed this point (Navratilova et al., 2011). Because they do not require theta rhythm, attractor dynamic models also do not require the modulation of theta rhythm frequency by running speed. In contrast to the interference models described above, the mechanism of path integration is a separate process superimposed on the network dynamics for grid firing (Burak and Fiete, 2006; Fuhs and Touretzky, 2006; McNaughton et al., 2006; Welinder and Fiete, 2008). The spacing in these models will depend upon the magnitude of velocity input, whereas the size of the firing fields will depend upon the pattern of synaptic connectivity (Navratilova et al., 2011). The size and spacing of grid cells in the data seem to change in parallel along the dorsal to ventral axis of medial entorhinal cortex (Hafting et al., 2005), but they are determined by two independent parameters of the attractor dynamics models. This contrasts with the interference models in which size and spacing are both determined by the magnitude of frequency change with velocity. In addition, because the grid pattern in attractor dynamic models requires particular patterns of synaptic connectivity, this might prevent these models from accounting for the flexible grid cell response in different tasks. Attractor models do not yet account for the immediate shift to context-dependent firing of entorhinal neurons shown in linear tasks such as the hairpin maze (Derdikman et al., 2009) or the spatial alternation task (Lipton et al., 2007), relative to the firing in the open field. In contrast, as shown in chapter 7, interference models can potentially account for context-dependent firing in linear tasks with modification in the form of phase reset.

Biophysical Models of Entorhinal Cortex

As described in the appendix, biophysical models can account for a range of intrinsic properties of entorhinal neurons, including membrane potential oscillations (White et al., 1998a; Fransén et al., 2004; Dudman and Nolan, 2009), membrane resonance (Heys et al., 2010), and persistent spiking (Fransén et al., 2002; Fransén et al., 2006). However, these detailed biophysical properties have not been fully incorporated into network models of grid cell firing properties. Instead, the initial models of grid cells described above used simplified representations of individual neurons. For example, the published attractor dynamic models of grid cells use continuous firing rate representations of individual neurons (Fuhs and Touretzky, 2006; McNaughton et al., 2006; Burak and Fiete, 2009). Current research has shown how grid cell firing properties can be obtained with networks of spiking neurons with more detailed conductance properties (Navratilova et al., 2011).

The initial interference models of grid cells represented rhythmic oscillations and spiking with cosine functions that vary in frequency but do not vary in amplitude. To fully understand these systems, we must use more physiologically detailed individual neuron models. An initial simulation has shown how networks with biophysically detailed spiking models of entorhinal neurons can simulate oscillators with frequencies that are controlled by velocity and interact to create grid cell firing fields (Zilli and Hasselmo, 2010). In this model, the effect of resonance currents on the spacing of firing fields could arise from differences in the slope of the relationship of firing frequency to depolarization. This could provide a different framework for understanding the influence of resonance currents on network spiking (Engel et al., 2008).

Hippocampal Place Cells Can Read and Store the Code

The phase of grid cell oscillations depends on location, but how can the oscillation phase be accessed by other neural structures for the encoding and retrieval of spatial location in episodic memory? As can be seen in the following figures, the spiking activity of grid cells occurs in repeated locations in the environment, giving some ambiguity about the location coded by an individual grid cell. However, the pattern of firing in multiple grid cells can be combined to provide an accurate representation of spatial location, and the readout of spatial location could be provided by a population of hippocampal place cells. This section of the chapter will address the generation of place cell responses based on grid cell input. Examples of place cells that can be generated by the model are shown in figures below.

The influence of grid cells on place cells has been modeled in several different ways (Fuhs and Touretzky, 2006; McNaughton et al., 2006; Rolls et al., 2006; Solstad et al., 2006; Gorchetchnikov and Grossberg, 2007; Fiete et al., 2008; Hasselmo, 2008c, 2009; Savelli and Knierim, 2010), but research has not converged on a single model. In general, the place cell representation arising from the grid cell representation in entorhinal cortex would depend upon the synaptic connectivity from grid cells to place cells, that is, from the entorhinal cortex to the hippocampus. Despite the multiple papers showing how place cells could be updated by grid cells, recent data suggests that grid cells might not be the only input capable of generating place cell firing in familiar environments. For example, place cells appear to develop more rapidly than grid cells in young animals (Langston et al., 2010; Wills et al., 2010), and place cells can be spared by medial septal infusions of muscimol that block theta rhythm and block grid cell firing (Brandon et al., 2011b; Koenig et al., 2011).

The creation of place cells depends upon the range of frequencies and firing field spacings described in the medial entorhinal cortex. If all the grid cells had the same spatial periodicity, then shifting them relative to each other (by shifting spatial phase) would still not uniquely define a single location in the environment. If two grid cells of the same spacing overlap in one location, then they will overlap in multiple other locations. However, if grid cells have different spacing, then their combination can uniquely identify a specific location in a small environment as shown in figure 3.18. (In a very large environment, the grid patterns in the example will eventually overlap, but the overlap could be reduced by grid cells with even larger firing fields and larger spacing.)

We can relate this to the analogy of the medial entorhinal cortex as an orchestra with instruments of different pitch that also play at different temporal intervals, with violins tending to play faster melodic lines and basses playing slower lines. Imagine that the violins play a quarter note every 2 beats, violas play a note every 3 beats, cellos play every 7 beats, and basses play every 12 notes. I admit this will sound strange, but the point is that they are playing quite actively, yet their playing will only converge at a very long interval. If a cymbal player is instructed to play a cymbal crash when all of the strings play a note simultaneously, then he would have to wait 84 beats before he would play. After 84 beats, all the stringed instruments hit a note at the same time because 84 is the least common multiple of the individual string

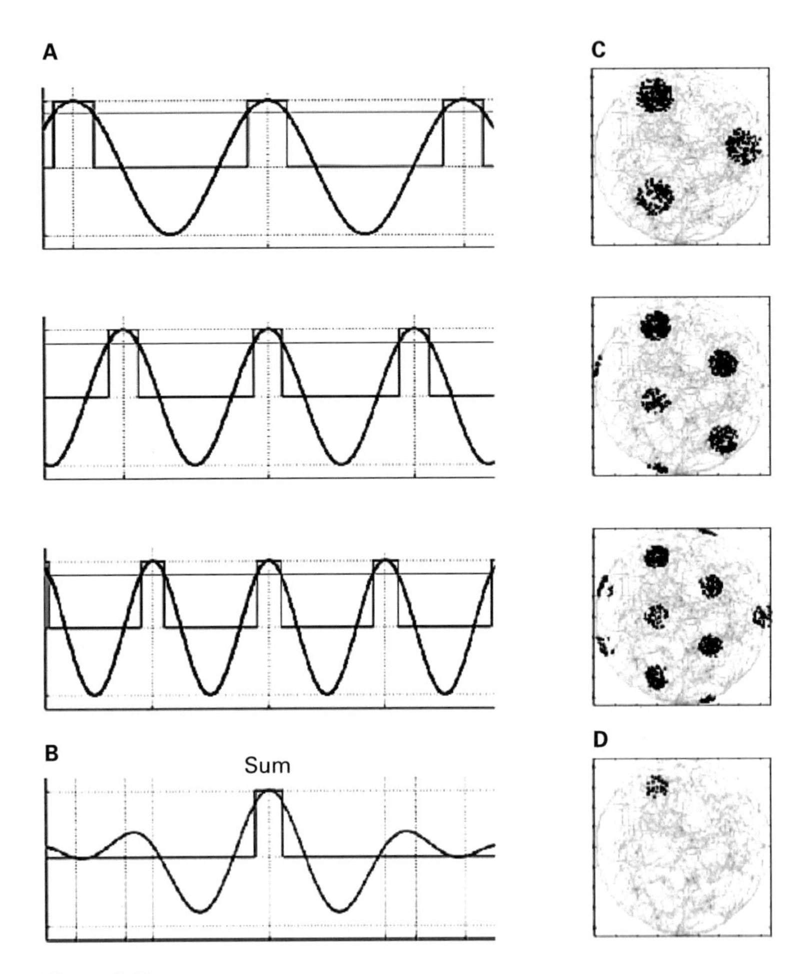

Figure 3.18
Example of how grid cells of three different spatial scales (top three rows) can be combined to generate a single place cell (bottom row) in a small environment (Erdem and Hasselmo, 2010). Left column shows neural activity induced at different locations along one dimension. Right column shows two-dimensional plot of firing patterns.

patterns (Burak and Fiete, 2006; Gorchetchnikov and Grossberg, 2007; Fiete et al., 2008; Welinder et al., 2008). In this example, the timing of the instruments could provide a code for temporal interval or, if the tempo of the instruments depends on velocity, they could code spatial location.

This is how the repeating code of grid cells could activate a very specific code of place cells that is activated only in a localized manner. This could form the association between the position along a spatiotemporal trajectory (e.g., number of beats) and the event or item at that position (the cymbal crash). Coverage of the trajectory can be obtained by having instruments play at different phases. For example, there could be 12 bass players, each of whom starts his or her repeating pattern on a different beat, and the same for the other instruments. The percussion could then play on the basis of specific combinations of individual instruments and code a cacophony of events at a series of specific times. If the hippocampus transcribers in the analogy effectively mark these in the score, then the full score can be played based on the starting phase of each stringed instrument and based on percussion players' having music that indicates their response based on specific combinations of string players.

In addition to differences in spacing, grid cells in more ventral regions of medial entorhinal cortex show larger sizes of firing fields in both the data (Brun et al., 2008) and the models. Based on this, the summation of these firing fields could result in different sizes of firing fields for hippocampal place cells. This was supported by further data from the Moser laboratory showing a progressive increase in the size of firing fields in more ventral regions of the hippocampus (Kjelstrup et al., 2008). As shown in figure 3.19, the larger firing fields show slower theta phase precession, consistent with a smaller frequency difference generating theta phase precession in the oscillatory interference model that was used to simulate the effect in the figure (Hasselmo, 2008a).

The larger field size in ventral hippocampus could provide a more general context signal—for example, coding one room versus another in the example of finding out where you left your car keys. The broader context signal in the ventral hippocampus has also been suggested to be more relevant to coding the phemomenon of fear or anxiety, based on the differential effects of lesions of the ventral hippocampus on fear or anxiety in the rat (Moser and Moser, 1998). This is consistent with the greater connectivity of ventral hippocampus with structures implicated in fear and anxiety such as the amygdala. This suggests that coding of larger spatial scales could be more relevant to the mechanisms of fear and anxiety (Hasselmo, 2008b).

In the model of episodic memory presented here, place cells are created by selection of random subsets of three grid cells' inputs and computation of the overlap in firing for these three grid cells. The place cell is assumed to fire anywhere these three grid cells fire simultaneously. This generates a spatial firing pattern for each place cell that is evaluated by taking the standard deviation of spiking location in the x and y

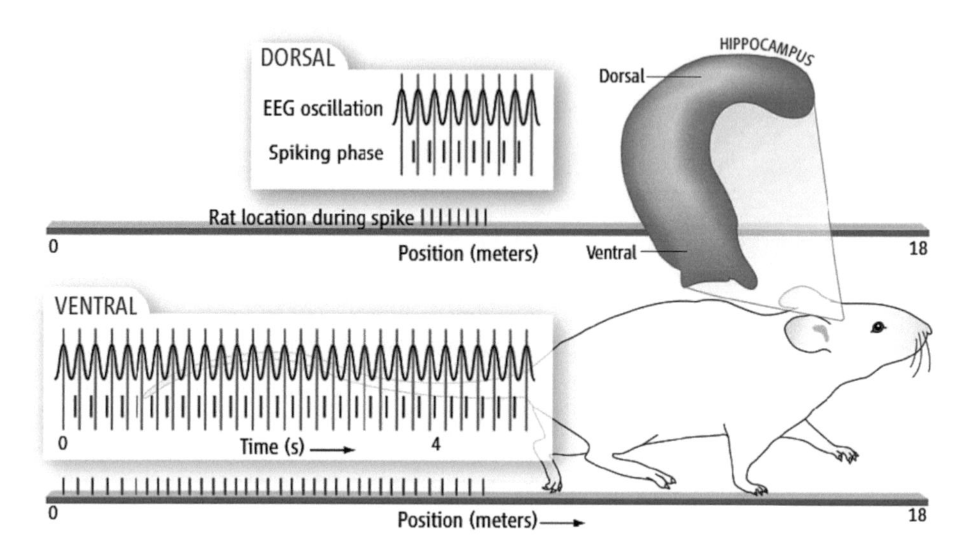

Figure 3.19
Schematic demonstration of data from the Moser lab (Kjelstrup et al., 2008) showing the difference in firing field size for place cells recorded in dorsal hippocampus (top) and ventral hippocampus (bottom). In the data, the larger firing fields are associated with slower theta phase precession of spikes relative to theta oscillations. The firing fields and theta precession were created using the oscillatory interference model (Hasselmo, 2008a, b; P. Huey/SCIENCE).

dimension. If the standard deviation of spiking location is smaller than a previously set parameter in both dimensions, then this place cell is selected for inclusion. Thus, the model specifically selects combinations of grid cells that overlap only in small regions of the environment. The input synapses from the randomly selected subset of three grid cells to this place cell are then strengthened. This ensures that the same place cells are reliably activated in the same location dependent on the pattern of grid cell spiking induced by the spatial phase of grid cells. Examples of place cell firing generated by the model are shown in figure 3.20.

The figure shows three stages of the system. Input from speed modulated head direction cells can drive entorhinal neurons with different frequency to velocity relationships thereby producing grid cells with different spatial scales (Giocomo et al., 2007). Grid cells with different spatial scales at different dorsal to ventral positions in the entorhinal cortex can drive place cells of different sizes in the hippocampus, with smaller firing fields in dorsal hippocampus and larger fields in ventral hippocampus (Hasselmo, 2008c; Kjelstrup et al., 2008). The difference in intrinsic frequency and spatial scaling forms the basis for the orchestra analogy of medial entorhinal cortex and hippocampus. The effective coding of space and time could depend upon responses

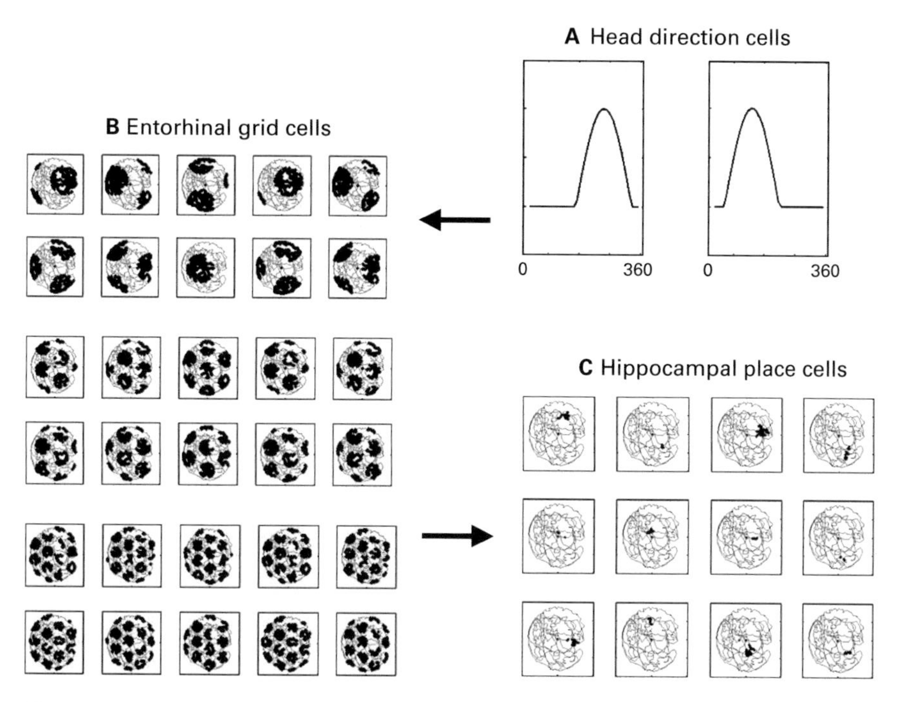

Figure 3.20
(A) Examples of the tuning properties of head direction cells used in the episodic memory model, showing activity versus head direction angle. (B) Examples of firing fields produced by 30 grid cells in the model during an extended period of rat exploration in a circular open field environment, selected from a total of 75 grid cells in the model. The difference in spacing and size of firing fields arises from differences in the slope of the frequency to velocity relationship. The difference in spatial phase offset results from differences in initial phase. (C) Examples of the firing fields in the circular environment of 12 place cells created in the model through random selection of inputs from groups of three grid cells (Hasselmo, 2009).

on a number of different spatial or temporal scales that correspond to different resonance properties of neurons. This model is described mathematically in the appendix, where the population of grid cell firing can be described as a grid cell vector, and the pattern of place cell firing can be described as a place cell vector. These can interact via a matrix of synaptic connections.

The examples in this chapter have illustrated how neurons can code continuous dimensions of space and time. The following chapter will describe how the interactions of these neurons can be used to code the spatiotemporal trajectory of an episodic memory.

4 Encoding and Retrieval of Episodic Trajectories

Now that we have the basic components of the model, we can finally address the central question: How does the model of episodic memory encode and retrieve a spatiotemporal trajectory? The previous chapter describes how spiking activity in the entorhinal cortex can encode the dimensions of space and time, through either the relative phase of spiking or the firing rate. The previous chapter also described how input from neurons coding speed and head direction can update the activity of grid cells and how the output of the entorhinal cortex can be read out by the spiking activity of hippocampal neurons.

In the framework presented here, episodic memory involves encoding a spatiotemporal trajectory and forming associations between events or items and specific positions on this trajectory. The encoding of the spatiotemporal trajectory itself involves coding of different positions along the trajectory that can be defined by location or by time or even by features of sensory input (Hasselmo, 2009). The process also involves a mechanism for linking between positions on the trajectory. The process of linking between positions on a trajectory could use a variety of techniques, including associations of trajectory position with velocity, associations of trajectory position with speed, or even the representation of evolution of time during the trajectory in the form of specific temporal intervals. This chapter will describe how these populations of neurons could be used for the processes of encoding spatiotemporal trajectories and, subsequently, how these circuits could mediate the process of retrieval. This model can be applied to human episodic memory as well as rat episodic memory. The computational details of the model are presented in the Appendix (Hasselmo, 2009).

Encoding a Trajectory

The encoding of a spatiotemporal trajectory involves the formation of associations between positions of the spatiotemporal trajectory so that one position can activate subsequent positions (or even preceding positions). The description here will focus on the use of phase coding in the persistent spiking model as this provides a framework that could be used for updating neural activity based on velocity or speed but also could use updating based on temporal intervals alone. In the case of velocity or speed, the grid cells activate place cells that activate cells coding velocity or speed and thereby further update the grid cell firing pattern. For updating based on temporal intervals, persistent spiking cells are activated by sensory features, and they fall into a characteristic spiking frequency that allows coding of temporal intervals as shown in the section on phase coding of time in chapter 3.

I will first describe a mechanism based on associations of position with velocity as presented in a recent paper (Hasselmo, 2009). In this framework, synaptic modification plays a role in encoding of a new trajectory. Encoding of associations between position and velocity can be performed by modification of synaptic connections between active hippocampal place cells and active head direction cells. This synaptic modification allows formation of associations between individual states (spatial locations) and the actions (movements) performed in those states.

During initial encoding of a trajectory, the activity of all cells in the network is determined by the actual velocity of the animal during movement. The sensory input about the actual running speed and head direction of an animal drives the activity of a population of cells sensitive to head direction and running speed that code velocity. Different speed-modulated head direction cells in the population will respond differently depending on their preferred direction, so that together they can accurately code velocity (as a head direction cell vector).

The velocity input from these cells updates the population of grid cells according to the oscillatory interference model of grid cells (creating a grid cell vector). As described in chapter 3, the velocity coded by an array of head direction cells updates the frequency and thereby the phase of an array of cells with different initial phases and different frequency to velocity relationships. These cells then activate different grid cells with different spatial phases and field spacing. The grid cells then activate a pattern of individual place cells in a population (a place cell vector). Note that the framework presented here could also use other variants of the grid cell model, such as the attractor model of grid cells, as long as grid cell firing is updated by head direction cells and speed, and place cell firing is influenced by grid cell output.

In this framework, initial encoding of a trajectory involves Hebbian strengthening of a set of synapses associating each location (coded by the population of place cells, also known as the place cell vector) with the associated action for movement performed at that location (coded by populations of cells coding speed and head direction, the head direction cell vector). The cells coding speed and head direction could be separate, but in this example we will assume we have speed-modulated head direction cells. Thus, a trajectory is stored by formation of associations via synaptic modification of the synaptic connections between the population of place cells and the

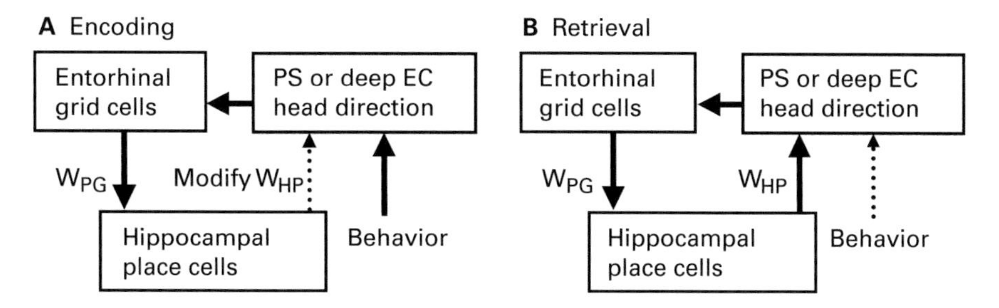

Figure 4.1

Proposed circuit for episodic encoding of trajectories. (A) During encoding, cells coding head direction and speed in the postsubiculum (PS) or deep layers of medial entorhinal cortex (deep EC) are driven by external behavior. Head direction and speed input updates grid cells in medial entorhinal cortex that drive place cell firing in the hippocampus. Associations between state (place) and action (speed and head direction) are formed by strengthening synapses between place cells and cells coding head direction and speed (dotted line: W_{HP}). (B) During retrieval, behavioral input does not drive the cells coding head direction and speed (dotted line). Place cells drive cells coding head direction and speed via previously modified synapses W_{HP}. Cells coding head direction and speed update grid cells that update place cells to complete the loop driving internal retrieval. Figure adapted from Hasselmo (2009).

population of head direction cells. As described in the Appendix, the mathematical model represents the synapses with a synaptic connectivity matrix. This synaptic modification during encoding of the trajectory completes the circuit shown in figure 4.1 that forms the basis for later retrieval of the trajectory. Note that during encoding the synapses being modified are not the predominant influence on postsynaptic activity. Instead, head direction activity is driven by behavior, and the place cells do not influence the head direction activity during encoding.

The synaptic connections that are strengthened between place cells and head direction cells could correspond to direct projections from region CA1 of the hippocampus to the postsubiculum (van Groen and Wyss, 1990) or to deep layers of entorhinal cortex (Naber et al., 2001), or to polysynaptic effects of projections from the hippocampus to the subiculum (Swanson et al., 1978; Amaral and Witter, 1989) as well as the postsubiculum (dorsal presubiculum) and parasubiculum (Naber and Witter, 1998) and from postsubiculum to layer III of the medial entorhinal cortex and from parasubiculum to layer II (Caballero-Bleda and Witter, 1993, 1994). This is just one possible mechanism for encoding of trajectories. The encoding of trajectories does not necessarily require associations between place cells and head direction cells. Alternate mechanisms could involve coding of temporal intervals in entorhinal cortex that can independently activate representations of place and action in the hippocampus and postsubiculum.

Retrieving a Trajectory

As shown in figure 4.1, during retrieval the dynamics differ from encoding. During encoding the head direction activity is driven by behavior whereas during retrieval the head direction activity is primarily driven by synaptic output from the hippocampus. This retrieval output involves place cell activity that spreads across synapses that were previously strengthened during encoding.

The cue for retrieval can take a number of forms. The cue can be the actual current location of the rat, coded by place cells updated by grid cells. The cue could also be environmental stimuli that have been associated with a particular pattern of place cells that can reactivate the place cell pattern (vector). Sensory stimuli are used to cue retrieval in some of the examples shown below, using the mathematical representation of associations between sensory stimuli representing items or events and places. Finally, a place cell cue that does not correspond to the actual current stimulus or location in the environment could be activated by internal mechanisms.

Whatever the source, the place cell cue can initiate retrieval of a trajectory by activating the associated action. Thus, the head direction activity can be retrieved from the current place cell activity. This forms one component of a functional loop for retrieval in the model. This loop retrieves trajectories via three stages: (1) The place cell vector activates the associated head direction cell vector via the synaptic connectivity matrix W_{HP}, (2) the head direction cell vector updates the phase of a population of entorhinal grid cells via the grid cell model, and (3) the grid cell vector updates the place cell vector via the synaptic connectivity matrix W_{PG}. The new place cell activity vector then activates the associated head direction pattern, and the loop continues driving retrieval of the trajectory. The retrieval loop is summarized in figure 4.1 and the mathematical details are provided in the Appendix.

The previous place cell activity spreads across the matrix W_{HP} to drive activity in the head direction cells. The grid cell model integrates this activity to update grid cell phase, potentially using interference of persistent spiking or oscillations or using attractor dynamics driven by head direction cells. Then the grid cell activity is transformed through a matrix W_{PG} representing the synaptic drive of grid cells on hippocampal place cells. Thus, the pattern of connectivity in the two synaptic matrices in the figure encodes the episodic memory for the trajectory in the model.

The effect of head direction on grid cell activity during retrieval requires that the head direction vector (coding a velocity vector corresponding to rat action) needs to maintain activity for the period Δt until a new place cell representation causes retrieval of the next head direction response. This is supported by graded levels of persistent firing that have been shown in intracellular recording from neurons in deep layers of entorhinal cortex slice preparations (Egorov et al., 2002) and intracellular recording in the postsubiculum (Yoshida and Hasselmo, 2009a). This provides a mechanism by which head direction cells in the deep layers of entorhinal cortex (Sargolini et al., 2006) and postsubiculum (Taube et al., 1990b; Sharp, 1996) may be involved in encoding and retrieval of trajectories.

Note that retrieval has different dynamics from encoding, as shown in figure 4.1 above. During encoding, the sensory input of velocity determines activity of head direction cells, which then drive grid cells and place cells. During retrieval, sensory input has less influence on the system, and head direction activity is determined primarily internally by previously modified synapses. Awareness of this internal difference in dynamical state allows differentiation of retrieved events from current sensory input. The difference in dynamics could be determined by modulatory or attentional influences on the postsubiculum or deep entorhinal cortex determining the relative influence of different synaptic inputs. Encoding and retrieval of single associations or items have been modeled as occurring on different phases of each theta cycle (Hasselmo et al., 2002a). A slower transition between encoding and retrieval could involve modulatory regulation of afferent versus feedback transmission by muscarinic acetylcholine receptors (Hasselmo, 2006). The possible role of acetylcholine and theta rhythm in regulating encoding and retrieval is discussed further in chapter 6.

Example of Trajectory Encoding and Retrieval

The model described here performs encoding and retrieval of complex spatial trajectories. Figure 4.2 illustrates the basic components of the trajectory retrieval, which includes grid cell phase, place cell activity, and head direction cell vectors (see appendix). In the figure, the actual trajectory run by the rat is shown in gray. This trajectory is from experimental data obtained in the Moser laboratory (Hafting et al., 2005). The rat forages along a meandering trajectory through an open field environment.

Part A of figure 4.2 shows the locations coded by the phase of grid cells in the entorhinal cortex during encoding (gray) and during retrieval (dashed black line). The trajectory starts in the upper right. The lines are based on decoding the two components of the phase in entorhinal cortex driven by head direction input. As described in chapter 3, if the network were receiving input from one cosine tuned head direction cell coding north–south, and another head direction cell coding east–west, then the phase could be mapped directly to the x and y Cartesian coordinates in the figure. However, for the grid cell model, the head direction input cells have preferred angles differing by 120 degrees, so these slanted coordinates need to be decoded back to Cartesian coordinates for the figure.

In part B of figure 4.2, the individual place cells activated by the pattern of grid cell activity at different positions of the trajectory are shown as open circles. During encoding, each of these place cells is associated with the pattern of head direction cell

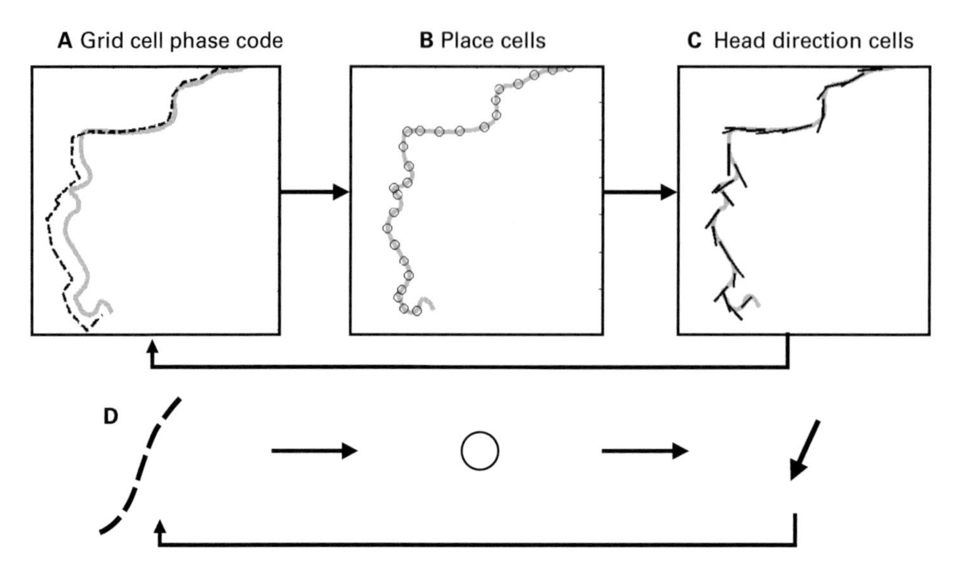

Figure 4.2
Encoding and retrieval of a continuous trajectory. (A) The continuous set of locations coded by the relative phase of modeled grid cells during encoding are plotted in gray, and the locations coded during retrieval are plotted as a dashed black line (Hasselmo, 2009). During retrieval, the relative phase shifts according to the head direction vector until it activates new place cells that alter the head direction vector and change the direction of phase shift. (B) The sequentially activated place cell representations are shown as open circles on the gray line representing the actual trajectory. (C) Examples of head direction activity vectors associated with specific locations along the trajectory during encoding are shown as straight, black line segments. During retrieval, place cells activate corresponding head direction vectors that drive the update of grid cell phase. (D) An example of retrieval at one location is shown. A place cell (circle) activates a head direction vector (arrow) that shifts the relative phase code (dashed line) of grid cells in the direction coded by the head direction vector.

activity shown as a vector in part C. During retrieval, activation of these individual place cells within the place cell vector reactivates the associated head direction cell vector.

In part C, the straight line segments are vectors representing the pattern of speed-modulated head direction cell activity induced by the behavioral input during encoding, and induced by place cell activity during retrieval, starting in the upper right. For example, if the place cells activate head direction cells with preferred direction west, then the vector points to the west (as in the upper right of C), or if the place cells activate head direction cells coding south, then the vector points to the south.

In simulations, during retrieval, the network can retrieve the full sequence from a cue consisting of an initial pattern of place cell activity along the trajectory. The initial place cell activity (one circle in 4.2B) activates the head direction cell activity (line in 4.2C) associated with that pattern of place cell activity. The head direction vector drives the update of the phase vector driving the grid cells (4.2A) for a period of time. This occurs through the influence of the head direction activity on the oscillation frequencies of velocity-controlled oscillators that drive grid cells. This causes the velocity controlled oscillators to show a continuous change in the phase representing location (4.2A) that moves in the direction coded by the head direction vector (C).

The change in the phase vector of velocity-controlled oscillators changes the pattern of activity in the grid cell vector until it eventually activates a new place cell representation (4.2B). Activation of a new place cell activates the associated head direction cell activity vector (4.2C), and this new head direction vector drives further update of the phase of velocity controlled oscillators that update the grid cell vector representing location. Thus, a recursive process allows place cell activity (state) to drive retrieval of the next action (speed modulated head direction) to update the phase of velocity controlled oscillators. This updates the grid cell vector to activate the next place cell pattern (state) in a repetitive manner along the length of the trajectory.

The network retrieves complex spatial trajectories with accuracy dependent upon the resolution of place cell representation. For lower numbers of place cells, the network shows progressively poorer performance, and the internal representation of place deviates more from the actual trajectory (Hasselmo, 2009).

The network shown above can be interpreted as a chain of associations between place, head direction, and grid cells, but the network does not necessarily have to use this circuit. There are objections against chaining models based on experimental data showing that participants can effectively retrieve the end of a sequence after missing one or more items, or they can retrieve the wrong order of items in a sequence (Lashley, 1951; Henson, 1998; Terrace, 2005). These objections do not necessarily apply to a spatiotemporal trajectory because the items are not elements of the trajectory itself. However, there are alternative mechanisms that could be used instead of the one described above, or that could be used together with the one described above. In one alternative mechanism, the network could have input that activates evolution of spiking based on time alone, as described in chapter 3, or based on running speed.The circuit dynamics of these alternate models are shown for retrieval in figure 4.3. In figure 4.3A, the cue activates theoretical entorhinal cells that fire at regular time intervals (time cells), and their synchronous firing could activate theoretical time cells coding specific time intervals from the cue that could trigger other cue cells (or cells coding time interval could directly activate the cue cells). The time cells could arise from oscillatory interference in the manner shown in figure 3.11.

Another alternate framework (figure 4.3B) has the running speed of an animal updating cells that code arc length intervals along the trajectory (see appendix). Arc

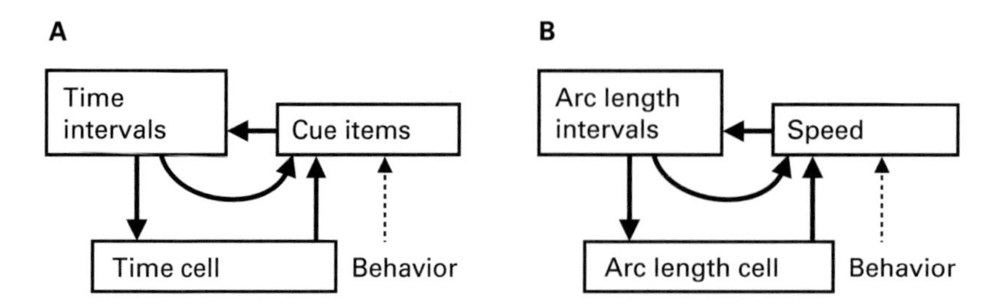

Figure 4.3
(A) Alternate circuit for episodic memory. During retrieval in this circuit, cue items activate theoretical cells coding time intervals that can update further cue items. (B) Another alternate circuit during retrieval. Cells coding arc length of a trajectory undergo recurrent interactions with cells coding running speed.

length refers to the measurement of the path length along an irregular trajectory, similar to what an odometer or pedometer would measure as one travels along a winding trajectory, as opposed to the direct two-dimensional Euclidean distance "as the crow flies" from a starting point to a given location. Arc length cells are still theoretical, as data has not differentiated cells coding Euclidean distance versus arc length. Cells coding arc length can be generated by oscillatory interference driven by the running speed of an animal. These cells could then either directly or indirectly retrieve associated running speed.

Associations with Items or Events

Episodic memories usually include encounters with events or items at particular positions along the spatiotemporal trajectory. The example episode of my morning at Boston University in chapter 1 has events familiar to most people, involving speaking with different people at different locations. How can these types of events be encoded? The simulation of episodic memory described here includes formation of associations between the place cell population and neuron population activity representing the stimulus features of individual objects or events experienced at specific positions on a trajectory. Individual objects that did not move were assigned locations near the trajectory, and the object vector was active when the agent was within a specific distance of the object location. For the events in the example in chapter 1, the location as well as the time window were defined. During encoding, Hebbian modification of synapses formed associations between the place cell vector and the stimulus feature vector (the neural activity coding items in the lateral entorhinal cortex). The retrieval of items shown in figures below resulted from the place cell vector activating the stimulus feature vector.

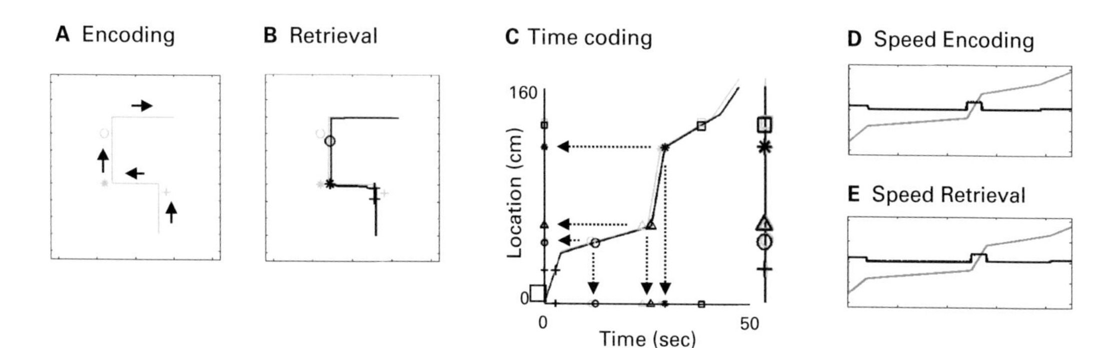

Figure 4.4
Example of retrieval of events or items along simple trajectories. (A) A four-segment trajectory is shown as a gray line during encoding. The individual events or items experienced as sensory stimuli along this trajectory are shown as three different symbols (plus sign, asterisk, circle) at different locations and times. (B) The location retrieved by grid cell phase is plotted as a thick black line that matches the gray line showing location during encoding. Events or items are plotted as black symbols according to grid cell phase at time of retrieval, showing retrieval that matches location and time along the trajectory during encoding (plus, asterisk, and circle symbols plotted in black). (C) Encoding and retrieval of a separate one-dimensional trajectory (line at right) with five items at different locations during different running speeds (changing from 10 cm/sec to 1 cm/sec to 20 cm/sec to 2 cm/sec to 5 cm/sec). Spatial location on the trajectory (y axis) is plotted versus time (x axis). The circle and triangle are close in location, but slow running speed (1 cm/sec) means they are encoded and retrieved at a long time interval. The triangle and asterisk are far apart in location, but fast running speed (20 cm/sec) means they are encoded and retrieved at a short time interval. (D–E) The plot of running speed (black lines) during change in location (gray line) shows that running speed during encoding (D) is replicated during retrieval (E). Figure adapted from Hasselmo (2009).

Figure 4.4 shows the encoding and retrieval of a simple trajectory with individual items (or events) at individual points near the trajectory. In this simulation, individual items are located near specific points along the trajectory as shown by the location of the gray symbols in part A. The trajectory first passes the gray plus sign, then the gray asterisk, then the gray circle. During encoding, the pattern of place cell activity is associated with patterns representing the individual items. During retrieval, the trajectory progressively activates place cell activity at different positions along the trajectory as shown in part B. Individual place cells associated with specific items then activate vectors representing the individual items. The locations where individual item representations are activated are shown with black symbols directly on the trajectory. Note that during retrieval, the network effectively retrieves first the black plus sign (near the first turn), then the black asterisk (at the second turn), and finally the black circle.

Because the network includes a representation of running speed, this allows the network to encode and retrieve trajectories based on temporal intervals as well as spatial location. As shown for a different trajectory in figure 4.4C, items or events can be retrieved based on the temporal intervals at which they were encountered. In this different trajectory example, five items are encountered and encoded during movement in one dimension. During retrieval, the circuit retrieves the trajectory and the position of the items on this trajectory. The speeds and distances used in this simulation correspond to standard running speeds of a rat encountering items in an environment, but they could be adjusted to represent human movement speeds. Two items (circle and triangle) encountered at a slow movement velocity (1 centimeter per second) are encoded and retrieved at a long temporal interval (12 seconds, x axis) despite being close in space (12 centimeters, y axis). In contrast, two items (triangle and asterisk) encountered at high velocity (20 centimeters per second) are encoded and retrieved at a short temporal interval (3–4 seconds, x axis) despite being far apart in space (60 centimeters, y axis). As shown in figure 4.4D and 4.4E, the model encodes the different velocities at different positions (going from 10 centimeters per second to 1 centimeter per second to 20 centimeters per second to 2 centimeters per second to 5 centimeters per second) and effectively replicates these different velocities during retrieval.

Theoretical Time Cells and Arc Length Cells

Figure 4.5 shows a trajectory that causes problems for encoding based purely on a chain of associations between locations and actions. When there is ambiguity due to overlap in the location (state) representations, the network can miss segments of the trajectory or become trapped in loops. In figure 4.5B, the network shows retrieval which misses a large loop in the upper right portion of the trajectory. This occurs because at the point of trajectory overlap, place cells are associated with two different actions (go north and go west) on the trajectory. In figure 4.5B, the place cells activate the action (go west) associated with the later portion of the trajectory and this causes retrieval to skip ahead to the last segment of the trajectory, missing the loop in the upper right and two of the items.

The retrieval of trajectories with self-crossing sections or overlapping sections can be greatly enhanced by combining the mechanism based on place cells with one of two other theoretical types of cells: (1) cells that code time intervals alone or (2) cells that code the arc length of the trajectory (Hasselmo, 2007). The generation of these theoretical cell types is described in figure 3.11, figure 7.7 and appendix p. 266. These types of theoretical cells can be obtained using the variant of the model described above in figure 4.3 that use different inputs to update frequency. If head direction input is removed and the frequency is only modulated by running speed in the model,

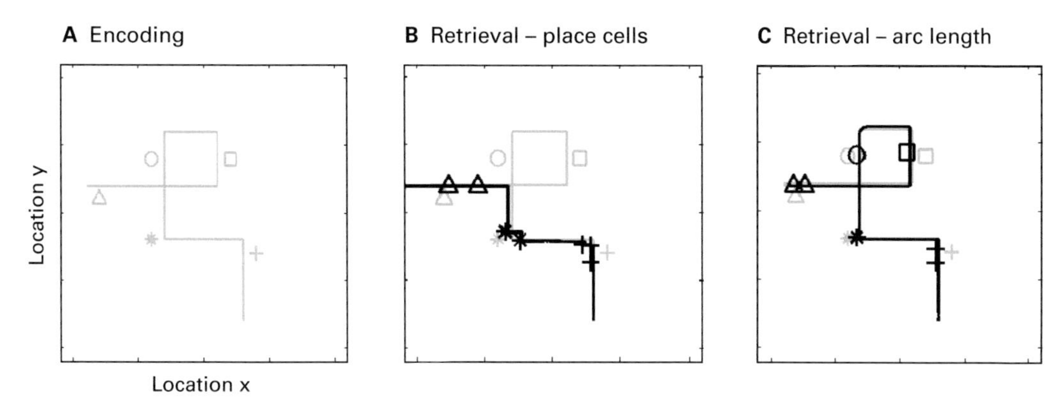

Figure 4.5
Self-crossing trajectory. (A) Gray lines show the self-crossing trajectory that will be encoded, along with the location of individual items at different positions along the trajectory. (B) With place cells only, the retrieved trajectory (thick black line) follows the actual trajectory (gray) until it encounters the self-crossing section. At this time, the place cells activate head direction cells going the wrong direction, causing the retrieval to miss a segment of the trajectory. (C) With theoretical arc length cells, the network retrieves all components of the trajectory in order. The grid cell phase representation of location (black line) effectively follows the full self-crossing trajectory. At the self-crossing location, theoretical arc length cells maintain context-dependent separation of the pathways because they respond differently based on the one-dimensional measure of trajectory arc length from starting point instead of two-dimensional spatial coordinates. Adapted from (Hasselmo, 2009).

then cells will respond on the basis of the specific arc length (travel distance) along the trajectory from a given reference location. In contrast, if both head direction and speed input are removed, then cells will maintain fixed differences in frequency that will cause interference at fixed time intervals (see figures 3.11 and 7.7).

Both types of theoretical cells can be obtained in simulations by assuming that a specific item or location activates persistent spiking neurons in the entorhinal cortex. Persistent spiking neurons with fixed frequency will interfere with a baseline frequency at a specific time interval from their onset as described in chapter 3. Neurons involved in arc length could fire at a rhythmic rate that is modulated by speed but not by head direction. The modulation by speed drives neurons at frequencies that differ from the baseline frequency of other persistent spiking cells. The neurons receiving convergent input from these persistent spiking cells will fire dependent upon when the shift in frequency causes constructive interference with the baseline cells.

Lesions of the hippocampus impair the ability to learn and disambiguate nonspatial sequences such as a sequence of odors in a sequential discrimination task (Agster et al., 2002) and impair the ability to retrieve information about the order of stimuli

(Fortin et al., 2002; Kesner et al., 2002), indicating the role of the hippocampus in encoding the temporal position of an event.

As described above, the mechanism for arc length cells resembles the model of grid cells, except that the interaction involves only two oscillations and the modulation of frequency depends on running speed only rather than speed-modulated head direction. Note that each arc length cell will fire at an arc length dependent upon its initial phase. The threshold and frequency of arc length cells were set in the simulations so that they would fire once during the trajectories studied here.

Simulations of arc length cells (Hasselmo, 2007) are able to replicate a number of features of the physiological data from the hippocampal formation, including the phenomenon of context-dependent "splitter" cells that fire selectively for right or left turn trials in continuous spatial alternation (Wood et al., 2000; Lee et al., 2006), the forward shifting of these splitter cells toward goal locations (Lee et al., 2006), and the context-dependent firing of neurons in a delayed nonmatch to position task (Griffin et al., 2007) or plus maze tasks (Ferbinteanu and Shapiro, 2003; Shapiro et al., 2006; Smith and Mizumori, 2006a). A similar function could be obtained from cells with frequencies fixed in time that can code pure temporal intervals (Hasselmo et al., 2007). As shown in figure 4.3, a circuit composed of arc length cells, speed modulated cells, and cells with frequencies modulated by speed can model sequential activation of arc length cells. Analogous to the connection between place cells and head direction cells, a connectivity matrix links a population of arc length cells with the associated running speed at each position along the trajectory (figure 4.3B). The analogous connection in the time interval circuit connects neurons coding specific time intervals with neurons representing cue items that can activate a separate set of neurons coding time intervals from the appearance of that cue (the same could be obtained by having the circuit coding time intervals connect with itself).

Note that the arc length cell could be associated with the mean speed on a segment of the trajectory rather than the instantaneous speed as speed can vary considerably along a segment, but the arc length depends on mean speed and time interval. Once an arc length cell has activated the appropriate mean speed during retrieval, the neurons representing the mean speed should modulate the frequency of another set of neurons, potentially the persistent firing cells in layer III of entorhinal cortex (Tahvildari et al., 2007). The change in frequency of these cells can update the phase of the interference giving rise to arc length cells. In simulations, in addition to the associations with speed, arc length cells are combined with the place cell activity and represented together by the vector of place cell activity so they together form associations with speed modulated head direction cells. This forms associations between arc length cells and the action at the corresponding position on the trajectory. Arc length cells are also associated with stimuli representing items or events.

During encoding, arc length cells are created at regular intervals of arc length when the magnitude of interference crosses a threshold. In these simulations, the threshold and frequency shift of arc length cells were set so they would be active only once on the trajectory. At the start of retrieval, the arc length phases are reset to a cue phase and arc length activity is computed based on retrieved speed rather than external speed. This causes sequential activation of arc length cells to drive speed-modulated head direction and thereby drive the correct update of grid cell phase and place cell firing over time.

Figure 4.5C shows how this mechanism overcomes the problem of self-crossing trajectories. This mechanism uses arc length cells that provide a context-dependent signal for different portions of the trajectory, allowing differential representation of the same location depending on the preceding spatial locations. As described above, the arc length cells respond on the basis of the arc length along the trajectory rather than actual spatial position, thereby disambiguating overlapping locations occurring at different arc lengths along the trajectory. Each time a new arc length cell is activated, the arc cell activates the correct head direction cells and drives the grid cell phase to the correct spatial location. Figure 4.5C shows how retrieval of self-crossing trajectories is enhanced by inclusion of arc length cells. At the overlapping segment, the arc length cells activated for the early part of the trajectory are distinct from those coding the later segment, so they guide the grid cell phase through the overlapping segment and allow activation of the grid cell phase and place cells representing the loop in the upper right corner of the trajectory.

Remembering Where You Parked Your Car

One of the stated goals of this model was to address aspects of episodic memory associated with the use of episodic memory in day-to-day life. In particular, episodic memory is commonly described as the memory mechanism that allows you to perform familiar everyday tasks such as remembering where you parked your car in a parking lot. The episodic memory example I gave in the first chapter starts with parking my car in the Warren Towers parking garage at Boston University. Almost every day that I return to my parked car, I think about the mechanisms for retrieval of this episodic memory.

Figure 4.6 illustrates how the model presented here can be used to encode and retrieve common types of episodic memories, such as the location of a parked car. The trajectories in figure 4.6 illustrate the example of parking a car in the same parking lot on two different days. On the first day (cue yesterday), the trajectory in gray represents the car driving through the entrance (plus sign) and parking in a location to the left (circle). The trajectory continues with the driver walking into the building

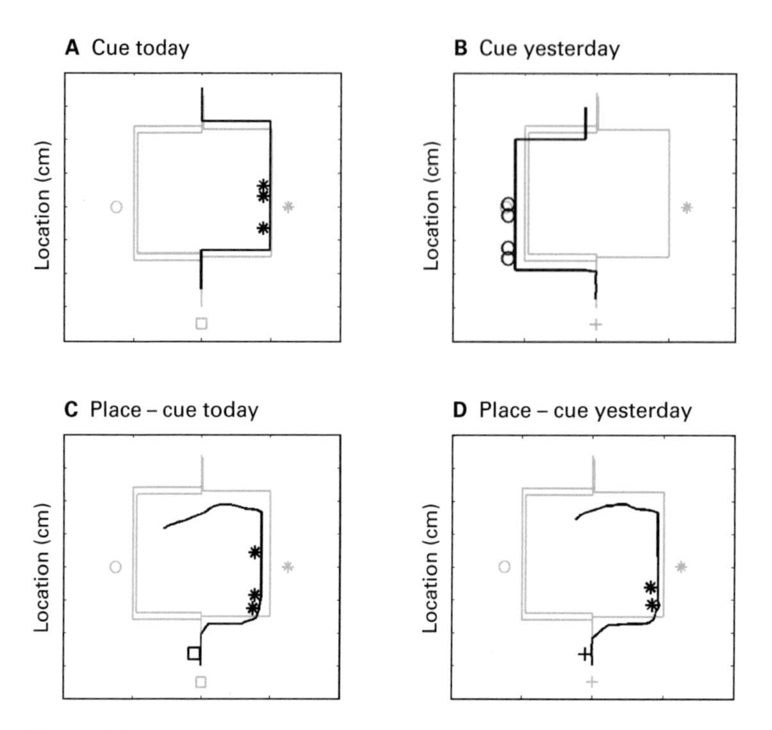

Figure 4.6
Episodic memory for parking location. The trajectory in gray shows the encoding of two consecutive days of parking in the same parking lot. (A and B) Arc length cells allow selective retrieval of either day. (A) When cued with a stimulus (square symbol) specific to only the second day (cue today), the arc length cells effectively retrieve the correct second-day parking location (right side, asterisk symbol). (B) When cued with a stimulus (plus sign) specific to only the first day (cue yesterday), the arc length cells effectively retrieve the correct first-day parking location (left side, circle symbol). (C and D) Place cells only retrieve the second day. (C) When cued with the location at the time of the second-day entrance to the parking lot (square symbol, cue today), the place cells effectively retrieve the second-day parking lot location (right side, asterisk symbol). (D) When cued with the location at the time of the first-day entrance to the parking lot (plus sign, cue yesterday), the place cells still retrieve the second-day parking lot location (right side, asterisk symbol) because of weakening of place to action associations with the first-day trajectory. Adapted from Hasselmo (2009).

entrance (top) and later returning to the car on the left and driving back out the parking lot entrance from the left. On the second day, the trajectory represents the car driving through the same entrance but with a different cue indicating the second day (square symbol, cue today). This could indicate a different event or stimulus on that day. The trajectory then goes to a parking location on the right (asterisk). The trajectory continues with the driver walking into the building entrance (top).

Retrieval of the episodic memory is tested on the second day as the driver comes back out of the building. In figure 4.6A, the cue for memory retrieval is a specific event or stimulus (square symbol) that took place during the drive to the entrance on the second day (cue today). This activates reset of arc length cells that guide correct retrieval of the trajectory to the right side parking location (asterisks).

Alternately, the arc length cells can mediate retrieval of the parking location on the previous day (cue yesterday) as shown in figure 4.6B. This uses a cue for memory retrieval in the form of a stimulus or event (plus sign) that took place during the drive to the entrance on the first day (yesterday). The first day cue activates reset of arc length cells that then guide retrieval of the trajectory to the left side parking location (circle).

Can the system work with only place cells? No, with place cells alone the network will only retrieve the more recent trajectory. As shown in figure 4.6C, a network without arc length cells but with place cells can be cued with an event from the current date at the entrance of the parking lot and can retrieve the recent episode (square symbol, cue today) of parking on the right (asterisk). However, a retrieval cue using an event at the entrance on the first day (plus sign, cue yesterday) does not retrieve the left side parking location (circle). Instead, the trajectory retrieval follows the recent trajectory to the right side (asterisk) as shown in figure 4.6D. This occurs because the recent modification of synapses between place cells at the entrance and speed modulated head direction cells coding the right turn reduces the strength of previously modified synapses associated with the previous left turn. Thus, the entrance location is associated with the recent trajectory and cannot retrieve the older trajectory. Pure place cell to action associations represent temporal recency by differences in strength. In contrast, the arc length cells in this model allow distinct encoding and retrieval of the location in which a car was parked on a previous day, as long as a distinct cue can be accessed to appropriately cue the reset of arc length cells.

The sensory cueing of trajectory retrieval is demonstrated further in figure 4.7A. This figure shows examples of a trajectory passing multiple items. During retrieval, presentation of individual items as cues (such as the 'x', the 'o,' and the diamond) can activate the associated phase and arc length cell activity, initiating retrieval of the remainder of the trajectory and subsequent items. As shown in figure 4.7B, the arc length cell representation allows successful retrieval of one trajectory even if it has strong overlap with a different trajectory. The retrieval of pathways with different

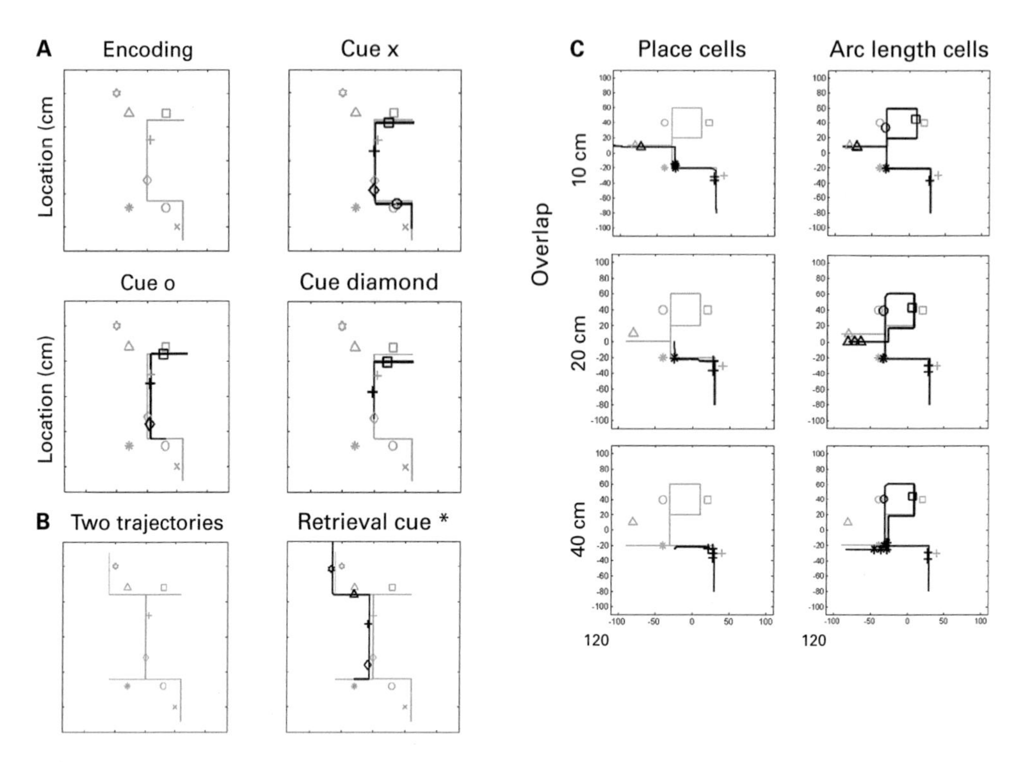

Figure 4.7

(A) Encoding and retrieval of a trajectory (gray line) passing multiple items (gray symbols). Sensory input vectors representing individual items (cue items 'x', 'o,' or diamond) can cue initiation of retrieval at different points along the trajectory, after which the remainder of the trajectory and associated items are retrieved. (B) The left side shows two overlapping trajectories learned by the network. The right side shows successful retrieval of one of these overlapping trajectories cued by the item at the asterisk. (C) Retrieval with different amounts of overlap of trajectories. Each row shows unsuccessful performance with 400 place cells and successful performance with 50 arc length cells. Top row has overlap of 10 cm, middle is 20 cm and bottom is 40 cm overlap. Adapted from Hasselmo (2009).

lengths of overlap has been addressed in figure 4.7C, which shows that arc length coding is effective for retrieving trajectories that overlap for extended segments of 10 centimeters, 20 centimeters, or 40 centimeters, whereas place cell coding breaks down for all types of overlap.

In these examples, retrieval was cued by a specific stimulus feature vector. This was done by using bidirectional Hebbian modification of synapses during encoding to form bidirectional associations between the stimulus feature vector associated with an item and the place cell vector (including arc length cells or time cells). In addition, a matrix coded associations between individual arc length activity and the vector of oscillation phases in the entorhinal grid cells. These bidirectional connections allowed an individual stimulus feature vector during retrieval to activate the associated pattern of place cells and arc length cells, and the arc length cell representation to activate the associated grid cell oscillation phase. This allowed cueing of retrieval from a specific stimulus or position along the trajectory. In this manner, you could cue retrieval of a segment in the middle of a spatiotemporal trajectory based on the cue, allowing retrieval of an isolated segment in the middle of an episodic memory. Retrieval of episodic memory often involves a specific item or position calling to mind the associated segment of an episode.

Grid cells and place cells show stable firing over multiple different exposures to an environment, even with long delays between exposures (Hafting et al., 2005). This requires associations with sensory input. The velocity input alone only maintains the stability of firing response during the time the velocity is present. To maintain stability of firing across different exposure sessions, sensory cues are necessary to set the phase of firing to the same starting point. In the simulations presented here, the sensory cues representing specific items are associated with specific positions on the trajectory, and then these sensory cues can cue retrieval of that position on the trajectory.

The encoding and replay of trajectories shown here provides potential mechanisms based on coding of time, arc length or space for replay phenomena in neurophysiological data. The circuit was used to simulate (Hasselmo, 2008) the temporally structured replay of spiking activity during REM sleep (Louie and Wilson, 2001), and could simulate the replay at choice points during behavior (Johnston and Redish, 2007), or with an increase in oscillation frequency or strength of synaptic connectivity could simulate the replay of previously experienced trajectories during sharp wave ripples (Skaggs and McNaughton, 1996; Lee and Wilson, 2002; Diba and Buzsaki, 2007).

This chapter demonstrated a model for encoding and retrieval of segments of spatiotemporal trajectories as a mechanism for episodic memory. This process requires a number of different types of associations. In particular, it requires associations between the code for space and time and the coding of actions, items, or events. The code for space and time comes in the form of place cells, arc length cells, and time cells. These are associated with actions in the form of speed cells and head direction cells. The formation of associations constitutes an important component of the process of encoding and retrieving spatiotemporal trajectories. The model uses bidirectional associations between the code of space and time with cells coding features of individual items or events at particular locations and times. These associations allow a spatiotemporal trajectory to cue retrieval of an item or event, and they allow an item or event to cue a spatiotemporal trajectory leading to other items and events.

The hippocampus is proposed to mediate the formation of these associations between a spatiotemporal trajectory and items and events. The following chapter will address mechanisms for the formation of associations between the spatiotemporal trajectory and items or events.

5 Linking Events and Episodes

Events are an essential part of episodic memory. How do we associate locations and times along a spatiotemporal trajectory with specific items or events? I remember my dog chasing the cat at the top of our street one winter morning, or I remember meeting my graduate student on the street outside my office. How are these associations formed?

The previous chapters described the neural mechanisms for encoding an episodic memory, with an emphasis on how neurons of entorhinal cortex could encode spatial location and time along a spatiotemporal trajectory. However, not much physiological detail was provided about how segments of the spatiotemporal trajectory are associated with specific items or events. Also, the model includes the formation of associations between segments of the trajectory and descriptions of action such as the speed and head direction at that position that could cue subsequent segments of the trajectory. This chapter will focus on the role of the circuitry of the hippocampus in the formation of associations between segments of the trajectory and the actions, items, or events in an episode.

A specific example of an episodic memory is useful for describing the function of the regions as I review the anatomy. I will use an old episodic memory of taking care of my daughter in our apartment in Eliot House when she was about two years old. Like other children that age, she would wake up early and wake her parents. I remember one morning walking down the hallway of our apartment to the kitchen, cutting some honeydew melon for her, and putting the melon in a bowl on the coffee table in our living room. She ate some melon, and then we started to read a book together. As we read, our Siamese cat, Louis, started to bat at the bowl and knocked it onto the floor. My daughter yelled the cat's name and then I picked up the melon. This episodic memory involves specific features of time and place that will be here described as "context," as well as items such as the cat and the melon and the event of the cat spilling the melon. These provide the basic elements of the episodic memory example for describing hippocampal function.

Hippocampal Circuits for Associations

As described in chapter 1, lesion and imaging data indicate that my initial encoding of the associations in this memory took place in the cortical structures of the medial temporal lobe. The connections of the medial temporal lobe in the human are summarized in figure 5.1, along with the corresponding regions in the rat. This figure shows the position of the hippocampal formation in the human brain, where it receives convergent input from a broad range of sensory association cortices. This input arrives in cortical areas including the parahippocampal cortex, the perirhinal cortex, and the entorhinal cortex. The entorhinal cortex projects into different subregions of the hippocampal formation, including the dentate gyrus, region CA3, region CA1, and the subiculum. Regions CA1 and the subiculum project back to the entorhinal cortex and parahippocampal cortices (Amaral and Witter, 1989).

The anatomy shown in chapter 2 focused on the topography within individual regions, using the analogy of an orchestra. In this chapter, viewing these regions as a cross section gives a different perspective, showing the complex layered structure of these regions that proves important when discussing their physiological properties. The layering is visible in the anatomical cross-section in figure 5.2B. As noted above, when looking at a cross section of the human brain, the hippocampus leaps out as having a particularly ornate structure, like the fine carvings around the door of a medieval church, or the wooden scroll on a violin. The distinctive curvilinear structure of the hippocampus gives an immediate sense of structured and orderly interactions between the different components. I will describe the layered structure that makes the hippocampus distinct from other cortical regions.

The coding of items like the cat and the melon in my episodic memory involve many levels of feature extraction in the cortex. In an anatomical cross section, most of the cortex on the outer surface of the cerebral hemispheres contains neurons arranged in a manner that appears diffuse to the untrained eye. However, the trained eye of an anatomist can categorize six layers in these regions referred to as neocortex. The untrained eye first detects the smaller cell bodies sandwiched in the middle layer of the neocortex, known as layer IV. This layer is referred to as the granular layer. Most neocortical regions receive their first sensory input via synapses arriving from the thalamus in layer IV. The sight of my cat and the sound of my daughter's voice first entered my neocortex from the thalamus. The layers above the granular layer are called supragranular and receive input from layer IV and send output to other neocortical areas higher in the hierarchy of sensory processing. For example, the initial processing of borders and colors in my primary visual cortex spread to higher cortical areas extracting features that ultimately mediated recognition of my cat. Layers below the granular layer are called infragranular and send feedback to the thalamus and to cortical areas lower in the hierarchy of sensory processing, allowing a feedback process that enhances recognition.

Identifying a cat does not require the hippocampal formation, as semantic knowledge does not require the hippocampus. For example, the recognition of individual items in a category of animals or faces appears to involve areas of the inferotemporal cortex in primates (Hasselmo et al., 1989; Riches et al., 1991; van Turrenout et al., 2000). However, identifying the specific episode of seeing the cat depends upon the connections from the association cortices to areas around the hippocampus, starting with perirhinal and postrhinal cortex (postrhinal is known as parahippocampal gyrus in humans). These neocortical areas look similar to the rest of the neocortex and may code information about how recently an item was seen by responding more strongly to novel items (Riches et al., 1991; van Turennout et al., 2000; Sohal and Hasselmo, 2000; Bogacz and Brown, 2003). For example, seeing my cat the first time that morning would have evoked a larger response than seeing it a second or third time. The parahippocampal and perirhinal regions project to the medial and lateral entorhinal cortex, respectively.

The entorhinal cortex starts to look different from the neocortex because it has little thalamic input and no cell bodies in layer IV, which appears as a gap in the density of cell bodies that is known as the lamina dissecans. This may reflect the fact that the entorhinal cortex mediates encoding of information that has undergone cognitive processing in other cortical regions, not sensory information from the thalamus. As described in chapters 2 and 6, the persistent spiking of entorhinal neurons may maintain working memory for novel information (Hasselmo and Stern, 2006), so that I could remember the sight of the cat and the melon long enough to form synaptic associations.

The hippocampus is even more distinct from the neocortex in that the diffuse distribution of cell bodies in different layers of the neocortex is replaced by tight layering of cell bodies that gives the hippocampus its ornate, roulade structure. Perhaps the most distinct hippocampal subregion is the dentate gyrus, which contains tightly packed cell bodies of dentate granule cells in a curving arc known as stratum granulosum.

The entorhinal cortex sends divergent connections to the dentate gyrus. In the rat, about 250,000 neurons in entorhinal cortex project to about 1 million neurons in the dentate gyrus (Amaral et al., 1990). This divergent connectivity inspired theories proposing that the dentate gyrus reduces the overlap between different input patterns (Marr, 1971; McNaughton and Morris, 1987). These connections might be important for separating the many different episodes in which I saw our Siamese cat Louis in the living room of our apartment. Each of those would involve the same inferotemporal activity, but the divergence of connections could allow a separate set of dentate gyrus neurons to be activated for each episode.

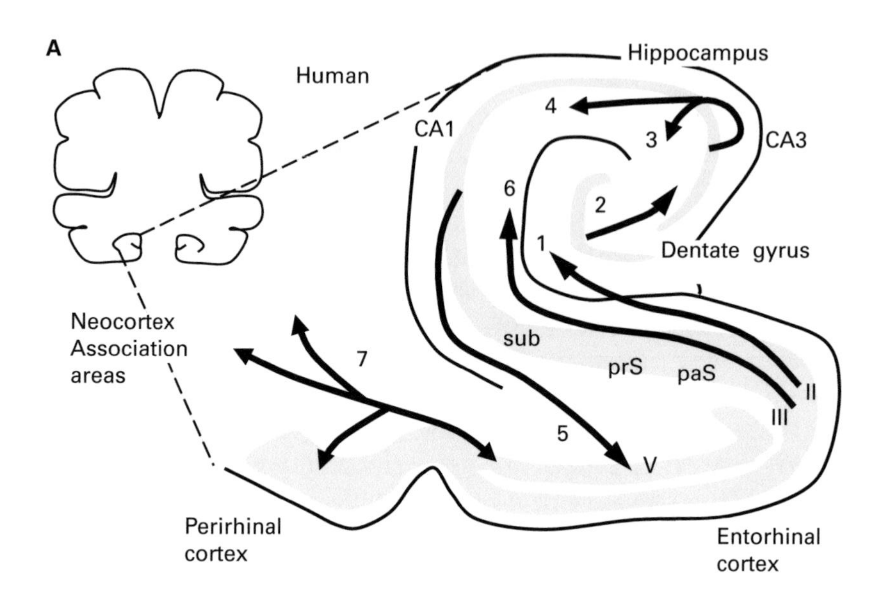

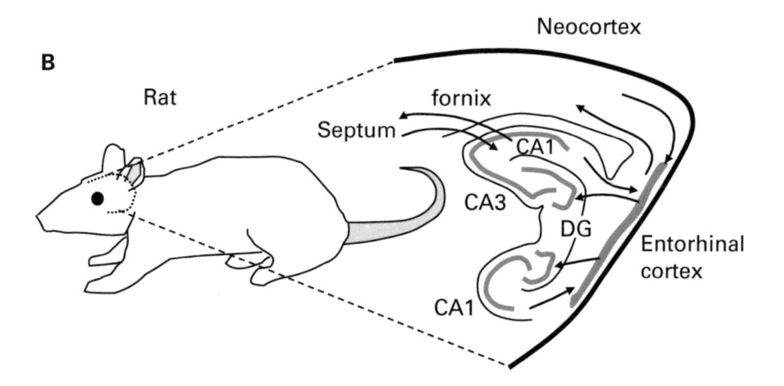

Figure 5.1

(A) Anatomy of the hippocampal formation in humans. Sensory input and motor actions are processed in a range of multimodal association cortices that send convergent connections to the entorhinal cortex (EC) via the perirhinal cortex and parahippocampal gyrus. Layers II and III of EC projects to subregions of the hippocampus and output from the hippocampus projects to deep layers of EC. Connections within the hippocampus include (1) perforant path inputs from EC layer II that synapse in the dentate gyrus (DG) and CA3, (2) mossy fibers projecting from the dentate gyrus that synapse on region CA3 pyramidal cells, (3) the longitudinal association fibers making excitatory recurrent connections between pyramidal cells in region CA3, (4) the Schaffer collaterals connecting pyramidal cells of region CA3 to pyramidal cells of region CA1, (5) excitatory feedback projections from region CA1 and subiculum (sub) to deep layers of EC, (6) direct

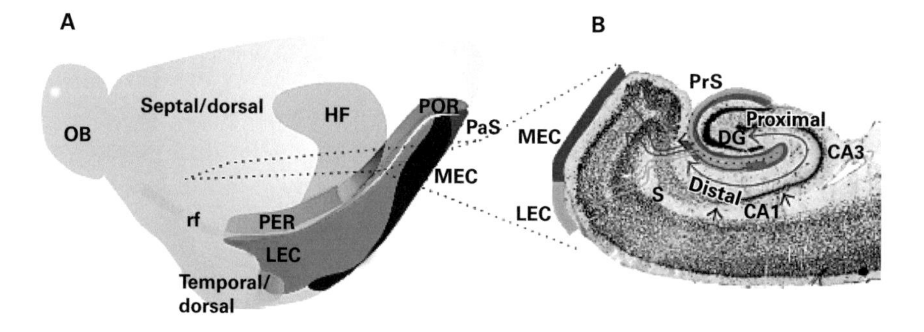

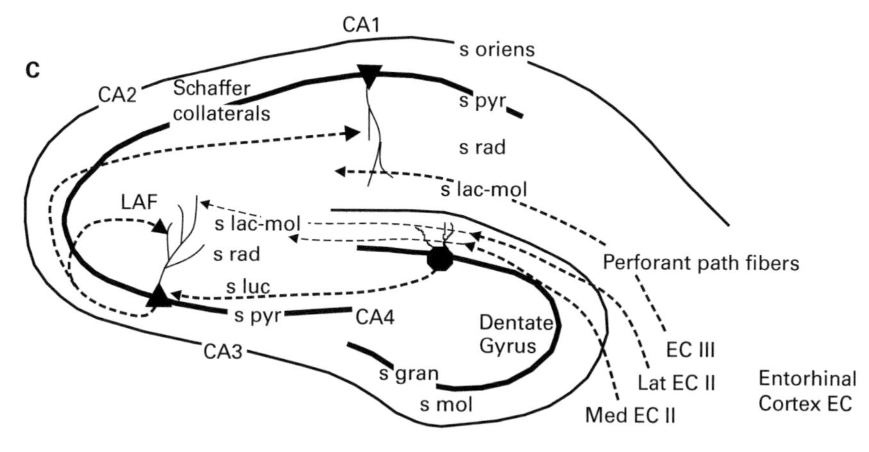

Figure 5.2
(A) Horizontal cross section of the entorhinal cortex and hippocampal formation showing (B) the lateral entorhinal cortex (LEC) and medial entorhinal cortex (MEC) the dentate gyrus (DG), region CA3, region CA1, subiculum (S), presubiculum (prS) (figure reprinted with permission from Canto et al., 2008). (C) Cross section showing synaptic connections (dashed lines). Perforant path input from entorhinal cortex layer II terminates in stratum lacunosum-moleculare (s lac-mol) of CA3 and in stratum moleculare (s mol) on granule cells in dentate gyrus with cell bodies in stratum granulosum (s gran). Entorhinal cortex layer III input synapses in s lac-mol of CA1. Longitudinal association fibers (LAF) of CA3 terminate in stratum radiatum (s rad), and Schaffer collaterals terminate in s rad of CA1. Dentate gyrus mossy fibers terminate in stratum lucidum (s luc) of CA3. Other abbreviations are stratum pyramidale (s pyr) and stratum oriens (s oriens).

Figure 5.1
(continued)
perforant path projections from EC layer III to region CA1, and (7) bidirectional connections between EC and higher order association cortices, including perirhinal cortex and parahippocampal gyrus. Connections not shown include projections from EC layer II to region CA3, from presubiculum (prS) and parasubiculum (paS) to EC, and from the subiculum (sub) to subcortical structures influencing neurons in medial septum. The diagram does not show modulatory cholinergic and GABAergic input from the medial septum to the hippocampus described in chapter 6. (B) Corresponding regions in the rat including connections with septum.

The separation of different episodes could involve contextual convergence from lateral and medial entorhinal cortex on different layers in the dentate gyrus molecular layer (stratum moleculare). In the inner molecular layer, adjacent to the cell bodies, the dendrites of granule cells receive synapses from inhibitory neurons and excitatory synapses from mossy cells in regions CA3 and CA4. Further out, in the middle molecular layer, the granule cells receive input from cells in the medial entorhinal cortex, coding location of events. Furthest out, in the outer molecular layer, the distal ends of the granule cell dendrites receive synaptic inputs from the lateral entorhinal cortex that may code items. In this manner, the information about my cat arriving in the outer molecular layer might be gated in the middle molecular layer by the spatial context of the coffee table in the living room, allowing spiking activity only in granule cells receiving input about that conjunction of cat and location occurring in that episode. The cells in lateral entorhinal cortex providing this most distal input also have the most broadly distributed connections with all regions of the neocortex (Insausti et al., 1997), and these cells are the first to degenerate with tangles in Alzheimer's disease (Braak and Braak, 1991), suggesting some selective sensitivity to the disease, perhaps due to the memory demands of coding single items or individuals occurring in multiple different episodes (Hasselmo, 1994).

As described in chapter 2, in addition to its striking laminar structure, the dentate gyrus also has topography of connections along its length (Witter et al., 2000) from the septal end (near the septum) to the temporal end (near the temporal pole in primates). More dorsal areas of entorhinal cortex, coding more selective spatial locations with higher oscillation frequencies, project to more septal areas of dentate gyrus. Thus, the location of my cat on the coffee table might have been coded near the septal end, whereas the location in the living room in our old apartment might have been coded near the temporal end.

In contrast to the dentate gyrus, the rest of the hippocampal cell layers are continuous with the entorhinal cortex, as if they were a gutter molded along the edge of the neocortex, collecting memories like rain running from the rooftop of the cerebral cortex. As can be seen in figures 5.1A and 5.1B, the cell body layers of entorhinal cortex transition into the parasubiculum and the presubiculum, where head direction cells were first discovered, and eventually into the subiculum. The subiculum contains a single diffuse layer of cell bodies that morphs into region CA1, which has tighter packing of cell bodies. Surprisingly, despite the cell layers' being continuous with entorhinal cortex, the flow of information appears to go the opposite way, with feedforward connections from CA1 to subiculum, presubiculum, and parasubiculum and on to entorhinal cortex.

In CA1, the cell body layer is known as stratum pyramidale, named after the pyramid shaped cell bodies tightly packed in this layer (Ramon y Cajal, 1911). This layer is where place cells were first discovered and have been most studied (O'Keefe and Burgess, 2005). These cells send their dendritic branches up and down into the adjacent layers that contain fewer cell bodies. In the most distant layer from the cell bodies, stratum lacunosum-moleculare, the CA1 pyramidal cell distal dendrites receive excitatory synaptic inputs from layer III of the entorhinal cortex, as shown in figure 5.2. Near the border with subiculum, region CA1 receives possible item information from lateral entorhinal cortex, and more distant from the subiculum, region CA1 receives spatial information from medial entorhinal cortex, including input from grid cells. Proximal to the cell bodies, the dendrites of CA1 pyramidal cells receive excitatory synaptic inputs from CA3 in stratum radiatum. Below the cell body layer, fibers from the medial septum arrive in the stratum oriens.

The cell body layer of CA1 is continuous with the tightly packed cell body layer of CA2 and CA3. The proximal dendrites of CA3 pyramidal cells receive powerful synaptic input from the dentate gyrus, where axons form large clusters of synapses that appear like moss, inspiring the name of mossy fibers (in stratum lucidum). These synapses have been called detonator synapses (McNaughton and Morris, 1987) and might strongly drive neurons for coding of specific items in an episode such as the cat and the melon. Further out, the CA3 pyramidal cells receive excitatory recurrent synaptic inputs from broadly distributed longitudinal association fibers that arise from other CA3 cells and make connections along extensive portions of the septal to temporal length of region CA3 (Amaral and Witter, 1989). In models, these recurrent connections are proposed to form associations between items within an episode, such as associating the cat and the melon. The distal dendrites of CA3 cells in stratum lacunosum-moleculare receive input from the entorhinal cortex, but these cells receive input from layer II of the entorhinal cortex, in contrast to the layer III input to CA1.

The distinctive and highly segregated pattern of connectivity within the hippocampal formation suggests specific functional roles for the pattern of connectivity. Some of the possible functional roles of this connectivity are described in the models of individual subregions below, but the functional role of the intriguing laminar segregation of hippocampal synaptic connections is still not fully understood.

Function of Hippocampal Subregions

Many different models attribute specific functional properties to individual subregions of the hippocampus. This description will focus initially on individual subregions of the hippocampus, separately addressing models for each subregion. Only after addressing the function of individual subregions will the function of full models of the hippocampus be described. Some common hypotheses for function of individual subregions are summarized in figure 5.3.

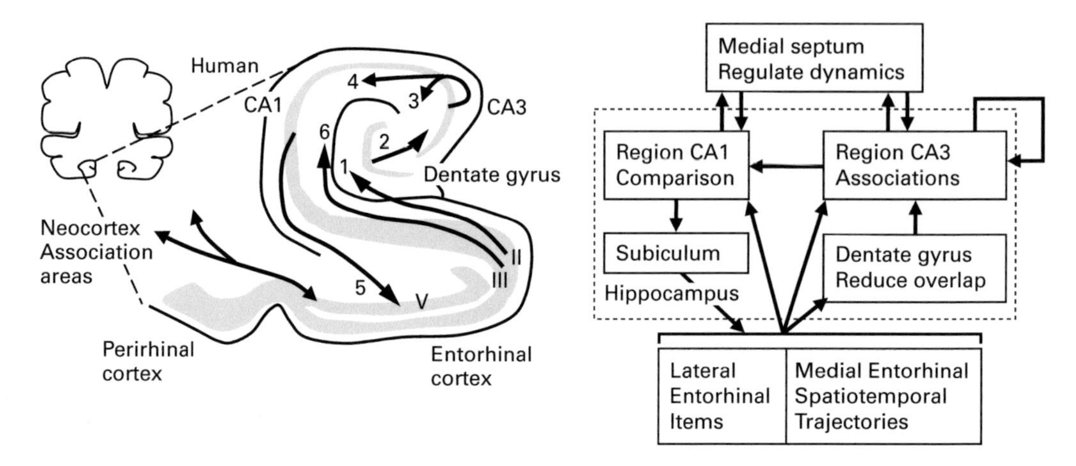

Figure 5.3
Hippocampal subregions on left are modeled on right with populations of units (rectangles). Arrows represent synaptic connectivity. The excitatory recurrent connections in region CA3 may encode associations between neurons in region CA3. The dentate gyrus is proposed to reduce overlap of input patterns to enhance the formation of associations in CA3. Connections from region CA3 to region CA1 may mediate the formation of associations and may allow the dentate gyrus and CA3 to regulate the formation of new representations in CA1. This may involve a comparison process in CA1 between entorhinal input and CA3 input that may regulate the encoding versus retrieval dynamics of the hippocampus. The encoding and retrieval dynamics may be regulated by theta rhythmic GABAergic and cholinergic input from the medial septum.

Region CA3 Associative Memory

Region CA3 provides the central locus for associations in many models of the hippocampus. Region CA3 contains broadly distributed excitatory recurrent connections, termed the longitudinal association pathway (Amaral and Witter, 1989) as shown in figure 5.4. These connections distribute widely along the longitudinal extent of the hippocampus (i.e., from the septal end to the temporal end) and provide a percentage connectivity which is larger than that of most other brain regions. (However, the 4% connectivity between CA3 neurons is still very modest compared to the 100% connectivity of many computational models.) The synapses of these fibers release the excitatory transmitter glutamate, which activates postsynaptic AMPA and NMDA receptors.

Many modelers have proposed that region CA3 may function as an associative memory. This was originally proposed by Marr (Marr, 1971) and extended by many subsequent models (McNaughton and Morris, 1987; Treves and Rolls, 1992; Hasselmo and Wyble, 1995; Levy, 1996). These models encode distributed patterns of activity

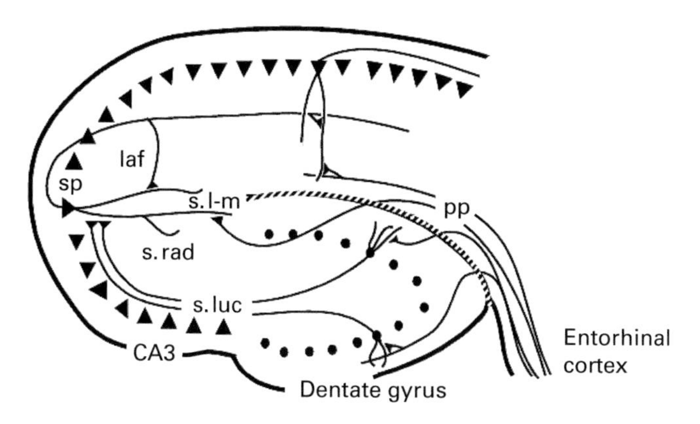

Figure 5.4
Connectivity of region CA3 addressed in models. The perforant path (pp) connections from entorhinal cortex layer II connect to the distal dendrites of region CA3 pyramidal cells in stratum lacunosum-moleculare (s. l-m) and contact granule cells of the dentate gyrus that connect to proximal dendrites of region CA3 pyramidal cells in stratum lucidum (s. luc). The longitudinal association fibers (laf) provide broadly distributed excitatory recurrent connections between different pyramidal cells within region CA3 that synapse in stratum radiatum (s. rad) on middle portions of CA3 dendrites.

along the length of region CA3 that can subsequently be retrieved with incomplete cues containing only a subset of the units active in the initial pattern. The retrieval of the linked components of an episode has been referred to as pattern completion. For example, region CA3 may form associations between a pattern consisting of region CA3 neurons with activity representing a specific position on a spatiotemporal trajectory (my living room) and region CA3 neurons with activity representing a specific item encountered on the trajectory (my cat). Or they may associate features within a pattern from a point in time including two items such as the cat and the melon.

In these models of region CA3, the pattern of neuronal activity is proposed to represent multimodal sensory information converging from association cortices onto entorhinal cortex (Amaral and Witter, 1989; Hasselmo and Wyble, 1997; Lipton and Eichenbaum, 2008). The input from entorhinal cortex enters region CA3 either indirectly through the dentate gyrus or directly at synapses of the perforant pathway in stratum lacunosum-moleculare of region CA3.

This convergence of multimodal sensory information on region CA3 means that strengthening of synapses here could provide associative links between distinct sensory stimuli without any strong prior associative link (such as the cat and the melon). In addition, in region CA3, the capacity for rapidly inducing large changes in size of synaptic potentials (long-term potentiation; LTP) suggests that this region can more rapidly encode associations than many other regions. Rapid mechanisms of synaptic modification would be required for me to encode the episode of the cat and the melon during a single period lasting several seconds in a manner that allows me to recall the event over a decade later.

The pattern of activity arriving in region CA3 via the mossy fibers from the dentate gyrus can represent a particular pattern of sensory stimuli encoded at a specific time and place. The mossy fiber inputs appear to be very strong, activating a pattern of spiking activity within the population of region CA3 pyramidal cells (McNaughton and Morris, 1987). In the episodic memory example presented here, the input pattern could correspond to place cells and other neurons coding position on a spatiotemporal trajectory near the coffee table, as well as neurons responding selectively to the cat and the melon in that context as shown in figure 5.5A.

In a model of CA3, the pattern of spiking activity in my CA3 pyramidal cells could induce Hebbian modification of the excitatory synapses of the recurrent connections from the longitudinal association pathway that synapse in stratum radiatum of region CA3. Hebbian modification means strengthening of synapses that have both pre- and postsynaptic activity. Experimental data show that synapses between neurons in

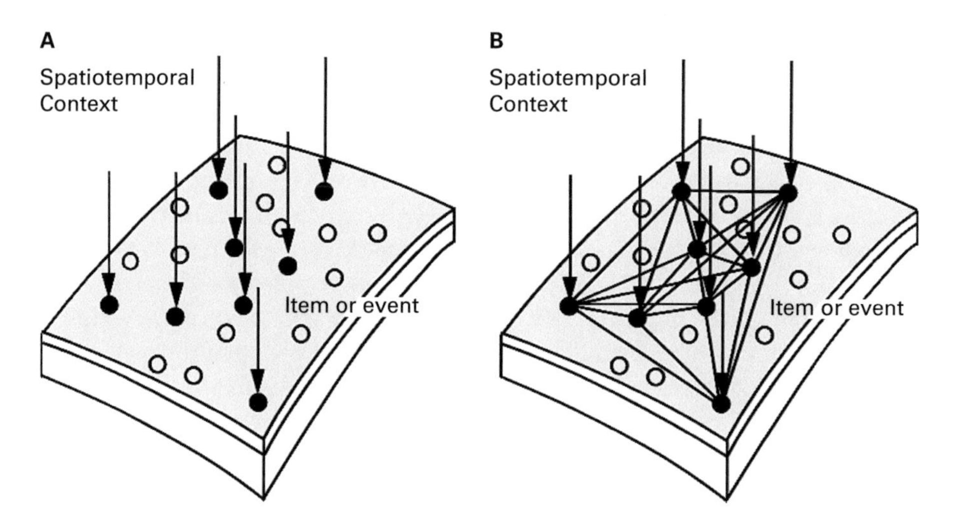

Figure 5.5
(A) Afferent input evokes a pattern of activity in region CA3. Filled circles represent active neurons in the network. The pattern of input includes a representation of spatiotemporal context that might include place cells coding my living room. The pattern of input may also include features of specific items such as my cat and the melon. (B) Hebbian synaptic modification strengthens synapses between active neurons. The black lines between the neurons represent excitatory synaptic connections that have increased the size of their synaptic potentials, making them more likely to induce postsynaptic spiking activity.

region CA3 undergo synaptic modification that has Hebbian properties dependent on NMDA receptors (Zalutsky and Nicoll, 1990; Bi and Poo, 1998).

The Hebbian synaptic modification can encode the initial association between the coffee table and the cat by increasing the strength of synapses between neurons in the full pattern as shown in figure 5.5B.

After encoding, a retrieval cue could evoke activity in a subset of neurons activated by the encoded pattern. For example, my wife asked me later what happened in the living room that morning. This activates the semantic representation of the living room in my neocortex along with the specific context (that morning). This evokes the pattern of activity in region CA3 coding the spatiotemporal context of the living room on that morning as shown by the arrows pointing at filled circles in figure 5.6A. The activity in the units coding spatiotemporal context spreads across the strengthened excitatory recurrent connections (lines) to cause spiking in neurons representing the full pattern of activity as shown by the open circles that become filled circles on the right in figure 5.6B. This is called pattern completion. The full pattern includes other elements of the episodic memory—for example, the neurons representing the event of seeing my cat knocking the bowl of honeydew melon onto the carpet.

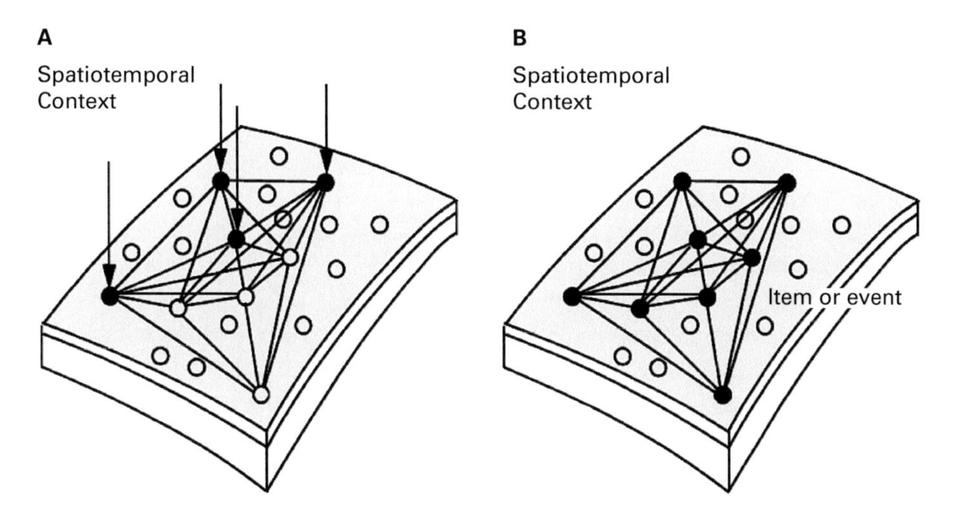

Figure 5.6
(A) A retrieval cue such as a position on a spatiotemporal trajectory evokes activity in a subset of interconnected neurons in region CA3. The filled circles represent neurons that generate spikes in response to the afferent input of the retrieval cue. The open circles represent neurons that do not respond with spiking activity to the retrieval cue. (B) Activity spreads along excitatory recurrent connections within region CA3 to evoke spiking activity (filled circles) in all neurons that were part of the originally learned input pattern. In this manner, the position within a spatiotemporal trajectory might retrieve a specific item or event associated with that position.

This provides a model for retrieval of an individual item or event at a particular spatiotemporal position in an episodic memory. This model of hippocampal function in region CA3 has been used to model performance in specific behavioral tasks used to study memory, such as cued recall (paired associates) and free recall (Hasselmo and Wyble, 1997). For example, in a free recall experiment, the subject follows a path to a particular experimental testing room and then views a series of words in that room with a particular experimenter. When subjects are asked what words were on the list, they can use the position on a specific spatiotemporal trajectory as a cue for memories of specific words stored during the experiment.

The mathematical representation of associative memory is described in the section of the appendix on Associative Memory Function. Most associative memory models utilize the mathematical notation of linear algebra to describe associative networks. The models represent the activity of populations of neurons with one-dimensional arrays of numbers (termed vectors), and the pattern of synaptic connectivity between the various neurons in the population are represented with two-dimensional matrices of numbers, in which each element represents the strength of one synapse. Most network models utilize two major descriptive equations: (1) a learning rule and (2) an activation rule.

Learning Rule

The learning rule describes how synaptic modification in response to different input patterns determines the pattern of synaptic connectivity. In particular, most models utilize a Hebbian learning rule, changing synaptic strength in proportion to pre and postsynaptic activity. An example of the function of such a learning rule is shown in figure 5.5 above, where strengthened synapses are represented by black lines. Only the synapses between active neurons are strengthened.

Physiological data from experiments on LTP demonstrate that the excitatory synapses arising from region CA3 of the hippocampus show changes in synaptic strength which depends upon the conjunction of pre- and postsynaptic activity (Zalutsky and Nicoll, 1990), and studies in organotypic cultures of CA3 cells show that this Hebbian synaptic modification requires that presynaptic pulses precede postsynaptic pulses by a narrow time window of less than about 40 milliseconds (Bi and Poo, 1998; Bi and Poo, 2001) a phenomenon referred to as spike timing dependent synaptic plasticity (STDP). The Hebbian properties of synaptic modification are necessary to maintain selectivity for the memory. If synapses were strengthened based on presynaptic or postsynaptic activity alone, then activating neurons coding my cat might strengthen all prior associations with that cat, causing me to retrieve some other prior association. For example, if strengthening were based on presynaptic activity alone, I might have strengthened an association of my cat playing with a toy and would confuse this association with the episode in the living room that morning.

Activation Rule

The activation rule describes how the activity of neurons in the network at one point in time depends upon the activity at previous points in time. Thus, the activation rule shows how afferent input (arrows) initially induces activity in specific neurons in figure 5.6. The activation rule also describes how the presynaptic activity spreads across synapses to influence the postsynaptic activity. Thus, the activation rule describes how presynaptic activity causes spiking in cells that induces synaptic transmission that induces synaptic potentials in other neurons, causing them to generate spiking activity.

An initial theory about region CA3 of the hippocampus was laid forth by Marr (Marr, 1971), and a clearer review of this potential function of region CA3 is provided by McNaughton and Morris (McNaughton and Morris, 1987). These articles describe region CA3 associative memory function as a single-step matrix multiplication process, drawing on early models of associative memory function (Anderson, 1972; Kohonen, 1972). Most of the examples in the McNaughton and Morris article use simple single-step dynamics for both the learning rule and the activation rule as shown in the section of the appendix on Associative Memory Function.

These simple models illustrate the basic properties of associative memory function, but in real brain function neurons are continuously receiving input and continuously sending output. Their function is not discretized into single steps of input and output with no preceding or subsequent activity. Realistic models must deal with ongoing dynamics in the network. The dynamics of associative memory models run into problems with interference between overlapping memories. Overlapping memories can result in the problem of interference during encoding described in chapter 6 and below and can also result in the problem of interference during retrieval described here.

Interference during Retrieval

All associative memory models encounter the problem of interference during retrieval as shown in figure 5.7. In this example, two separate episodic events share a number of active neurons. Because of this, a retrieval cue that activates some of the overlapping neurons will activate both of the stored memories simultaneously. For example, I have many memories that occurred in my living room at Eliot House, but when remembering the specific episode of the cat and the melon, I was able to avoid retrieval of other episodes that occurred in the same location.

The problem of interference during retrieval can be partly addressed in simple associative memories by including inhibition in the activation rule for the network. For example, in an early model of associative memory function in the hippocampus, feedforward inhibition was proposed to be activated in proportion with the number of active units in an input pattern, and this would prevent activation of the weaker of the two retrieved memories (McNaughton and Morris, 1987). However, the problem

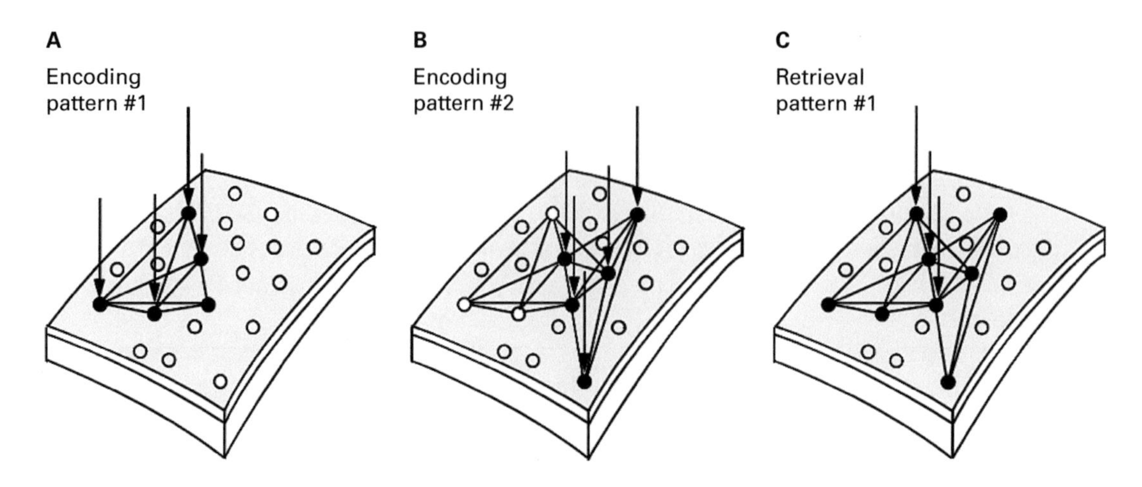

Figure 5.7
Example of interference during retrieval between patterns in region CA3. (A) Encoding of pattern 1 strengthens connections between five neurons. (B) Encoding of pattern 2 strengthens connections between five neurons including three that overlap with pattern 1. (C) If a retrieval cue (arrows) meant to activate pattern 1 contains overlapping elements, then pattern 2 can be activated as well, resulting in interference between the retrieved patterns (retrieval of both).

of interference during retrieval can be addressed better with the more complex retrieval dynamics of attractor neural networks as described next.

Attractor Dynamics

Single-step associative memories do not provide realistic dynamics for describing the function of region CA3. For example, in a recurrent network, the pattern of activity elicited after one step will spread along the recurrent connections again, eliciting further activity. This cycle will repeat itself, causing positive feedback effects on excitatory activity which will cause the network to explode in activity as shown in figure 5.8A. This explosion might occur in cortical networks in the form of seizure activity and could underlie the greater propensity of region CA3 for seizure activity. The feedback excitation must be controlled for effective memory function.

Modeling work has addressed mechanisms for controlling the level of activity in a network model with excitatory recurrent connections, and for storing memories as individual stable fixed points in the dynamics of the network, referred to as attractor states in an attractor network (Amit, 1988). These attractor models were initially described as general models of memory function without reference to specific brain regions (Hopfield, 1982; Cohen and Grossberg, 1983; Hopfield, 1984; Amit, 1988), but later work focused on using attractor networks to model the function of hippocampal region CA3 (Treves and Rolls, 1992, 1994; Hasselmo et al., 1995a). Models of attractor

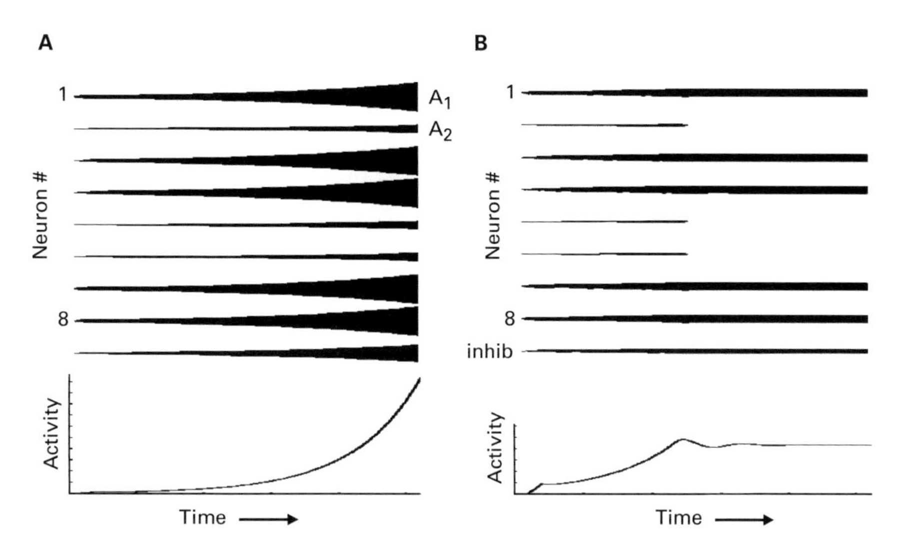

Figure 5.8
(A) Explosive feedback in a recurrent network. The top traces show the activity of individual neurons represented by thickness of black lines with time plotted horizontally. Note that excitatory feedback can lead to exponential growth of individual neuron activity (marked A_1), as well as spread of activity between different neurons within the network (marked A_2). The bottom trace shows an example of the activity of an individual neuron within the network during this explosive activity. (B) Attractor dynamics allow the network to approach stable activity states with subsets of neurons active at reasonable activity levels. In this example, feedback inhibition (inhib) provides this stability, allowing maintenance of self-sustained activity in the five active neurons and inhibition of the other three. Bottom trace grows to a stable level.

states require some means of preventing the explosive growth of activity in the network shown in figure 5.8. Different models have approached this using different techniques.

Limit on Maximum Firing Rate Most early attractor network models used an artificial upper limit on the firing of neurons to help control excitatory feedback (Hopfield, 1982; Hopfield, 1984; Amit, 1988). For example, many of these models use sigmoid input–output functions to describe how an individual neuron changes its firing rate depending on the level of synaptic input. As synaptic input increases to very large levels, these sigmoid functions approach a maximum level of firing rate output that cannot be surpassed. This prevents the activity of individual neurons from surpassing the maximum. As described in the section on single-cell properties, real neurons do reach a maximum firing rate when strongly depolarized. However, this maximum firing rate is far above the firing rate observed during normal function. Thus, the use of an artificial maximum on neuronal output is not a realistic way of obtaining attractor dynamics.

Even with an artificial maximum on the output of a neuron, activity can still spread laterally to other units in the network. Most early attractor network models avoid this by setting up detailed point-to-point inhibitory connectivity (Hopfield, 1982; Cohen and Grossberg, 1983; Hopfield, 1984; Amit, 1988; Menschik and Finkel, 1998). This is also unrealistic as real inhibitory interneurons receive convergent input from many neurons and send output to many neurons. Thus, a single excitatory neuron cannot selectively inhibit another excitatory neuron without diffusely influencing numerous other cells. Thus, models of region CA3 must use more realistic representations of inhibitory connectivity.

Inhibitory Subtraction A simple mechanism for controlling excitatory feedback uses subtractive inhibition (Wilson and Cowan, 1972; Gardner-Medwin, 1976; Hasselmo et al., 1995a). In the hippocampus, excitatory neurons (pyramidal cells) send convergent connections to inhibitory interneurons, and these interneurons send inhibitory connections back to all the excitatory neurons. The representation of this feedback inhibition in computational models is summarized in figure 5.9A.

For certain parameters, subtractive inhibition can balance excitatory feedback, to allow the network to approach a stable equilibrium value without explosive growth as shown in figure 5.9B and C. The term subtractive refers to the mathematical representation of inhibition with a reversal potential well below resting potential. Details of the model are presented in the appendix. Synapses from the inhibitory interneurons release gamma-amino-butyric acid (GABA), which activates both $GABA_A$ and $GABA_B$ receptors on pyramidal cells. The potassium currents activated by postsynaptic $GABA_B$ receptors have a reversal potential well below resting, so they have a subtractive effect for most values above resting potential. Models of this type can encode and retrieve multiple different patterns as different attractor states as shown in figure 5.10.

These models simplify the spiking of neurons into continuous firing rates. This representation can use the threshold linear input–output functions described in the Appendix, allowing analysis of the suprathreshold dynamics of equations using linear differential equations. However, these models do not explicitly simulate the timing of spikes in individual neurons.

Inhibitory Division Another means to control runaway activity is to divide the activity of individual units in the network by the total activity of the network. This mechanism normalizes activity levels in a manner that only allows a subset of units to be active. This technique has been used in some models of region CA3 (Treves and Rolls, 1992, 1994; Levy, 1996). This technique has been used for different representations

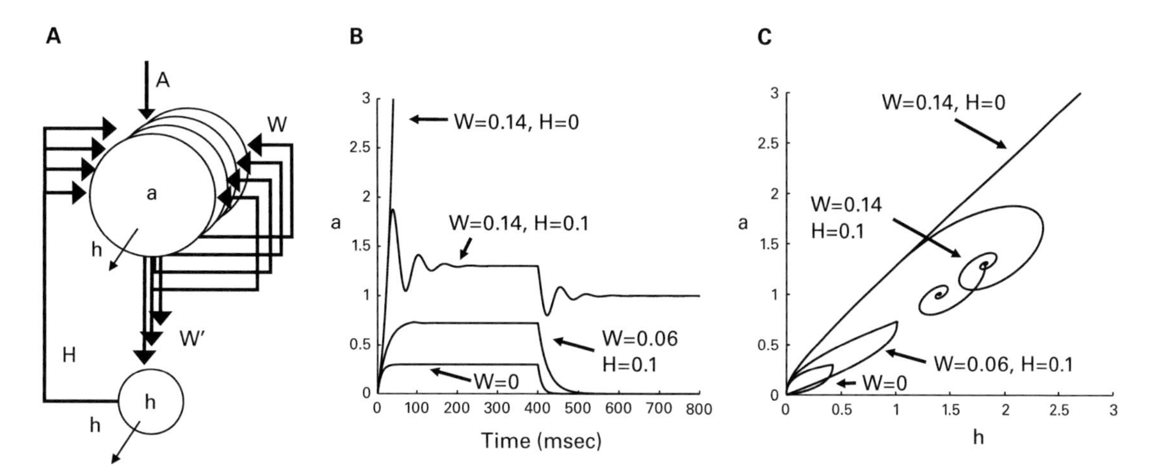

Figure 5.9
(A) Network with feedback inhibition controlling total activity. Excitatory neurons with activity *a* make excitatory recurrent connections with strength W and excitatory connections with strength W′ on an inhibitory interneuron. The inhibitory interneuron with activity h sends back distributed inhibitory connections with strength H. (B) Attractor dynamics in the network. The plot shows the activity (*a*) of a single neuron across time for different values of the synaptic parameters. With no excitatory feedback (W = 0), the network shows only a passive increase in activity when afferent input is present (the afferent input ends at 400 msec). With no feedback inhibition (W = 0.14, H = 0), the network shows explosive growth of activity (top). With a proper balance of excitatory and inhibitory feedback (W = 0.14, H = –0.1) the activity grows to a level that is sustained after 400 msec in the absence of afferent input. (C) In this graph, the excitatory activity (*a*) is plotted on the y axis versus inhibitory activity (*h*) on the x axis. For W = 0.14, H = –0.1, as *a* grows, *h* grows until it starts to bring activity *a* down again. The ongoing interaction of *a* and *h* causes network activity to spiral in toward a stable equilibrium point. This point is referred to as a fixed point attractor. Figure using model in Hasselmo et al. (1995a).

of neuronal output. Output can be represented as a continuous value which represents number of action potentials fired per unit time (rate code), or it can be represented by generation of spikes each time the neuron reaches threshold (integrate and fire).

This use of inhibition is still relatively abstract, in that a single inhibitory unit receives excitatory input from the full network and contacts every excitatory neuron in the network. However, the division effect could be similar to effects of shunting inhibition due to activation of $GABA_A$ receptors (McNaughton and Morris, 1987; Abbott, 1991). Activation of $GABA_A$ receptors increases membrane conductance to chloride, which has a reversal potential very close to the resting potential of the neuron. This results in the inhibition having an effect more similar to division than

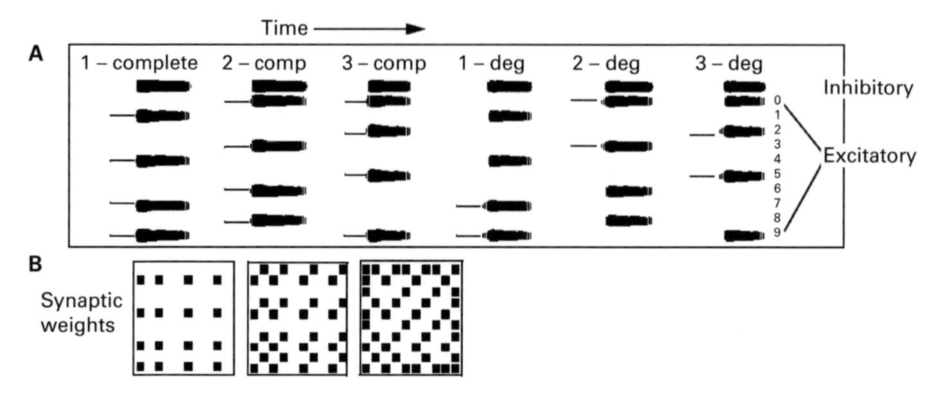

Figure 5.10
Encoding and retrieval of multiple patterns as fixed point attractors in a network with subtractive inhibition. (A) Activity of excitatory neurons (0–9) plotted as width of black lines with time plotted horizontally. The top trace shows activity of the single inhibitory interneuron. Three complete patterns are presented sequentially to the network (1-complete, 2-comp, 3-comp). Subsequently, degraded versions of these input patterns are presented. The thin horizontal lines mark the neurons receiving a cue input that then evokes the complete patterns marked 1-deg, 2-deg, and 3-deg. Looking closely, you will see that each pair of cue neurons activates the full pattern that previously contained the cue. (B) The pattern of synaptic connectivity is shown with black squares representing strengthened synapses after sequential encoding of each of the three input patterns. Adapted from Hasselmo et al. (1995a).

subtraction. This division is similar to that used by Minai and Levy in model networks storing sequences of activity in CA3 (Minai and Levy, 1993; Levy, 1996).

Most physicists performing analysis of attractor neural networks have emphasized computing the capacity of these networks, making the assumption that biological network function is optimized for long-term capacity. However, it is not possible to experimentally quantify the information capacity of long-term episodic memory in animals, nor is it clear that the system is optimized for storage capacity. In contrast, the capacity of short-term memory can be computed experimentally and has been modeled.

Using limits on the maximum spiking output of individual neurons is less realistic, but there is physiological evidence suggesting that synaptic transmission does show a frequency limitation—with strong decreases in excitatory transmission as synapses are activated at higher frequencies. This phenomenon has been utilized as a means of limiting excitatory feedback to obtain attractor dynamics in networks of spiking neurons (Fransén and Lansner, 1995). In more abstract models not concerned about the detailed dynamics of neuronal interaction, the total activity within the network can be limited by artificially imposing a "k-winners-take-all" scheme (O'Reilly and

McClelland, 1994; Norman and O'Reilly, 2003), in which only a set number of k neurons can be active.

Attractor network models have certain features appealing for modeling of episodic memory. These include the fact that the strong driving force of recurrent excitation ensures that even very weak retrieval cues can evoke full, active attractor states and the fact that initial conditions only choose which memory will be retrieved; they do not determine the final features of the retrieved pattern. In a single-step associative memory, the final memory retrieved with a retrieval cue activating two neurons might be only half the strength of that retrieved with a cue activating four neurons. With attractor dynamics, if the correct fixed point attractor is entered and sufficient time is allowed, the final fixed point attractor state will not differ dependent upon the starting retrieval cue.

Behavioral work has not explored whether initial conditions have a strong influence on the nature of retrieved episodic memories, but on an intuitive level retrieved memories reach a similar final state of detail. Attractor network models represent different elements of memory in a single pattern of activity distributed across different neurons (a spatial pattern). In most of these models, the progression of time results in activity settling on a single spatial pattern of activity, which can be described by a single point in vector space. Therefore, those networks are referred to as fixed point attractor networks. If the capacity of the attractor network has not been reached, then the attractor network can prevent interference during retrieval between the stored patterns.

Interference during Encoding

In addition to the potential for causing explosions of activity and causing interference during retrieval, excitatory recurrent connections also have the potential to interfere with the encoding of new patterns. The attractor dynamics described above can overcome the problem of interference during retrieval, based on the dynamics of their activation equation. However, they cannot overcome the problem of interference during encoding because this involves the dynamics of both the activation equation and the learning equation. Because of this, most associative memory models prevent the problem by using different dynamics during encoding of patterns, with clamping of activity to the desired input pattern (Kohonen, 1972; Hopfield, 1982; Amit, 1988). This requirement for different encoding and retrieval dynamics raises specific physiological issues that have been addressed in experiments and models described in more detail in chapter 6.

One proposed solution to this problem is that the mossy fibers could strongly activate a subset of CA3 pyramidal cells on the basis of the activity pattern in dentate gyrus. As mentioned above, this type of activation has been referred to as detonator synapses (McNaughton and Morris, 1987). This framework has been elaborated in a model in which the mossy fibers set the encoded pattern, and the direct perforant path input from entorhinal cortex acts as a weaker cue during retrieval (Treves and Rolls, 1992). A problem with this framework is that detonator synapses can ensure that a subset of neurons are active but cannot as effectively prevent activity in neurons receiving recurrent activation.

Another solution is to weaken the excitatory recurrent connections during encoding, preventing them from interfering with the new input pattern (Hasselmo et al., 1992; Hasselmo and Schnell, 1994; Hasselmo et al., 1995a). The problem of interference during encoding and this solution are elaborated extensively in chapter 6. These two mechanisms of reducing interference are not incompatible, and they are very likely to be implemented together by cholinergic modulation as acetylcholine has been shown to suppress excitatory recurrent connections in region CA3 (Hasselmo et al., 1995a) and also increases excitatory activity in the dentate gyrus (Foster and Deadwyler, 1992), thereby increasing mossy fiber input to region CA3.

Experimental Tests of Pattern Completion

As shown above, the models of associative memory function in region CA3 propose its role in pattern completion. In this framework, the pattern of spiking activity in region CA3, including both place cells and representation of other events, should result in Hebbian modification of the synapses of the longitudinal association fibers. This Hebbian synaptic modification should then allow an incomplete cue, such as a sensory cue associated with a specific location, to cue pattern completion that activates all of the sensory features initially encountered at that position. An attractor state should be able to take incomplete sensory cues and complete the full array of sensory features.

This hypothesis was initially proposed by Marr in 1971 and then discussed more extensively by researchers in the 1980s and 1990s (McNaughton and Morris, 1987; Treves and Rolls, 1992; Hasselmo et al., 1995a). In more recent years, specific experiments have been designed to test the hypothesis of pattern completion in region CA3. The notion of attractor dynamics was initially supported by data showing that the firing selectivity of place cells and head direction cells will rotate together as a unit (Knierim et al., 1995). However, when local and distal cues are rotated in opposite directions, place cells shift in different directions suggesting that attractor dynamics are not dominant (Knierim, 2002).

Another study testing pattern completion involved genetic manipulations allowing selective removal (conditional knockout) of the NR1 subunit of the NMDA receptor in region CA3 (Nakazawa et al., 2002). This involves the insertion of specialized sequences (called loxP sequences) into the DNA on each side of the DNA coding the NR1 subunit. In an adult animal, selective induction of the enzyme Cre recombinase by association with a gene primarily present in CA3 selectively cuts at the loxP site and removes the NR1 subunit in CA3 pyramidal cells and splices the DNA together at that point.

The mouse with the selective conditional knockout of NMDA receptors in region CA3 showed normal performance in the Morris water maze when the same cues were present on all sides of the task during both encoding and retrieval. When only one cue was present during retrieval, control mice could still perform the task. However, the NMDA CA3 knockout mice were impaired at the task, presumably due to loss of pattern completion of the sensory cues (Nakazawa et al., 2002). The CA3 knockout mice also showed reduced activity of place cells in the environment with reduced cues whereas control mice showed place cell activity more closely resembling the activity with all cues present. These experimental data are consistent with the theory of CA3's playing a role in pattern completion.

A similar pattern of impairment on the cue completion version of the Morris water maze was observed in animals with selective expression of tetanus toxin in region CA3, resulting in a shutdown of synaptic transmission arising from CA3 (Nakashiba et al., 2008).

Another study of completion involved initial learning of a square and a circular environment made of different materials in which place cells showed distinct responses (Wills et al., 2005). Spiking responses of place cells were then tested in an environment with a single set of walls that were moved between sessions to different intermediate configurations between a circle and a square (Wills et al., 2005). In this study, place cells abruptly and simultaneously switched their firing patterns to match the firing pattern previously associated with either the square or circle pattern, suggesting entry to one of two attractor states coding either the circle or the square.

Sequence Storage Models

The networks with stable fixed point attractors described above do not provide an easy representation of the temporal order of multiple different items, such as the interitem associations which appear to influence the order of retrieval of words in a free recall task. In a free recall task, once one word from a list is retrieved, there is a much higher probability of retrieving words that were adjacent on the list (Howard and Kahana, 2002).

These interitem and cross-temporal associations can be more effectively represented by storage of multiple patterns in a sequence rather than a single pattern. The storage of a sequence of activity patterns has been proposed as an alternative function of the excitatory recurrent collaterals in region CA3. In many models, this simply involves the formation of Hebbian associations between sequentially presented patterns (Marr, 1971; McNaughton and Morris, 1987; Jensen and Lisman, 1996b; Levy, 1996; Wallenstein and Hasselmo, 1997). These types of models are described in more detail in chapter 7 as different models of theta phase precession.

Sequence retrieval can also be modeled in other ways. For example, chapters 3 and 4 described how a sequence can be coded as a spatiotemporal trajectory, either with a circuit linking locations to actions that trigger the next location (Hasselmo and Brandon, 2008; Hasselmo, 2008, 2009) or using the oscillatory interference model to code a representation of intervals in time for associations with individual items (Hasselmo, 2007; Hasselmo et al., 2007).

Encoding of temporal order has also been simulated with the temporal context model that uses a gradual evolving context vector that is associated with individual items and events, with the change in the context vector dependent on the number of items experienced (Howard and Kahana, 2002; Howard et al., 2005).

Attractor dynamics can be used for sequence retrieval. For example, each pattern in a sequence can be an individual fixed point attractor. Rapid excitatory recurrent connections can push the network into each individual attractor, whereas slower excitatory connections evoke elements of the next attractor in the sequence (Kleinfeld, 1986). Individual attractor patterns can be turned off by fast feedback inhibition allowing the transition (Lisman, 1999). Alternately, the entire sequence of patterns can be stored as a limit cycle attractor, with later patterns evoking earlier patterns such that the sequence repeats itself (Hasselmo, 1999a).

Encoding of multiple overlapping sequences runs into the problem of interference between stored sequences. Theta rhythm oscillations induced by cholinergic and GABAergic input from the medial septum could assist in disambiguating overlapping sequences (Sohal and Hasselmo, 1998a, 1998b). More extensive description of sequence encoding and retrieval is provided in chapters 3 and 7.

Biophysical Models of Region CA3

Many of the hippocampal models described above use simplified continuous firing rate neurons to study the dynamics of the network. The next step in relating these mechanisms to the physiological data requires development of models with spiking neurons, and a further step involves modeling the network with detailed compartmental biophysical simulations representing a range of membrane currents.

Detailed compartmental biophysical simulations of hippocampal subregions have simulated physiological phenomena observed in slices, including the bursting properties of single CA3 neurons (Traub et al., 1991) and the theta frequency oscillations observed in isolated slice preparations in CA3 (Traub et al., 1992) and CA1 (Rotstein et al., 2005). Associative memory function has been obtained in networks with excitatory recurrent connections and Hebbian modification (Barkai et al., 1994; Wallenstein and Hasselmo, 1997a; Menschik and Finkel, 1998). However, approximations to attractor dynamics are difficult to obtain in spiking networks (Menschik and Finkel, 1998) because CA3 neurons have relatively low firing rates, and the synaptic potential decays to zero before a new spike is generated, so they do not produce a steady output influence on other neurons.

In terms of behavioral function, most spiking models of region CA3 have focused on the encoding and retrieval of sequences of spiking activity. This allows replication of the mechanisms of theta phase precession using spiking network models (Jensen and Lisman, 1996c; Tsodyks et al., 1996; Wallenstein and Hasselmo, 1997b; Thurley et al., 2008). These models will be described in more detail in chapter 7.

Dentate Gyrus

Most theories and models of hippocampal memory function include an important functional role for the dentate gyrus—that of pattern separation. In these models, the dentate gyrus serves to reduce the overlap between different patterns of activity stored within region CA3 of the hippocampus. The dentate gyrus activity then strongly activates region CA3 pyramidal cells via the mossy fibers. As described above, there are a number of problems associated with the overlap between stored patterns in an associative memory, including both interference during retrieval and interference during encoding. A simple means of reducing these types of interference between encoded patterns involves preprocessing the patterns to make their representation less overlapping when they activate region CA3. This reduces the overlap that causes interference during retrieval as shown in figure 5.7 and interference during encoding as described in chapter 6. This requires a mechanism for automatically reducing overlap between multiple sequentially presented patterns, and a mechanism for mapping the altered, less overlapping representations in region CA3 back to the original input patterns from association neocortex.

Divergence for Pattern Separation

The cells in the dentate gyrus greatly outnumber those in the entorhinal cortex. In the rat, there are about 250,000 entorhinal cortex cells and about 1,000,000 dentate gyrus cells (Amaral et al., 1990). This means that there is a great divergence of connections from entorhinal cortex to dentate gyrus. If the pattern of connections between the structures is random, this divergence alone will result in pattern separation (McNaughton and Morris, 1987; McNaughton and Nadel, 1990; O'Reilly and McClelland, 1994). An example of this process is shown in figure 5.11.

Competition between Patterns

The overlap between stored patterns can be further reduced by allowing modification of the perforant path inputs from entorhinal cortex to dentate gyrus. This results in self-organization of the dentate gyrus representation due to competition between encoded representations. Strengthening of entorhinal input to dentate gyrus tends to

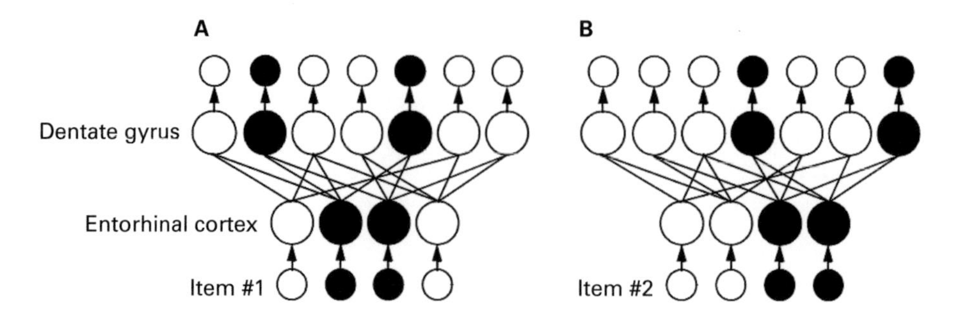

Figure 5.11
Example of passive separation of overlapping patterns due to random diverging connectivity. (A) Pattern 1 activates two entorhinal cortex neurons, but if the dentate gyrus threshold is set at 2 active inputs, only dentate gyrus neurons receiving input from both entorhinal neurons become active. (B) Pattern 2 activates two entorhinal cortex neurons, one of which was in pattern 1. However, the divergent connections with a threshold result in this second overlapping pattern only activating neurons in dentate gyrus which were not activated by pattern 1. Thus, the divergence results in no overlap in dentate gyrus.

cause greater overlap between patterns, but the separation of patterns is increased by weakening of entorhinal synapses in dentate gyrus based on postsynaptic activity without presynaptic activity, or presynaptic activity without postsynaptic activity (O'Reilly and McClelland, 1994).

By definition, episodic memories are encoded at one point in time. This means that any representation for the initial encoding of this memory must be formed on the basis of that single presentation. The afferent input could persist for a period of time (see discussion of the possible role of entorhinal cortex as a buffer), but it is not intermixed with the reactivation of other encoded memories. Thus, the encoded memories are not interleaved during learning. This presents a special problem for models of the dentate gyrus in episodic memory as most models of self-organization and competitive learning utilize interleaved learning of a large set of input patterns.

Sequential self-organization of representations in the dentate gyrus has been obtained by utilizing modification of inhibitory connections (Hasselmo and Wyble, 1997) as shown in figure 5.12. In this model, individual representations in dentate gyrus are initially activated due to random divergent input. At this point, excitatory feedforward connections undergo Hebbian modification to strengthen the excitatory input to that dentate representation. This would allow smaller entorhinal patterns to activate dentate gyrus. However, this excitatory modification is offset by enhancement of inhibitory feedback connections within the dentate gyrus. These can then prevent any patterns that do not strongly resemble the initial input pattern from activating the same representation.

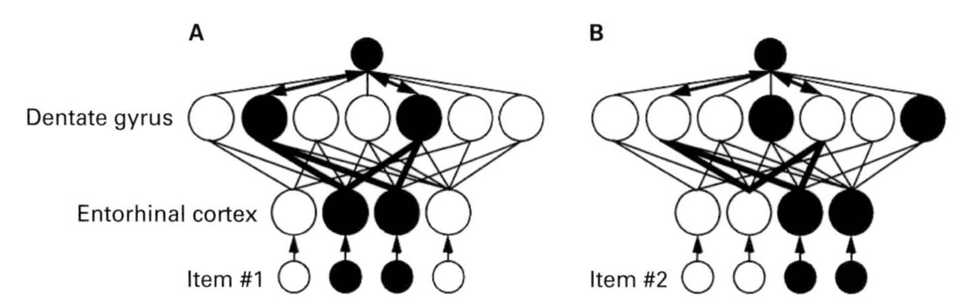

Figure 5.12
Example of sequential competition between stored patterns. Note that this model has modifiable excitatory input, modifiable feedback inhibition, and additional connections beyond those in the previous figure, which further prevent pattern 2 from activating the same dentate neurons as pattern 1. (A) Pattern 1 initially activates two dentate neurons, and excitatory entorhinal path input to these neurons is strengthened with Hebbian properties (thicker lines). In addition, synaptic connections to and from interneurons are enhanced (lines with arrows). (B) Note that pattern 2 could activate the same dentate representation due to the strengthened excitatory connections, but the modified feedback inhibition prevents this, ensuring that pattern 2 activates a separate representation in dentate gyrus. In this manner, modified inhibition can prevent sequentially presented patterns from developing overlapping representations in the dentate gyrus.

Other models of episodic memory have used self-organization of input to dentate gyrus but have utilized interleaved presentation of different patterns (Rolls et al., 1997; O'Reilly et al., 1998). These networks can obtain competitive self-organization without modification of inhibition, but the requirement of interleaved learning may be unrealistic for encoding of episodic memories unless these memories are repeatedly reactivated and utilized to refine the dentate gyrus representation (see discussion of McClelland et al., 1995, below).

Self-organization during interleaved presentation of input patterns has been studied extensively in models of the formation of feature detectors and topographic representations in the primary visual cortex (Erwin et al., 1995) and in more abstract models. Repeated interleaved presentation of patterns allows a longer term more gradual competition between stored patterns which results in formation of separate representations, that is, representations which can be used most accurately for retrieving the encoded patterns. This is important for semantic memory in neocortex.

Experimental Tests of Pattern Separation

Similar to the experimental analysis of pattern completion described above, experiments have analyzed the hypothesized role of the dentate gyrus in pattern separation. One study involved selective knockout of NMDA receptors in the dentate gyrus

(McHugh et al., 2007). Mice with this knockout showed normal contextual fear conditioning but were impaired in their ability to show fear conditioning selective for one but not the other of two similar contexts.

The role of dentate gyrus in spatial pattern separation was also analyzed with a task in which food was placed under an object on a sample trial, and then on a subsequent test trial an identical object with reward was placed in the same location and an identical object was placed in a position 15 to 105 centimeters away (Gilbert et al., 2001). Selective lesions of the dentate gyrus impaired the ability of rats to identify the same location unless the distractor was the maximal distance from the correct object.

In functional imaging work, techniques for registration of the fMRI activity onto anatomical scans have allowed more detailed analysis of different subregions activated during behavioral tasks. Using these techniques, Craig Stark and colleagues tested for activity in hippocampal subregions associated with pattern separation versus pattern completion (Bakker et al., 2008). Limitations on imaging resolution still forced them to combine the dentate gyrus with region CA3. They tested pattern separation by having participants encode visual stimuli (e.g., a rubber duck) and then presenting test stimuli that were exactly the same or had small differences (a rubber duck with slightly different shape and coloring). Participants were required to correctly identify these stimuli as similar but not identical (old). The dentate gyrus and CA3 showed differential activity between the similar (lure, correct rejection) and the old conditions (hit), suggesting a selective role of these regions in pattern separation.

One computational model of the dentate gyrus focuses on the trade-off between pattern separation and pattern completion on the feedforward connections (O'Reilly and McClelland, 1994). During encoding patterns must be separated as much as possible (as described above and in the next chapter) whereas during retrieval a cue must be able to complete the full correct episodic memory. Thus, the process of pattern separation implicitly includes a type of pattern completion phenomenon because those patterns that are not separated into distinct representations will activate the same unified representation. For example, you might tune up a network to separate the representation of two different faces. However, if the network is functioning properly, it should be able to recognize one of those faces based on partial cues if the person is wearing a hat or sunglasses. The partial cues activate the same representation that the full pattern evoked. Note that this involves activation of a single category representation via feedforward connections in contrast to the associative pattern completion of disparate features via excitatory feedback connections as in models of CA3.

Region CA1

The primary excitatory output from region CA3 involves synaptic projections via the Schaffer collaterals to region CA1. Region CA1 of the hippocampus has a pattern of connectivity that is dramatically different from the pattern of connectivity in region CA3 (Amaral and Witter, 1989). This includes three anatomical differences. The mossy fiber input from the dentate gyrus to region CA3 stops before it reaches region CA1, so stratum lucidum does not exist in region CA1. The cells of region CA1 have only very weak recurrent connections, so the vast majority of synapses in stratum radiatum of region CA1 arise from the Schaffer collateral pathway from region CA3. Finally, CA1 receives input from layer III of entorhinal cortex rather than layer II, and this input is segregated, with input from medial entorhinal cortex entering areas of CA1 proximal to CA3, and input from lateral entorhinal cortex entering CA1 more distal near the subiculum (Witter et al., 2000). This segregation contrasts with the input from layer II of medial and lateral entorhinal cortex that terminates in different layers of stratum lacunosum-moleculare in region CA3. Based on these striking anatomical differences, models have suggested different functions of CA1 and CA3, and physiological data have shown some subtle differences in the response of CA1 versus CA3.

Mapping Back to Neocortex

If the associations encoded in region CA3 require a transformation in the dentate gyrus to make them less overlapping, then there must be some means of reversing the transformation in order to link the encoded patterns back to the activity patterns associated with convergent input to entorhinal cortex. However, there are no direct connections from region CA3 to structures such as the entorhinal cortex. How can this link back to entorhinal cortex take place? Region CA1 appears well suited to this putative linking function (Hasselmo et al., 1995b; McClelland and Goddard, 1996; Hasselmo and Wyble, 1997). The lack of recurrent connections and the convergence of CA3 output with entorhinal input could allow region CA1 to perform Hebbian modification mapping CA3 activity to the associated activity in entorhinal cortex and to gate how strongly the CA3 output spreads back to influence entorhinal cortex.

Comparison Function

Region CA1 could play an important role in determining whether the reconstructed pattern in region CA3 satisfies criteria for linking to entorhinal cortex, based on a comparison of the region CA3 output with the current sensory input relayed through entorhinal cortex (Gray, 1982; Levy, 1989; Eichenbaum and Buckingham, 1990; Hasselmo and Schnell, 1994; Hasselmo, 1995; Hasselmo et al., 1995a; Vinogradova, 2001). The anatomical connectivity of region CA1 could allow a comparison function between the output from contextual convergence in CA3 and the separate representation of spatiotemporal position arriving from medial entorhinal cortex and the representation of individual items arriving from lateral entorhinal cortex. This convergence could allow evaluation of the memory. For example, during retrieval, the contextual convergence and pattern completion in CA3 might activate a memory that includes your colleague Erik, but the entorhinal input might not contain that information. This mismatch could indicate the retrieval of the wrong context and reset the dentate gyrus and CA3 to retrieve a different context. Alternately, if region CA3 and dentate have not previously encoded a particular context, the comparison function will also reveal a poor match that could push dynamics toward stronger encoding of novel experiences (Hasselmo and Schnell, 1994).

This matching function could be very important for ensuring that incorrect retrieval does not propagate back to entorhinal cortex. For example, if the input pattern has not been previously encoded, the immediate retrieval from region CA3 will not have any useful information. In addition, if there is severe interference of the new pattern with previously stored patterns, the immediate retrieval from region CA3 will contain additional undesired information which should be restricted from passing back to entorhinal cortex (Hasselmo and Wyble, 1997). The cognitive impairments in schizophrenia appear to include impairments in holding context to gate behavior, possibly due to reduced function of NMDA receptors in the hippocampus (Coyle et al., 2003). A model of impaired function in region CA1 simulated the reduced use of category context for gating episodic retrieval in schizophrenia (Siekmeier et al., 2006).

Region CA1 has the appropriate anatomical connectivity for this type of comparison function as shown in figure 5.13. The activity of region CA1 pyramidal cells reflects the convergence of Schaffer collateral input from region CA3 and perforant path input from entorhinal cortex (Katz et al., 2007). Intrinsic interactions within the dendritic tree or feedforward inhibition could ensure that CA1 pyramidal cells only fire if there is a match between CA3 input and entorhinal input. In this case, only matching patterns would activate region CA1 and only matching patterns could spread back to entorhinal cortex directly or via the subiculum.

As noted above, region CA3 receives lateral and medial input in different layers, potentially contacting all CA3 cells, whereas the lateral and medial input to region CA1 contacts different populations of CA1 cells. Thus, region CA3 has a stronger merging of the spatiotemporal context and item input, so they cannot be compared independently. In addition, even if region CA3 performs a comparison function, all output must pass through the potential comparison function of region CA1 as well. Having a separate comparison stage might be important in that it can allow region CA3 to settle into an attractor state or sequence based on contextual convergence without immediately vetoing this state on the basis of input from entorhinal cortex.

The theory of a comparison function in region CA1 has been described verbally in a number of locations (Gray, 1982; Levy, 1989; Eichenbaum and Buckingham, 1990; Vinogradova, 2001), but only in a few publications have the actual mechanisms of such a comparison been described in detail and analyzed mathematically or simulated (Hasselmo and Schnell, 1994; Hasselmo et al., 1995a; Hasselmo and Wyble, 1997; Katz

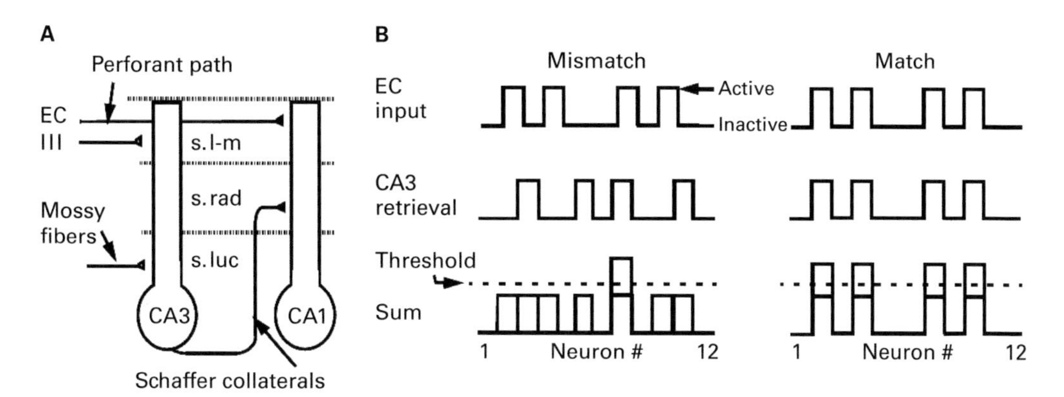

Figure 5.13
(A) Summary of region CA1 inputs. Input from entorhinal cortex (EC) layer III synapses in stratum lacunosum-moleculare (s. l-m). The Schaffer collaterals from region CA3 synapse in stratum radiatum (s. rad). (B) Overview of matching function. Top two lines represent activity of synaptic input to specific neurons from different sources (EC input and CA3 retrieval). Bottom line represents the sum of input effects on postsynaptic activity for each neuron relative to threshold. Mismatch: for unfamiliar patterns, the retrieval from region CA3 does not match with EC input, and postsynaptic activity is below threshold for most neurons. Match: for familiar patterns, the retrieval from region CA3 matches EC input and brings neurons in the pattern above threshold. s. luc, stratum lucidum.

et al., 2007). In some simulations, the sum of region CA1 activity was utilized to regulate levels of modulatory input from the medial septum to set the dynamics of encoding versus retrieval as described in chapter 6 (Hasselmo and Schnell, 1994; Hasselmo et al., 1995a). Regulation of cholinergic and GABAergic modulatory input from the septum could also determine the capacity for information to flow back to entorhinal cortex (see discussion of modulation in chapter 6). The matching function in region CA1 has also been used to generate context-dependent retrieval of sequences for guiding spatial alternation behavior as described in chapter 7 (Hasselmo and Eichenbaum, 2005; Katz et al., 2007). Region CA3 and CA1 have been proposed to have different roles in prediction based on prior experience, but simulations do not show strong differences consistent with the differences in anatomical connectivity (Treves, 2004) though incorporation of spike frequency accommodation enhances the prediction function.

Context-Dependent Remapping

Experimental studies have suggested different roles for region CA1 versus region CA3 consistent with a trade-off between separation and completion. Some unit recording studies have compared the neural activity in region CA3 versus region CA1 when features in the environment are changed. In a study in the lab of Jim Knierim (Lee et al., 2004), rats ran on a circular track with both local features on the track and distal features visible on the walls. Then the track was rotated different amounts relative to the walls. Region CA1 neurons showed a difference in firing between the two conditions that was proportional to the rotation, with less similar firing for larger rotations. In contrast, consistent with its greater recurrent connectivity, region CA3 appeared to show more completion, maintaining more similar firing patterns in the two conditions. However, this effect appears to contrast with another comparison of CA3 and CA1, in which neurons were compared in different rooms (Leutgeb et al., 2004). In this second study, CA3 neurons showed very different activity (global remapping) in different rooms regardless of similarity of environments whereas CA1 neurons showed firing dependent on the similarity of the environments.

These apparently disparate results were addressed in a further study looking at neural activity measured by activation of the Arc protein (Guzowski et al., 2004; Vazdarjanova and Guzowski, 2004). In that study, the amount of completion in CA3 depended on the level of environment similarity, with separation of very different environments and completion of more similar environments. This is consistent with a proposed trade-off of separation and completion for spatial coding (O'Reilly and McClelland, 1994; Guzowski et al., 2004), in which similar patterns are made more similar, but if patterns are sufficiently different, their differences are emphasized. For episodic memories, the spatiotemporal context has to be sufficiently different to trigger the differentiation mechanism for separate episodic memories occurring in the same spatial environment.

Full Hippocampal Models

The functions of individual regions described above have been combined in full network simulations of episodic memory in the hippocampal formation. Chapters 1 through 4 described a full network simulation of entorhinal cortex and hippocampus performing encoding and retrieval of episodic trajectories. In contrast, this section will focus on earlier models of the encoding and retrieval of discrete items to replicate experimental data on human verbal memory tasks such as free recall and cued recall (Hasselmo and Wyble, 1997) as well as recognition memory (O'Reilly et al., 1998; Hasselmo and Wyble, 1997). All of these simulations focus on encoding and retrieval of temporally discrete, binary patterns of activity (Hasselmo and Wyble, 1997; O'Reilly et al., 1998; Rolls et al., 1998). These models included multiple different hippocampal subregions but did not address the role of entorhinal cortex in coding spatiotemporal trajectories.

Modeling the effects of hippocampal lesions does not help constrain the structure of hippocampal simulations, as models of lesion effects require blocking or removing the structure. Therefore, simulations of hippocampal lesions have only been performed in models containing some representation of neocortical networks as well, such as the models focused on temporally graded retrograde amnesia (Alvarez et al., 1994; Hasselmo et al., 1996; Shen and McNaughton, 1996). In contrast, drug effects on memory function help to constrain the structure of models as the cellular effects blocked by drugs are very well described at the level of cellular physiology (Hasselmo and Wyble, 1997). In addition, the quantitative details of behavioral data can be utilized to constrain the structure of hippocampal models (O'Reilly et al., 1998).

Modeling Free Recall

In free recall experiments, participants are presented with a list of words during encoding. In a separate retrieval phase, they are asked what words were on the list and retrieve them in arbitrary order. No specific cue elicits the retrieval of each word. Instead, the general experimental context must serve as a cue for retrieval of each word. Thus, simulation of free recall mediated by the hippocampus requires a representation of context (Hasselmo and Wyble, 1997). The memory for individual words then takes the form of associations between this episodic memory of the context and the episodic representation of the individual word items.

A full hippocampal circuit was used to model the free recall of lists of words (Hasselmo and Wyble, 1997). In the model, input was presented to both superficial and deep layers of entorhinal cortex in the form of single binary vectors representing the context (the room and the time) and separate binary vectors represented individual words. This pattern entered the simulated dentate gyrus, which formed distinct representations of the sequentially presented input patterns. Hebbian modification of recurrent synapses in region CA3 formed fixed point attractors for context and items, and formed associations between the attractor created by experimental context and the episodic representation of individual words (items). Hebbian modification of the Schaffer collaterals associated the pattern of CA3 activity with a pattern of activity in CA1 caused by convergent input from CA3 and entorhinal cortex, and Hebbian modification also associated the CA1 pattern with the pattern of input activity in the entorhinal cortex (Hasselmo and Wyble, 1997).

During retrieval, presentation of the experimental context alone in superficial entorhinal cortex caused activation of the dentate gyrus context and activated the context attractor state in CA3. The context activity spread across previously strengthened recurrent connections to cause activity in multiple different attractors representing individual words (items). These item attractors competed via feedback inhibition, allowing only one item attractor to predominate. This CA3 attractor caused activity in CA1, and when there was sufficient convergence with the context input from entorhinal cortex, the activity would flow back to deep layers of entorhinal cortex to generate retrieval of individual words in free recall. After one item had been active for a period, a wave of inhibition would shut off all attractors (possibly analogous to the oscillatory inhibition associated with theta rhythm oscillations). Subsequently, activation of the context attractor state again caused activity to spread to multiple item attractors, but the previously retrieved item had sufficient residual spike frequency accommodation that it could not be retrieved again, and a different word could be retrieved in CA3 and spread back to deep layers of entorhinal cortex.

Modeling Recognition

Recognition memory for individual words has been proposed to involve two different processes: (1) explicit remembering of the episode when the word was encoded and (2) a feeling of knowing the word—that is, a vague sense of familiarity (Tulving, 1983). The distinction is sometimes referred to as remember versus know, though more recently has been referred to as recollection versus familiarity. Both types of memory can be used to perform a recognition memory task, but the explicit remembering has been proposed to be mediated by the hippocampus (Fortin et al., 2004; Eichenbaum et al., 2007) whereas the feeling of knowing may be a neocortical process. Extensive research in the field has focused on determining whether recognition memory involves two processes of recollection versus familiarity (Yonelinas, 2001), versus a single process with a variable threshold (Wixted and Stretch, 2004). Addressing the full range of data on this topic would require an entire book, so I will focus on simulations of the potential neural mechanisms.

Simulations of hippocampal function in recognition memory have addressed the explicit recollection of the episode when a word was encoded (Hasselmo and Wyble, 1997; O'Reilly et al., 1998; Norman and O'Reilly, 2003). In the Hasselmo and Wyble model, the process of recognition memory involves the reverse of the interaction used for free recall. In free recall, the subject is presented with the context and must retrieve individual items. In contrast, in testing of recognition, the network is presented with individual items, which activate the item attractors. The activity must then flow across strengthened connections to activate the attractor representing the context. Because the context is present for an extended period of time during encoding in this model, it develops strong recurrent connections. These strong recurrent connections allow the context attractor state to be activated very easily. Thus, presentation of individual items can activate the context attractor with relatively weak connections even if the context attractor could not activate that same item for free recall. This asymmetry in the strength of attractors simulates experimental data showing differences in number of items that can be elicited in free recall, which is small relative to the number of the same items that can be accurately recognized. This difference in strength also made the free recall mechanism much more sensitive to simulations of the effects of the acetylcholine receptor blocker scopolamine in the network, simulating experimental data showing reduced free recall of items (Ghoneim and Mewaldt, 1977) while having no significant effect on recognition of items (Hasselmo and Wyble, 1997). Simulation of the effects of scopolamine is discussed further in chapter 6.

The feeling of knowing of a word (familiarity) was not modeled in the Hasselmo and Wyble model, but this mechanism was explicitly modeled in a later paper by Norman and O'Reilly (Norman and O'Reilly, 2003) that explicitly compared the familiarity mechanism with the recollection mechanism described above. The familiarity mechanism might occur in areas outside of the hippocampus such as the perirhinal cortex or entorhinal cortex. Recordings of unit activity in these regions demonstrate neurons that show a strong reduction in activity for stimuli that were presented previously (Riches et al., 1991; Fahy et al., 1993; Sohal and Hasselmo, 2000). Similar reductions also appear in neocortical areas such as the inferotemporal cortex (Rolls et al., 1989). The reduction in activity for repeated stimuli has also been shown in human functional imaging studies (van Turennout et al., 2000; van Turennout et al., 2003). The mechanism for this response reduction is not clear, but it may involve competitive self-organization of the network (Sohal and Hasselmo, 2000) or alternately could involve anti-Hebbian modification of inputs to these neurons (Bogacz and Brown, 2003). In the simulations of two different mechanisms of recognition memory (Norman and O'Reilly, 2003), judgments based on familiarity depended upon the reduction of activity in neocortical areas whereas judgments based on recollection used internal hippocampal retrieval mechanisms based on associations in region CA3.

Relational Memory

Sequence retrieval has greater flexibility than fixed point attractors because it provides for cross-temporal associations between different elements of a sequence. However, in current models, it does not allow for significant flexibility in the particular patterns retrieved. Overlapping sequences can be encoded and retrieved (Levy, 1996; Wallenstein and Hasselmo, 1997), but the particular sequence retrieved usually depends upon earlier elements of the sequence.

Extensive data suggest that the hippocampus is particularly important for flexible retrieval of relational information about recent events (Cohen and Eichenbaum, 1993). That is, rather than being locked into a particular stereotypical pattern or sequence of patterns, the hippocampus appears to mediate flexible access to causal relationships within a large network of interitem associations. In the overall framework presented here, the entorhinal cortex would also be important for this function, consistent with impairments of relational memory function after entorhinal cortex lesions in monkeys (Buckmaster et al., 2004).

The relational memory framework corresponds to a case of multiple overlapping sequences encoded in region CA3, which can be accessed not in a rigid manner specific to individual sequences but with flexible transitions across the elements of the network, picking out different components in different sequences. This flexible retrieval of different events within an episode or different relational features could be obtained if the recurrent connections of region CA3 set up an associative network, within which specific subsequences of activity could be evoked depending upon the conditional features of the retrieval cue. Alternately, the episodic memory circuit described in chapter 4 can be used, with directional input guiding an update of the spatial representation in medial entorhinal cortex that drives the representation of place in the hippocampus. Prefrontal cortex could guide retrieval with directional cues.

This moves beyond trajectory retrieval based on associations between positions and actions, incorporating retrieval of trajectories that had never been experienced, based on the guided activation of particular internal representations of actions. For example, if one wishes to combine individual sequences that were not previously experienced together, the prefrontal cortex could cue activation of an action mediating the transition from one previously experienced trajectory onto a different trajectory. The retrieval of relational information from the hippocampus could involve the mechanisms of forward planning described in chapter 7 on memory-guided behavior.

6 Drug Effects on the Dynamics of Encoding and Retrieval

I have conducted experiments in which undergraduates were given drugs that transiently impaired their episodic memory function. Naturally, these studies were fully approved by our institutional review board for human studies. In these studies, physicians administered the commonly used drug scopolamine via intramuscular injection, and we saw clear impairments of the encoding of paired associates (Atri et al., 2004).

How do drugs affect episodic memory function? Studies of drug effects provide a large and useful body of data linking memory behavior to specific neurophysiological mechanisms at a cellular level. This chapter will address the question of how drugs affect episodic memory, first reviewing the extensive behavioral data, then addressing the relevant physiological data, then presenting a model linking the cellular physiological data to the behavioral effects of drugs.

Drug Effects on Human Memory

For thousands of years, humans have known that certain plant compounds will affect memory for events, and this knowledge has been refined as the specific chemical compounds in these plants have been isolated and tested both clinically and experimentally. Human memory function can be influenced by a number of different drugs. These drug effects can be used to further analyze the mechanisms of episodic memory in human participants. In contrast to lesions, which may completely remove the circuits necessary for a particular function, drug effects alter the function of neural circuits in a specific manner that can be modeled by simulating the effects of the drug on cellular physiology. This provides an opportunity to address the role of specific neural mechanisms for memory function. On the other hand, most drugs are administered systemically—that is, throughout the circulatory system of a human. Thus, despite the fact that drug effects are selective for a specific set of neuronal receptors, they are often not necessarily localized to a specific region of the nervous system. The following sections will introduce some drugs with strong effects on memory function.

Acetylcholine Blockers

Acetylcholine is a neurochemical released at a number of different locations in the nervous system. For example, when you type on a keyboard, the action potentials from neurons in your spinal cord spread along axons to the synapses on your muscles and cause release of acetylcholine, which activates nicotinic receptors that trigger the muscles to contract. After you eat a meal, parasympathetic synapses in your digestive system release acetylcholine that activates muscarinic receptors that enhance the action of intestinal muscles. Acetylcholine also plays an important role in the brain. Neurons containing acetylcholine in the brainstem regulate the flow of information into the brain during waking by releasing acetylcholine in the thalamus and other structures. Cholinergic neurons in the basal forebrain send projections that release acetylcholine throughout the cerebral cortex, where acetylcholine can influence cortical nicotinic and muscarinic receptors. The effects in the cortex appear to play an important role in encoding of memory.

The drug scopolamine blocks the muscarinic subtype of acetylcholine receptors. Scopolamine impairs the encoding of memory (Ghoneim and Mewaldt, 1975; Hasselmo, 1995 provides a review). The behavioral effects of plants containing scopolamine such as jimson weed has been known for centuries. At high doses, these plants cause hallucinations and delirium. The drug scopolamine was extracted from these plants in 1880 and is commonly used in a variety of clinical applications. In particular, anesthesiologists use scopolamine for its ability to keep the respiratory tract dry and noted that patients often do not recall episodes that occur while they are under the influence of the drug. This led to experimental studies showing that scopolamine impairs memory on a range of different tasks, including detection of novel objects or faces in a growing list or learning the locations of objects in a schematic house (Flicker et al., 1990) or learning the position of chess pieces (Liljequiest and Matilda, 1979).

Several studies have shown that scopolamine selectively impairs encoding while sparing retrieval. For example, scopolamine impairs the encoding of lists of words for free recall (Ghoneim and Mewaldt, 1975, 1977). In control conditions, a subject given 120 words could remember about 45 words when asked to engage in free recall of all the words on the list. However, if the same subject is given scopolamine before encoding the list of 120 words, then his or her later free recall is severely impaired, showing an average of only 5 words recalled. In contrast, if the subject learns the list of words before administration of scopolamine, then the drug has no effect on the recall of words (Ghoneim and Mewaldt, 1975, 1977). A similar effect appears with paired associate learning (Atri et al., 2004). A subject encoding a list of 16 pairs of related words before scopolamine injection can recall about 10 of the second words when cued with the first word in the pair after scopolamine. However, if the subject learns the pairs after injection of scopolamine, he or she only remembers about 6 of the associated words (Atri et al., 2004). In monkeys, scopolamine impairs the encoding of lists of objects for subsequent recognition, but does not impair recognition of items learned before drug administration (Aigner et al., 1991).

These data indicate an important role for acetylcholine in the encoding of new information within neural circuits. The impaired encoding of new memories caused by scopolamine can be offset by the drug physostigmine (Ghoneim and Mewaldt, 1977), which increases acetylcholine levels by blocking the breakdown of acetylcholine by acetylcholinesterase. Clinically, these data on the role of acetylcholine in memory function have been linked to data showing loss of cholinergic enzymes in cortical structures in postmortem tissue from Alzheimer's patients (Perry et al., 1977). This suggests that some of the memory impairments in Alzheimer's disease may be linked to the loss of cholinergic innervation. Current treatments for Alzheimer's disease include drugs such as donepezil (with the trade name Aricept) that inhibit the breakdown of acetylcholine by inhibiting acetylcholinesterase.

Behavioral data demonstrate that scopolamine is selective for aspects of episodic memory. Scopolamine impairs episodic memories for items stored at a specific time and place, but has little effect on performance in tasks testing semantic memory (memory for general facts; Caine et al., 1981; Broks et al., 1988) or procedural memory (memory for habitual actions; Nissen et al., 1987). However, there is some evidence that scopolamine impairs mechanisms of working memory for novel stimuli by reducing cellular mechanisms for active maintenance described below. Scopolamine reduces parahippocampal fMRI activity observed during the delay period of a delayed match to sample task (Schon et al., 2005). Loss of active maintenance could underlie the impairment of delayed matching function in humans caused by scopolamine (Green et al., 2005), as well as the impairments of encoding in monkeys observed with localized infusions of scopolamine (Tang et al., 1997). This hypothesis is also consistent with evidence that medial temporal lesions selectively impair working memory for new conjunctions of complex nonverbalizable visual stimuli (Olson et al., 2006).

Inhibition Enhancers

Another class of drugs that cause a striking impairment of episodic memory is the benzodiazepines and other sedatives. The amnesic effects of benzodiazepines are so potent that they have obtained a colloquial designation as "date rape" drugs. This arose from well-publicized cases where a rapist slipped a benzodiazepine into the drink of an unsuspecting female and then took advantage of her sedated state with the plan that the amnesia would prevent her from filing charges due to difficulty in remembering the event. The term is particularly strongly associated with one benzodiazepine with the trade name Rohypnol, which is not available in the United States for this reason. However, the amnesic effect is similar for a wide range of benzodiazepines that are commonly prescribed and used in the United States, such as Valium or Xanax (Ghoneim and Mewaldt, 1977; Frith et al., 1984; Richardson et al., 1984). Clinically, benzodiazepines such as midazolam (Versed) are commonly used to calm patients in preparation for administration of anesthesia and commonly cause complete or partial amnesia for the episodes occurring after administration of the drug. Anesthesiologists see the amnesic effects of drugs such as midazolam and scopolamine as an important adjunct to analgesia.

The benzodiazepines are a class of drugs with similar chemical structures that all increase the current through the channel of the $GABA_A$ receptor, which is one type of receptor activated by the inhibitory transmitter GABA. The effects of benzodiazepines on memory are very similar to the effects of scopolamine, despite their very different action on different sets of neuronal receptors. In fact, most studies have been unable to show significant differences between the behavioral effects of these drugs (Richardson et al., 1984), though one study suggested that benzodiazepines impair encoding of visual paired associates whereas scopolamine impairs delayed matching (Robbins et al., 1997). One common pathway that could underlie the similarity of effects concerns the dual modulatory input from the medial septum to the hippocampus, which includes both cholinergic and GABAergic fibers. Both cholinergic antagonists and benzodiazepines alter the hippocampal theta rhythm (McNaughton and Coop, 1991; McNaughton et al., 2007). Thus, the similar amnesic effect of both these drugs could result from physiological disruption of hippocampal theta rhythm. As mentioned above, the drug muscimol when infused into the medial septum causes a blockade of theta rhythm oscillations and grid cell firing patterns (Brandon et al., 2011b). Muscimol causes increased currents through the $GABA_A$ receptor that are also increased by benzodiazepines, suggesting that the benzodiazepines might be impairing memory via disruption of oscillations and grid cell firing patterns.

The effects of drugs on episodic memory provide a useful source of information for understanding the cellular and circuit dynamics important for episodic memory. However, most of the studies of drug effects in humans involve administration of the drug orally or by injection, so the drug spreads throughout the body and the location of the cells or receptors being influenced by the drug is not apparent from the human data. More selective techniques used in animals help localize the drug effects on episodic memory function.

Acetylcholine and Animal Memory

Extensive studies in animals support the role of acetylcholine in encoding of episodic memories and indicate a specific role for acetylcholine within the hippocampus and parahippocampal cortices including the entorhinal cortex and perirhinal cortex. For detailed description of the behavioral tasks used in these studies, see chapter 1.

The effects of scopolamine on episodic memory have been tested in animals by using the variant of the water maze task in which the hidden platform is moved at the start of each testing day, requiring the animal to remember where the platform is on a given day. In rats, scopolamine impairs performance on this task, most likely by impairing the encoding of new platform location (Whishaw, 1985; Buresova et al., 1986). Similarly, performance in the eight-arm radial maze is significantly impaired with scopolamine administration or by fornix lesions (Cassel and Kelche, 1989) which reduce the cholinergic innervation of the hippocampus. This impairment may be due to the disruption of encoding for previously visited arms. When a delay is imposed between sample and test phases of the eight-arm radial maze, the impairment in performance caused by scopolamine increases (Bolhuis et al., 1988). Similar to human and nonhuman-primate work with scopolamine, this impairment on maze tasks is only observed when scopolamine is given prior to the encoding period and not if given during the delay between sample and test phases (Buresova et al., 1986). Thus, memory tasks requiring the association of events with a specific time (sample vs. test) and a specific place (arms visited or platform location) appear to be impaired with the blockade of muscarinic receptors.

Localized infusions of cholinergic receptor blockers into specific anatomical structures demonstrate the importance of cholinergic receptors for particular aspects of memory tasks. Localized infusions of scopolamine into parahippocampal structures impair the encoding of information for subsequent recognition in both monkeys (Tang et al., 1997) and rats (Winters and Bussey, 2005). The studies in monkeys used tasks in which animals are exposed to one or multiple sample stimuli during encoding and are subsequently tested on their delayed recognition of these sample stimuli and rejection of other stimuli which were not presented during the sample phase. Systemic injections of scopolamine impair encoding but not retrieval in this task (Aigner et al., 1991). Local infusions into perirhinal cortex in monkeys impair encoding for subsequent recognition whereas infusions into dentate gyrus or inferotemporal cortex do not (Tang et al., 1997). Recognition memory in rats can be measured by the increased time that rats normally spend sniffing and exploring novel items versus familiar items. Local scopolamine infusions into perirhinal cortex reduce the difference in sniffing time for novel objects, but these infusions do not impair spatial alternation, suggesting task specificity (Winters and Bussey, 2005). These effects might be due to impairments of mechanisms for active maintenance of information for encoding into episodic memory.

Support for the effect of scopolamine on active maintenance of information for encoding also comes from studies on trace conditioning, in which animals hear a tone and then several seconds after the tone ends a shock is presented. Despite this long trace interval between tone and shock during initial exposure, rats learn the association and on later days they freeze when they hear the tone. Infusion of drugs that block muscarinic receptors into perirhinal or entorhinal cortex before the initial exposure to tone followed by shock reduces the effect of trace conditioning on rat freezing behavior (Bang and Brown, 2009; Esclassan et al., 2009), but does not impair conditioning when the tone overlaps with the shock (delay conditioning). This suggests that blockade of acetylcholine receptors blocks the mechanisms for active maintenance of the tone memory during the seconds-long trace interval between the presentation of the tone and the shock.

Local application of cholinergic antagonists into other regions also causes selective memory impairments. Infusions of scopolamine into region CA3 cause selective impairments of encoding but not retrieval of spatial trajectories in the Hebb-Williams maze (Rogers and Kesner, 2003). Infusions of scopolamine into the hippocampus during learning of a new platform location on each day of the Morris water maze impaired performance on the second trial, supporting an impairment of episodic retrieval (Blokland et al., 1992). Infusions of scopolamine into the medial septum actually increased acetylcholine release in the hippocampus but impaired spatial learning in the Morris water maze, suggesting that perturbations of acetylcholine levels can alter normal hippocampal function (Elvander et al., 2004). Infusions of carbachol into the medial septum, which increase levels of hippocampal acetylcholine, also impair memory (Bunce et al., 2004; Elvander et al., 2004), possibly by interfering with consolidation of previously stored memories (Bunce et al., 2004) as described below in the section on consolidation.

Anatomical localization of cholinergic function can also be studied with selective lesions of cholinergic neurons caused by localized injections of the toxin saporin conjugated with antibodies selective for cholinergic neurons. Retrograde transport of the saporin results in a selective lesion of cholinergic neurons innervating the structure that was injected. Selective lesions of the cholinergic innervation of the entorhinal cortex in rats cause impairments in delayed nonmatch to sample for novel but not familiar odor stimuli (McGaughy et al., 2005). Similarly, selective lesions of the cholinergic innervation of perirhinal cortex in monkeys cause impairments in visual delayed match to sample performance (Turchi et al., 2005). Selective cholinergic lesions of the medial septum do not cause impairments as strong as complete medial septal lesions, suggesting that the role of this cholinergic innervation in spatial memory encoding can be substituted by GABAergic innervation from the medial septum (Pang and Nocera, 1999). Combining cholinergic lesions with lesions of the GABAergic innervation causes much stronger impairments (Pang et al., 2001). These data indicate an important role of the combined cholinergic and GABAergic innervation of the hippocampus from the medial septum. The models presented here provide a framework for understanding the role of this modulatory input from the medial septum.

Anatomy of Acetylcholine

Despite the fact that acetylcholine plays an important role in regulating body function via the peripheral nervous system, drugs that affect acetylcholine receptors in the peripheral nervous system without crossing the blood-brain barrier do not influence

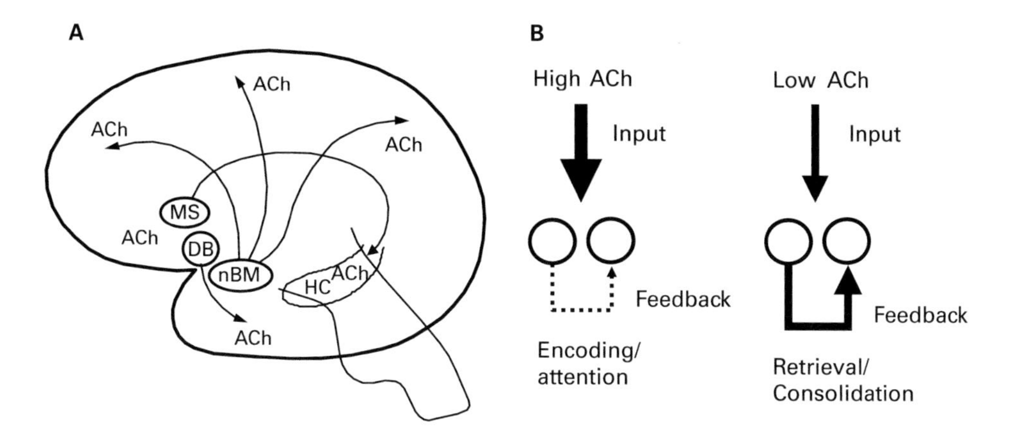

Figure 6.1
(A) Origin of fibers releasing acetylcholine (ACh) in cortical structures. Cholinergic innervation of the hippocampus (HC) arises from the medial septum (MS). Cholinergic innervation of parahippocampal structures arises from the MS and vertical limb of the diagonal band (DB). Cholinergic innervation of most of the neocortex arises from the nucleus basalis of Meynert (nBM). (B) Summary of the overall effect of cholinergic modulation in cortical structures. High levels of acetylcholine enhance the spiking response to afferent input and reduce excitatory feedback in the cortex. Lower levels of acetylcholine result in less influence of afferent input and stronger influence of excitatory feedback (adapted from Hasselmo, 2006).

memory. Therefore, the review of cholinergic anatomy will focus on the innervation in the central nervous system and, especially, the cortical structures implicated in memory function.

As shown in figure 6.1A, the hippocampus receives broadly distributed cholinergic innervation from the medial septum (Mesulam et al., 1983; Descarries et al., 1997), and parahippocampal structures such as the entorhinal cortex also receive modulatory input from the medial septum as well as the vertical limb of the diagonal band of Broca (Alonso and Kohler, 1984). In the model presented here, the cholinergic modulation is modeled as dependent on volume transmission (Umbriaco et al., 1995; Descarries et al., 1997), in which acetylcholine is released in a diffuse manner to cause relatively uniform and long-lasting activation of acetylcholine receptors throughout the structure. Anatomical data support volume transmission. For example, in the hippocampus, only seven percent of the axonal varicosities of cholinergic fibers are associated with postsynaptic densities (Umbriaco et al., 1995), suggesting that acetylcholine spreads diffusely, whereas all GABAergic varicosities in that study showed synaptic specializations. The hippocampus may regulate the level of cholinergic modulation by feedback connections to the medial septum (McLennan and Miller, 1974; Hasselmo and Schnell, 1994; Dragoi et al., 1999). Initial studies of acetylcholine levels in the hippocampus used microdialysis measurements that can only be taken over several minutes, and these results showed slow changes in tonic levels of acetylcholine, with high levels during active exploration (Marrosu et al., 1995), sustained attention (Sarter et al., 2005), and theta rhythm oscillations (Monmaur et al., 1997). However, recent studies using amperometry with higher temporal resolution show fast, phasic changes in acetylcholine levels in prefrontal cortex after sensory stimulus events (Parikh et al., 2007).

The hippocampus receives GABAergic inputs from the medial septum which appear to selectively target inhibitory interneurons in the hippocampus (Freund and Antal, 1988; Toth et al., 1997). These inputs play an important role in driving theta rhythm oscillations in the hippocampus which appear selectively during active exploration of the environment in the rat hippocampus (Green and Arduini, 1954; Buzsaki et al., 1983) and entorhinal cortex (Chrobak and Buzsaki, 1994) and in parahippocampal cortices of human during performance of maze tasks and working memory tasks (Kahana et al., 1999; Raghavachari et al., 2001; Rizzuto et al., 2003; Rizzuto et al., 2006).

Physiological Effects of Acetylcholine

The behavioral studies described above show that memory can be influenced by drugs that affect acetylcholine. What are the cellular effects of acetylcholine that are influenced by these drugs? Extensive physiological data show the effects of acetylcholine within the hippocampus and parahippocampal structures. The model of episodic memory presented here provides a theoretical framework for understanding the role of these physiological effects in episodic memory function.

As summarized in figure 6.1B, acetylcholine may enhance the encoding of memory by enhancing the influence of feedforward afferent input to the parahippocampal cortices and the hippocampus, making cortical circuits respond to features of the external world while decreasing excitatory feedback activity mediating retrieval and consolidation. This change in dynamics results from effects at different subtypes of acetylcholine receptors that include both nicotinic and muscarinic receptors.

Nicotinic Enhancement of Afferent Input

The drug nicotine has received considerable negative press because of its role in addiction to cigarettes. However, cognitive data clearly show that nicotine enhances the encoding of new episodic memories (Buccafusco et al., 2005; Levin et al., 2006). The nicotinic enhancement of memory function may partly result from nicotinic enhancement of excitatory synaptic potentials in the hippocampus caused by afferent input to hippocampal region CA3 from the dentate gyrus (Radcliffe et al., 1999) and from entorhinal cortex (Giocomo and Hasselmo, 2005). Nicotine does not enhance recurrent excitation. Similarly, in thalamocortical slice preparations of somatosensory cortex (Gil et al., 1997), activation of nicotinic receptors enhances thalamic input to cortex but not excitatory feedback synapses in the cortex. Nicotinic enhancement of glutamatergic transmission has also been shown at the medial dorsal thalamic input to prefrontal cortex (Gioanni et al., 1999).

Muscarinic Enhancement of Spiking Activity

Acetylcholine can also enhance encoding by increasing the spiking response of cortical pyramidal cells to afferent input (Patil and Hasselmo, 1999; Hasselmo and McGaughy, 2004). This effect results from activation of postsynaptic muscarinic M1 receptors that cause depolarization of pyramidal cells by suppressing a leak potassium current (Cole and Nicoll, 1984; Barkai and Hasselmo, 1994). These muscarinic receptors also reduce spike frequency accommodation by reducing a calcium-activated potassium current (Madison and Nicoll, 1984; Barkai and Hasselmo, 1994) as shown in figures 6.2 and A.19.

Muscarinic Presynaptic Inhibition

The enhancement of response to cortical input described above seems consistent with the enhancement of memory encoding. In contrast, it might initially seem puzzling that acetylcholine also reduces excitatory feedback within cortical circuits due to presynaptic inhibition of glutamate release from excitatory feedback synapses. Early physiological studies showed that acetylcholine causes presynaptic inhibition of excitatory glutamatergic synaptic potentials in the middle molecular layer of the dentate gyrus (Yamamoto and Kawai, 1967; Kahle and Cotman, 1989) and in stratum radiatum in region CA1 (Hounsgaard, 1978; Valentino and Dingledine, 1981). The selectivity

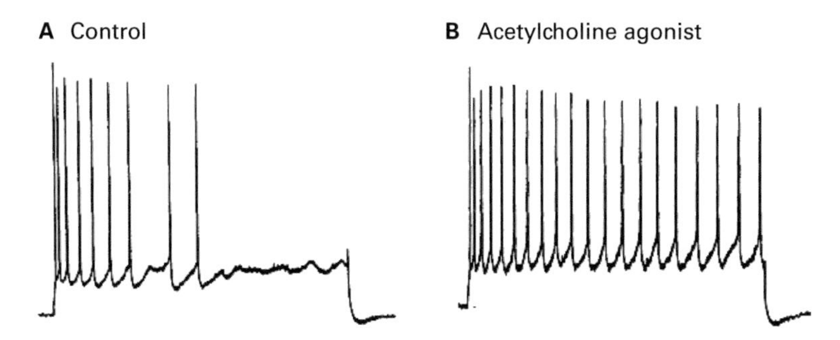

Figure 6.2
Spike frequency accommodation during a one-second-long current injection in control conditions (A) is reduced by activation of muscarinic receptors by the acetylcholine agonist carbachol (B), allowing more sustained spiking activity in response to depolarizing input (Barkai and Hasselmo, 1994).

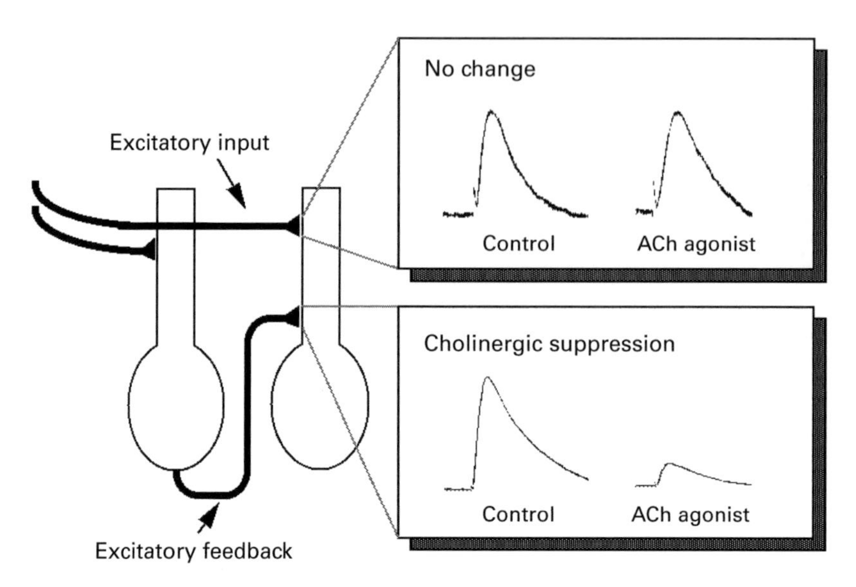

Figure 6.3
Selective presynaptic inhibition by acetylcholine receptors. Synaptic potentials can be elicited by stimulation of excitatory input fibers (top) and excitatory feedback (bottom). Activation of muscarinic acetylcholine receptors by the acetylcholine agonist carbachol (ACh agonist) causes no change compared to control conditions of synaptic potentials evoked by stimulation of afferent excitatory input (top). In contrast, the bottom traces show how the ACh agonist carbachol causes strong presynaptic inhibition of synaptic potentials at excitatory feedback synapses (ACh agonist) due to reduction of glutamate release compared to control conditions (Hasselmo and Bower, 1992; Hasselmo et al., 1995a).

of cholinergic presynaptic inhibition for excitatory feedback synapses was first shown by my work in olfactory cortex. As shown in figure 6.3, activation of acetylcholine receptors in the olfactory cortex causes selective presynaptic inhibition of synaptic potentials activated by stimulation of excitatory feedback connections in layer Ib while having a much weaker effect on synaptic potentials induced by stimulation of excitatory input fibers from the olfactory bulb in layer Ia (Hasselmo and Bower, 1992). The selective effect on excitatory feedback but not afferent input has also been shown in region CA3 of hippocampus, where muscarinic receptors cause presynaptic inhibition of excitatory glutamatergic transmission at recurrent feedback connections of the longitudinal association fibers in stratum radiatum, but not at afferent fiber excitatory input synapses in stratum lucidum (Hasselmo et al., 1995a; Vogt and Regehr, 2001) or stratum lacunosum-moleculare (Kremin and Hasselmo, 2007).

Acetylcholine has a similar selective effect in hippocampal region CA1, causing presynaptic inhibition of excitatory synaptic potentials in stratum radiatum of region

CA1 but causing smaller reductions in stratum lacunosum-moleculare, where entorhinal cortex layer III input terminates (Hasselmo and Schnell, 1994). Presynaptic inhibition appears to be stronger for synapses with AMPA receptors versus silent synapses in the hippocampus (de Sevilla et al., 2002) consistent with physiological evidence that presynaptic inhibition is stronger for recently potentiated synapses in the olfactory cortex (Linster et al., 2003). Presynaptic inhibition also appears in areas providing hippocampal feedback to the neocortex such as the subiculum (Kunitake et al., 2004). The selective sensitivity of glutamatergic synapses on different pathways to presynaptic inhibition is summarized in figure 6.4. More recent work has shown that presynaptic inhibition results from activation of the M4 subtype of muscarinic receptors, allowing selective enhancement of cholinergic presynaptic inhibition by drugs that enhance M4 receptor activity (Shirey et al., 2008), and selective blockade of cholinergic presynaptic inhibition with knockout of the M4 receptor (Dasari and Gulledge, 2011).

Given the behavioral data described above that showed a role of acetylcholine in the encoding of new memories, the evidence for cholinergic presynaptic inhibition initially appeared somewhat paradoxical. Why would a substance important for

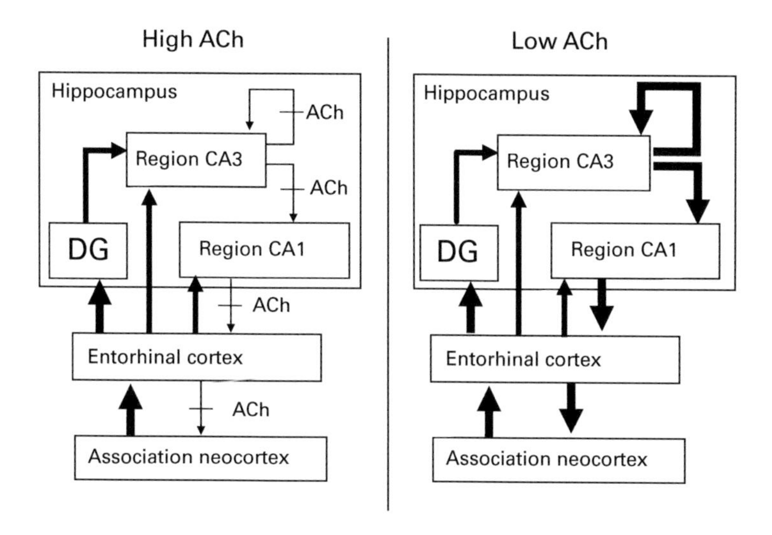

Figure 6.4

Locus of presynaptic inhibition. High ACh: High levels of acetylcholine (ACh) cause presynaptic inhibition (thin arrows) of excitatory recurrent connections in CA3 as well as connections from region CA3 to region CA1 and feedback connections between neocortical structures. Other connections are less affected (thick arrows). Low ACh: Low levels of ACh allow stronger synaptic transmission (thick arrows) at the previously inhibited synaptic connections. DG, dentate gyrus. Adapted from Hasselmo (1999).

encoding cause inhibition of synaptic transmission? Computational modeling of these effects in the framework of associative memory function provided a resolution for this paradox. As described later in this chapter, these models provide a clear picture for why acetylcholine should suppress excitatory feedback relative to afferent input.

The initial descriptions of selective cholinergic presynaptic inhibition in the olfactory cortex (Hasselmo and Bower, 1992) and hippocampus (Hasselmo and Schnell, 1994) inspired a number of subsequent studies to search for similar selectivity in neocortical structures. Acetylcholine was shown to suppress feedback but not afferent input synaptic potentials in the primary somatosensory cortex (Hasselmo and Cekic, 1996). In slice preparations including both thalamus and cortex, thalamic input showed nicotinic enhancement of synaptic potentials, whereas excitatory feedback showed muscarinic presynaptic inhibition (Gil et al., 1997). Acetylcholine was also shown to cause selective presynaptic inhibition of excitatory feedback synapses that reduce excitatory spread in the primary visual cortex (Kimura and Baughman, 1997; Kimura, 2000).

Enhancement of Synaptic Modification

Acetylcholine can also enhance encoding by enhancing synaptic modification. Acetylcholine enhances LTP (long-term potentiation) in many areas, including the hippocampus (Burgard and Sarvey, 1990; Huerta and Lisman, 1995; Adams et al., 2004), entorhinal cortex (Cheong et al., 2001), and primary olfactory cortex (Hasselmo and Barkai, 1995; Patil et al., 1998). In region CA1, induction of LTP depends on phase relative to spontaneous oscillatory activity (Huerta and Lisman, 1995; Hyman et al., 2003). Stimulation of the medial septum enhances LTP induction in vivo (Ovsepian et al., 2004), and scopolamine blocks the LTP enhancement associated with medial septal activity (Leung et al., 2003). Recent studies also demonstrate nicotinic enhancement of LTP (Buccafusco et al., 2005).

Modulation of Inhibition and Theta Rhythm Oscillations

Acetylcholine may also enhance encoding through its role in increasing theta rhythm oscillations within the hippocampal formation (Bland and Oddie, 2001; Siok et al., 2006). As described in chapter 2, theta rhythm appears to play an important role in encoding of memory (Winson, 1978; Givens and Olton, 1994). Encoding in tasks is enhanced when stimuli are presented during periods of theta rhythmicity in delay conditioning (Seager et al., 2002) and trace conditioning (Griffin et al., 2004). Theta rhythm is reduced by cholinergic lesions (Lee et al., 1994) and blocked by combined lesions of the cholinergic and GABAergic input from the medial septum (Yoder and Pang, 2005). Cholinergic neurons show theta rhythmic firing, which could provide rhythmic modulation of neuronal function in the hippocampus (Brazhnik and Fox, 1999).

Interneurons play an important role in theta rhythm (Fox et al., 1986; Csicsvari et al., 1999; Klausberger et al., 2003). Cholinergic modulation directly depolarizes many hippocampal interneurons (Chapman and Lacaille, 1999; McQuiston and Madison, 1999; Alkondon and Albuquerque, 2001), which could enhance their activity during theta rhythm. Muscarinic receptors also cause presynaptic inhibition of GABA release (Pitler and Alger, 1992). The combination of depolarization and presynaptic inhibition of interneurons appears paradoxical, but computational modeling demonstrates that these combined effects reduce background activity while heightening the response to suprathreshold sensory stimuli (Patil and Hasselmo, 1999). Cholinergic modulation also increases the rhythmicity of some interneurons (Chapman and Lacaille, 1999). This cholinergic regulation of interneuron rhythmicity could contribute to regulating the encoding and retrieval dynamics of the hippocampus. In the hippocampus, muscarinic receptors selectively depolarize O-LM interneurons, but not non-O-LM cells (Lawrence et al., 2006). This could provide separate rhythmic timing of dendritic and somatic inhibition that could enhance separation of encoding and retrieval dynamics during theta rhythm oscillations (Hasselmo et al., 2002b; Kunec et al., 2005; Cutsuridis and Hasselmo, 2010) as described later in this chapter.

Enhancement of Persistent Spiking

Acetylcholine has been demonstrated to enhance the persistent spiking of individual cortical neurons, which could provide a mechanism for active maintenance of novel information. This effect has been shown in entorhinal cortex (Klink and Alonso, 1997a), as well as other regions. As shown in chapter 2, in standard control conditions, entorhinal neurons will respond to an intracellular depolarizing current injection by generating spiking activity during the current injection but will terminate spiking after the end of current injection. In contrast, during activation of muscarinic receptors with the cholinergic agonist carbachol, neurons respond to the same magnitude and duration of depolarizing current injection with an increased number of spikes, and when the current injection ends, they persist in spiking activity for an extended period of minutes (Klink and Alonso, 1997b; Egorov et al., 2002). This effect has also been described in other areas including perirhinal cortex (Leung et al., 2006) and prefrontal cortex (Haj-Dahmane and Andrade, 1998).

This persistent spiking provides an excellent mechanism for active maintenance of novel information both for short-term working memory and for encoding of information into long-term memory. Detailed computational simulations of the entorhinal cortex (Fransén et al., 2002) demonstrate how the cholinergic activation of intrinsic mechanisms for persistent spiking could underlie spiking activity during the delay period of delayed matching tasks in both rats (Young et al., 1997) and monkeys (Suzuki et al., 1997), as well as causing phenomena such as match and nonmatch enhancement and suppression which occur during these tasks. Modeling demonstrates how cholinergic modulation of a nonspecific cation current in entorhinal cortex may cause a regenerative cycle in which spiking causes voltage-sensitive calcium influx which further activates the nonspecific cation current to main persistent spiking (Fransén et al., 2002) as described further in the Appendix. These effects provide a mechanism for active maintenance of information for working memory. Modeling also demonstrates the cellular mechanisms for the neurons in deep layers of entorhinal cortex that maintain graded persistent spiking frequencies that can be shifted up and down by periods of depolarizing or hyperpolarizing current injection (Fransén et al., 2006).

These intrinsic mechanisms for persistent firing would be particularly important for working memory for novel stimuli (Hasselmo and Stern, 2006). Most previous models of working memory focus on persistent spiking maintained by excitatory recurrent connections in the prefrontal cortex (Durstewitz et al., 2000). However, novel stimuli would not match the pattern of previously strengthened synapses and therefore might not be maintained by recurrent excitation. As an alternative, working memory for novel stimuli might depend upon the intrinsic mechanisms for persistent spiking that are enhanced by acetylcholine (Hasselmo and Stern, 2006).

The role of acetylcholine in working memory for novel stimuli is consistent with data from experiments that were motivated by this hypothesis. Selective lesions of the cholinergic innervation of entorhinal cortex impair delayed nonmatch to sample performance for novel odors but not for familiar odors (McGaughy et al., 2005). Infusions of cholinergic antagonists impair trace conditioning but not delay conditioning, suggesting that acetylcholine is required to maintain information about the tone through the trace period so it can be associated with a shock (Bang and Brown, 2009; Esclassan et al., 2009). In contrast, scopolamine does not impair delay conditioning, in which a tone continues until the shock appears. In humans, scopolamine reduces parahippocampal fMRI activity observed during the delay period after presentation of novel sample stimuli in a delayed match to sample task (Schon et al., 2005). Loss of this persistent activity during the delay period could underlie the impairment of delayed matching function in humans caused by scopolamine (Robbins et al., 1997; Green et al., 2005).

Persistent spiking in perirhinal and entorhinal cortex may be important for encoding of information into long-term episodic memory, consistent with scopolamine causing a reduction of delay period activity correlated with subsequent memory in humans (Schon et al., 2005) and with local infusions of scopolamine causing impairments of encoding for subsequent recognition in monkeys (Tang et al., 1997). This hypothesis is also consistent with evidence that medial temporal lesions appear to selectively impair working memory for new conjunctions of complex nonverbalizable visual stimuli (Olson et al., 2006). The selective role in encoding of novel stimuli could explain the early work showing that medial temporal lobe lesions have little effect on delayed matching for small numbers of highly familiar stimuli (Correll and Scoville,

1965), but delayed matching for trial-unique stimuli is strongly impaired by fornix lesions (Gaffan, 1974) or lesions of perirhinal and entorhinal cortex (Meunier et al., 1993). In contrast, lesions of prefrontal cortex impair performance for small numbers of familiar stimuli, but not for trial-unique stimuli (Eacott et al., 1994). These selective effects may be due to the role of cellular persistent spiking mechanisms in parahippocampal cortices for maintenance of novel, trial-unique stimuli versus synaptic mechanisms of persistent spiking in prefrontal cortex for maintenance of familiar stimuli. Consistent with this, fMRI studies using the n-back working memory task show prefrontal activity with little medial temporal activity during working memory for familiar stimuli but show a dramatic increase of medial temporal lobe activity during performance of the same task with novel stimuli (Stern et al., 2001), highlighting the importance of medial temporal lobe structures in working memory for novel stimuli (Hasselmo and Stern, 2006).

Acetylcholine Enhances Encoding

Based on the physiological data described above, extensive modeling shows that the effects of acetylcholine set appropriate dynamics for the encoding of new episodic memories (Hasselmo et al., 1992; Hasselmo, 1993; Hasselmo and Bower, 1993; Hasselmo et al., 1996; Hasselmo and Wyble, 1997). Some of these effects are obvious. The muscarinic depolarization of pyramidal cells and the nicotinic enhancement of synaptic transmission clearly enhance the influence of external afferent input, allowing more accurate encoding of multiple features of an event or item. The muscarinic enhancement of synaptic modification serves to increase the long-term maintenance of associations—for example, between position on a spatiotemporal trajectory and an event.

However, it is not immediately clear why the same synapses that show muscarinic enhancement of long-term synaptic modification (Huerta and Lisman, 1995; Patil et al., 1998; Adams et al., 2004) should also show strong muscarinic presynaptic inhibition of glutamate release (Hasselmo and Bower, 1992; Hasselmo and Schnell, 1994). This presynaptic inhibition can reduce the magnitude of synaptic potentials to less than half their initial size. Why would one need to suppress transmission at the synapses undergoing modification during storage of a memory? Computational modeling allows understanding of this combination of effects, and the overall need for changes in dynamics during associative memory function.

Problem: Interference during Encoding

The problem of interference during encoding is an important issue for associative memory models of region CA3 as well as other regions of the hippocampal formation. Overcoming the problem of interference during encoding requires separate dynamics during encoding and retrieval, which is an essential feature of all the models of associative memory function, including the attractor dynamic models reviewed in chapter 5.

What happens if encoding and retrieval are not separated? This problem can be illustrated using the standard test of episodic memory known as the paired associates task (or cued recall task). In a paired associates task, participants are presented with a series of related word pairs, such as leather–holster. The first word in each pair is commonly labeled as word A and the second as word B. Then at the end of the list the subject is cued with the first word A in each pair (e.g., leather) and asked to generate the second word B (e.g., holster). This task is impaired by lesions of the hippocampus, suggesting that some process occurs in the hippocampus during the brief presentation of a single word pair that allows a subject to correctly respond to the cue.

Figure 6.5 shows the proposed mechanism for this associative function. The filled circles in figure 6.5A represent populations of neurons activated in region CA3 for the first context (context 1), the first word A (leather), and the second word B (holster). Sensory input of these words initially activates language areas of the neocortex, but the input also activates a selective episodic representation in the hippocampus.

Hebbian modification of synapses between spiking neurons causes strengthening of synapses as represented by thicker black lines in figure 6.5B. Synapses are strengthened between leather and holster as well as between context 1 and holster. Note that

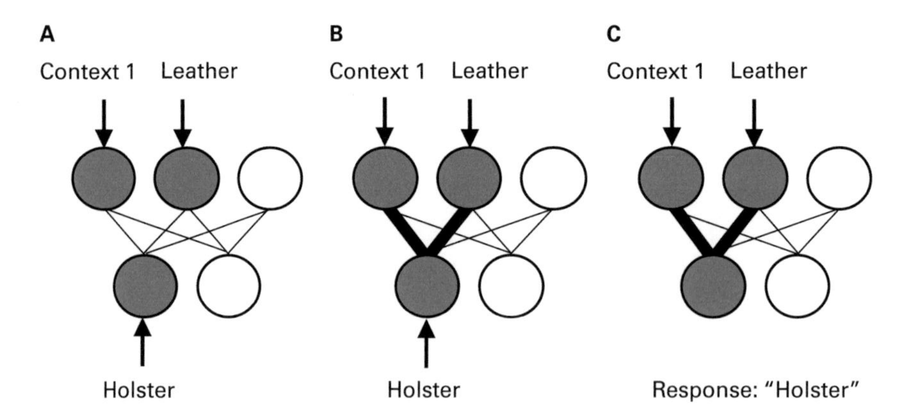

Figure 6.5
(A) During encoding of the A–B word pair, input to hippocampus (arrows) activates neurons (filled circles) coding the words leather and holster and context 1. (B) Hebbian synaptic modification strengthens synapses between the active neurons (thick black lines). (C) During retrieval, the cue word activates the population of neurons coding the word leather, and activity spreads across previously strengthened synapses (thick black lines) to activate the neurons coding the word holster.

the Hebbian properties of synaptic modification are important to ensure that only the connections with both presynaptic and postsynaptic activity are strengthened. If modification depended only on postsynaptic activity, then other associations with the neurons representing holster (such as "gun") could be strengthened. If modification depended only on presynaptic activity, then other associations from the neurons coding the word leather (such as "purse") could be strengthened.

After encoding, the network can simulate cued recall. When the subject is cued with the word leather in the first context, the spiking activity for this word and the context representation spreads across previously modified synapses to cause spiking in the population of neurons representing the word holster. This spiking activity then spreads back via entorhinal cortex to the language cortices to allow generation of the correct verbal response—saying the word "holster."

The learning of a single word pair is not a problem for associative memory models. However, difficulties occur when additional overlapping word pairs are learned. For example, many experiments test a second set of word pairs that share the first word A (e.g., leather) but associate it with a different second word designated with the letter C (e.g., boot).

Figure 6.6 illustrates the problem of interference during encoding. When the new word pair is heard, the word leather activates the same neurons but with a new context representation (context 2). The new word boot activates a different population of neurons.

Interference during encoding occurs because of retrieval of previously encoded associations. The spiking activity induced by the word leather causes synaptic transmission across previously strengthened synapses that causes spiking in the neurons that previously represented the word holster (filled circle on left). This could inhibit the activity caused by the word boot. In this example, if we apply the learning rule for Hebbian synaptic modification at the same time as the activation rule, this results in strengthening of an additional undesired connection (gray striped line) between the new context (context 2) and the word from the first context (holster). This is an incorrect association as the word holster did not appear in the new context.

During retrieval, when the first word leather is presented in the new context (context 2), the spiking activity in context 2 can cause synaptic transmission across the additional undesired strengthened synaptic connection. This can cause spiking activity in the neurons representing the incorrect word holster, resulting in the incorrect response. Thus, interference during encoding prevents accurate associative memory.

Solution: Suppress Retrieval during Encoding

The solution to the problem of interference during encoding is to suppress the retrieval of previously encoded associations during the encoding of new associations. This

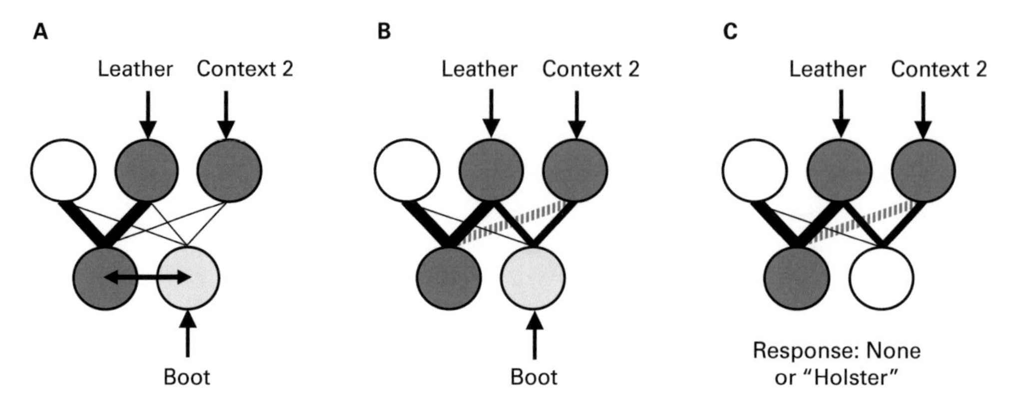

Figure 6.6
(A) During encoding of the A–C word pair, input to hippocampus (arrows) activates neurons coding context 2 and the words leather and boot. Interference during encoding results from the spread of activity across previously strengthened synapses (thick black lines) to activate the neurons coding the previous word holster (filled circle). (B) This causes strengthening of undesired synaptic connections between context 2 and the first word holster (thick striped gray line). In addition, activity of the word holster inhibits activity of the word boot, thereby reducing synaptic modification. (C) During retrieval, the cue word leather with context 2 causes spread of activity across the undesired synaptic connection (thick striped gray line), resulting in the activation of the incorrect response (holster) or no response due to response competition.

effect could be obtained by combining the enhancement of synaptic modification with the selective presynaptic inhibition of all excitatory recurrent synaptic transmission mediating retrieval during the new encoding. Note that the afferent input bringing new words and context into the network must not be suppressed, consistent with experimental data showing lack of presynaptic inhibition at afferent excitatory input synapses (Hasselmo and Bower, 1992; Hasselmo and Schnell, 1994; Hasselmo et al., 1995a). The suppression of the recurrent excitation (retrieval) during application of the learning rule (encoding) has been a standard feature of associative memory models since the first models were developed (Anderson, 1972; Kohonen, 1972; Hopfield, 1982), but no physiological mechanism was previously provided.

Acetylcholine causes a combination of physiological effects within the cortex that can overcome interference during encoding. The same synapses that show muscarinic enhancement of synaptic modification commonly show muscarinic presynaptic inhibition of glutamatergic synaptic transmission. In addition, the muscarinic presynaptic inhibition is selective for the synapses proposed to perform associative memory function in models of the hippocampus. Muscarinic presynaptic inhibition reduces transmission at the longitudinal association fibers in region CA3 (Hasselmo et al., 1995a) and the Schaffer collaterals input to region CA1 (Hounsgaard, 1978; Valentino and

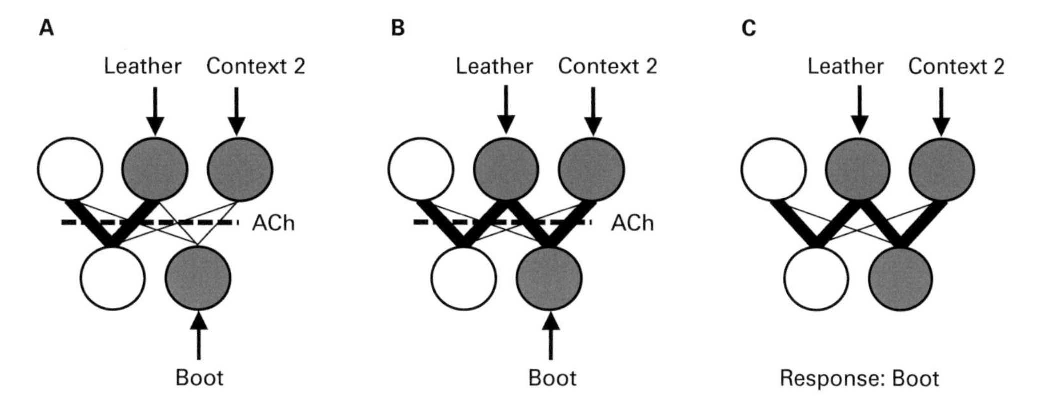

Figure 6.7
Encoding of the A–C word pair leather–boot in the presence of acetylcholine (ACh). (A) Cholinergic presynaptic inhibition reduces the spread of activity across all synapses including previously strengthened synapses (dashed line, ACh) but allows neurons to selectively respond to the new afferent input to the network (arrows). (B) The reduction of retrieval combined with the enhancement of synaptic modification by ACh allows strengthening of only the synaptic connections with the new word boot (thick black lines). (C) During retrieval after encoding with acetylcholine, the cue word leather in the second context can elicit the correct response (boot).

Dingledine, 1981; Hasselmo and Schnell, 1994) while having a much weaker effect on afferent fiber synapses bringing afferent input from entorhinal cortex to region CA1 (Hasselmo and Schnell, 1994) and to region CA3 (Hasselmo et al., 1995a; Kremin and Hasselmo, 2007) or dentate gyrus input to region CA3 (Hasselmo et al., 1995a; Vogt and Regehr, 2001). This pattern of effects was summarized in figure 6.4.

Implementing the effects of acetylcholine during encoding in the model prevents the retrieval of previously stored associations from interfering with the encoding of new associations. This is shown for the A–C pair in figure 6.7. Excitatory afferent input is not suppressed by acetylcholine, so spiking is still induced by the word leather and the word boot. However, presynaptic inhibition by acetylcholine at all excitatory recurrent connections between neurons in the model prevents retrieval due to synaptic transmission across previously strengthened synapses in figure 6.7A. Thus, there is no spiking in the neurons representing the word holster (no retrieval of prior associations).

The lack of interference from prior retrieval means that Hebbian synaptic modification selectively strengthens the associations between leather and context 2 and the C word boot in figure 6.7B. Even though acetylcholine causes presynaptic inhibition at these synapses that will reduce glutamate release, it does not completely prevent glutamate release. At the same time, acetylcholine also enhances NMDA currents (Markram and Segal, 1990) triggering synaptic modification between the neurons activated by afferent input. This direct enhancement of changes in synaptic strength ensures that new learning only occurs at the time that the other cholinergic effects are present to set appropriate dynamics for new encoding.

During retrieval after learning with acetylcholine, when the first word leather is presented in the new context (context 2), the spiking activity in context 2 causes synaptic transmission only across synapses that were only strengthened with the neurons representing the word boot. In this manner, the presence of acetylcholine during encoding enhances the accurate retrieval of the second overlapping association.

In further work, computational modeling showed how cholinergic modulation in region CA3 and CA1 could allow effective separation of the dynamics of encoding and retrieval for multiple different overlapping input patterns (Hasselmo and Schnell, 1994; Hasselmo et al., 1995a). The computational model included feedback regulation of cholinergic modulation based on the match between retrieval and afferent input. When an input pattern is novel, the overlap between retrieval and input is small, and acetylcholine levels stay high. This allows the suppression of previously encoded associations by cholinergic presynaptic inhibition of the longitudinal association pathway and Schaffer collaterals. The encoding of new information is enhanced by the increase in spiking response to afferent input and the cholinergic enhancement of long-term synaptic modification. When a pattern has been stored, the match between retrieval and input is strong, suppressing the levels of acetylcholine and allowing stronger synaptic transmission to mediate retrieval (Hasselmo et al., 1995a).

Experimental Test of Interference during Encoding

The impairment of encoding of episodic memories caused by the muscarinic receptor antagonist scopolamine can be understood based on the blockade of acetylcholine effects. Scopolamine prevents depolarization and persistent spiking and will allow spike frequency accommodation to reduce the spiking response to afferent input (Hasselmo and Wyble, 1997). In addition, scopolamine will reduce synaptic modification needed for encoding a spatiotemporal trajectory and its associations with events and items. This accounts for the selective impairment of encoding by scopolamine. In contrast, scopolamine will prevent presynaptic inhibition at previously strengthened synapses, allowing effective retrieval of previous associations.

Based on the example shown in figure 6.6, scopolamine should also cause greater proactive interference from previous paired associates (A–B) during encoding of overlapping paired associates (A–C) than during encoding of nonoverlapping paired associates (D–E). This prediction (Hasselmo and Wyble, 1997) motivated a study of the effects of scopolamine on cued recall. The study showed that scopolamine caused an impairment of the encoding of new paired associates that was stronger for overlapping

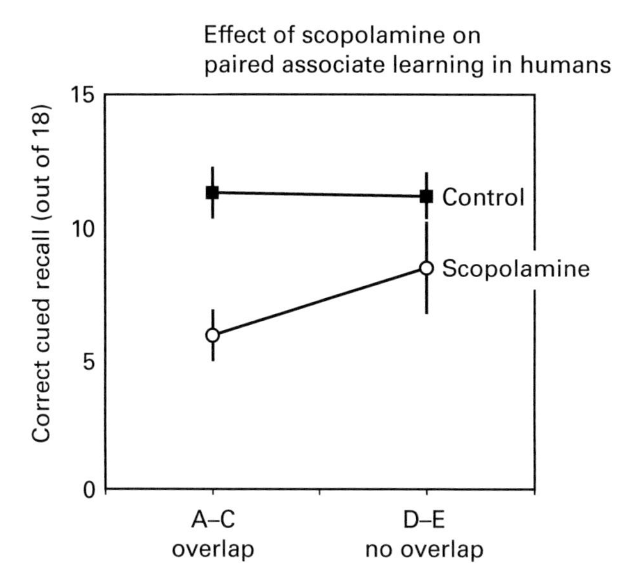

Figure 6.8
In control conditions, subjects correctly retrieve the second word for about 12 overlapping word pairs (A–C overlapping) and about 12 nonoverlapping pairs (D–E nonoverlapping). In contrast, the presence of scopolamine during encoding causes stronger impairment of cued recall for paired associates A–C that overlap with previously encoded pairs A–B, and causes a weaker impairment of cued recall for paired associated D–E that do not overlap (Atri et al., 2004). This supports the prediction of the model that blockade of acetylcholine receptors would enhance proactive interference during encoding (Hasselmo and Wyble, 1997).

paired associates than for nonoverlapping pairs (Atri et al., 2004) as shown in figure 6.8. A similar design was used to test overlapping odor associations in rats, showing that cholinergic receptor blockade or cholinergic lesions also caused enhanced proactive interference (De Rosa and Hasselmo, 2000; De Rosa et al., 2001).

Relationship to Other Studies

The physiology and modeling described here had an effect on a number of subsequent studies. As described above, the presentation of a functional role for stronger cholinergic modulation of feedback versus afferent input connections led to further physiological studies showing the selective presynaptic inhibition of feedback versus afferent input in thalamocortical slices of somatosensory cortex (Gil et al., 1997), slice preparations of auditory cortex (Metherate and Hsieh, 2004), and slice preparations of primary visual cortex (Kimura, 2000). The modeling indicates that the selective effects of acetylcholine should enhance the response to afferent input while reducing the spread of activity within cortical structures (Hasselmo et al., 1992; Linster and Hasselmo,

2001). Modeling also predicts that loss of cholinergic presynaptic inhibition should cause increased spread of activity, consistent with data showing an increase in background spiking activity in unit recordings of place cells found after lesions of the fornix that cut cholinergic input fibers (Shapiro et al., 1989) or after local infusion of cholinergic antagonists into the hippocampus (Brazhnik et al., 2003). In the primary visual cortex, acetylcholine reduces the extent of spatial integration in the visual field. For example, acetylcholine shifts the spatial tuning of visual cells to respond to shorter visual bar stimuli (Roberts et al., 2005). The enhancement of afferent input may also mediate attentional modulation of the responses of visual cortex neurons. The attentional responses of visual cortex neurons show enhancement by acetylcholine and are reduced by scopolamine (Herrero et al., 2008). The model predictions have also been tested in human visual areas, where increases in acetylcholine levels induced by the acetylcholinesterase blocker donepezil (Aricept) cause a reduction in the spread of activity as measured by fMRI (Silver et al., 2008).

The medial septum could also serve to regulate feedback to entorhinal cortex, with state changes mediated by both acetylcholine and GABAergic modulation. The flow of retrieved information back to entorhinal cortex requires transmission at both the Schaffer collaterals and the connections back to entorhinal cortex. As described above, cholinergic muscarinic receptors cause presynaptic inhibition of synaptic potentials in region CA1 (Hounsgaard, 1978; Valentino and Dingledine, 1981; Hasselmo and Schnell, 1994). Stimulation of the medial septum and local infusion of acetylcholine decreases the magnitude of synaptic potentials in stratum radiatum of region CA1 (Rovira et al., 1983). Sensory activation causes reductions in evoked synaptic potentials in CA1, and these reductions are blocked by scopolamine (Herreras et al., 1988). As noted above, a matching function in region CA1 could use current retrieval activity to regulate the balance of encoding and retrieval dynamics. This could involve the output from the region CA1 comparison function regulating the cholinergic and GABAergic modulation from the medial septum (Hasselmo and Schnell, 1994; Hasselmo and Wyble, 1997).

Detailed biophysical simulations of the primary olfactory (piriform) cortex were used to test this mechanism of cholinergic modulation for encoding of overlapping patterns of activity induced by odors (Barkai et al., 1994). The model utilized detailed biophysical simulations of individual neurons, with currents including the fast sodium and delayed rectifier currents underlying action potential generation, the AHP current and M current underlying spike frequency accommodation and the A current. Realistic synaptic potentials were modeled with dual exponential time courses. Patterns were stored via Hebbian synaptic modification of excitatory feedback connections, and then retrieval was tested with incomplete pattern cues consisting of subsets of the neurons in each pattern as shown in figure 6.9. Cholinergic modulation was modeled by reducing the maximum conductance of excitatory synaptic potentials, enhancing the

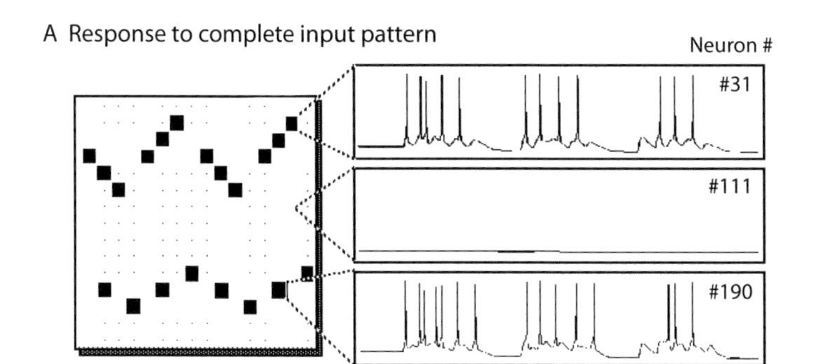

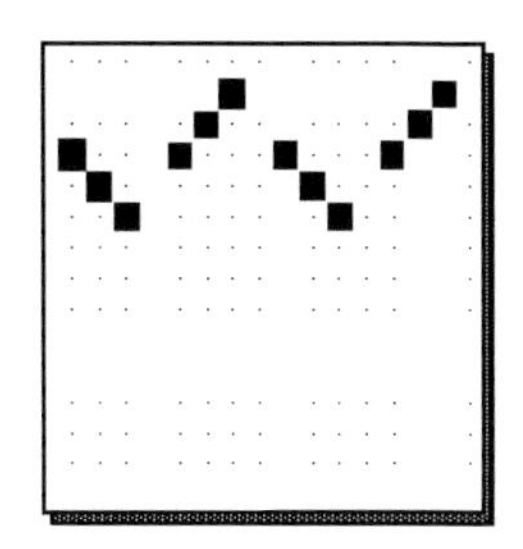

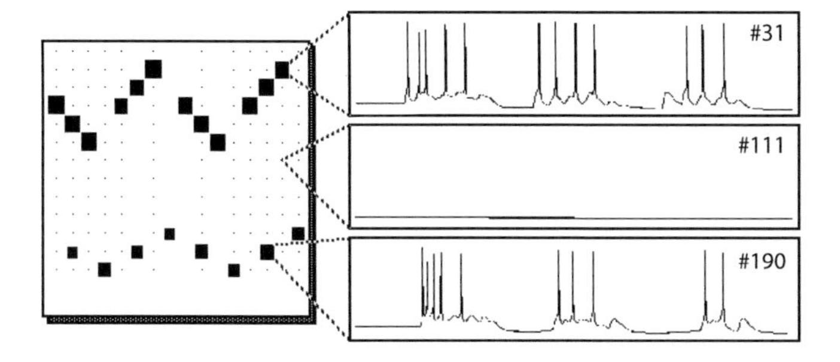

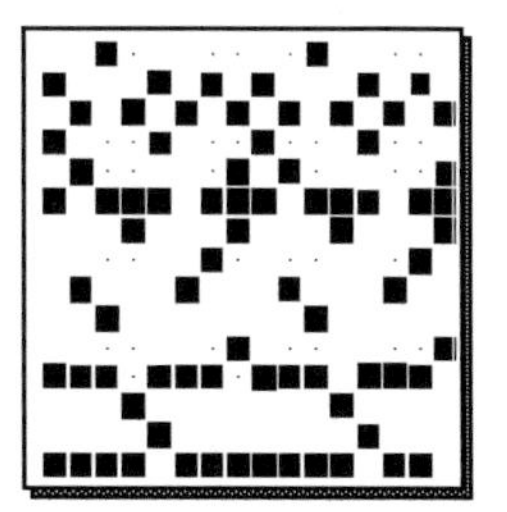

Figure 6.9

(A) Biophysical simulation of spiking response to afferent input. Size of black squares indicates amount of spiking activity on left. Example membrane potential traces are shown on right. (B) With no synaptic modification (no learning), a degraded input pattern only activates a subset of neurons. (C) After learning with acetylcholine (ACh), the network effectively completes missing components of the input pattern. (D) After learning without ACh for multiple overlapping patterns, proactive interference results in simultaneous retrieval of elements of all the multiple different input patterns.

induction of Hebbian modification and suppressing spike frequency accommodation with reductions in the AHP current and M current. As shown in figure 6.9, this biophysical network demonstrated that cholinergic modulation could reduce interference during encoding, preventing the stored patterns from becoming merged together into a single broadly distributed pattern (Barkai et al., 1994). Cholinergic reduction of spike frequency accommodation allowed stronger activity during encoding whereas increased spike frequency accommodation during retrieval prevented the strong excitatory feedback connections from causing excess activity.

All the models show that the effects of interference during encoding are cumulative, building up as more overlapping memories are stored. Figure 6.6 showed how an additional synapse is strengthened for each unit of an overlapping pattern. For

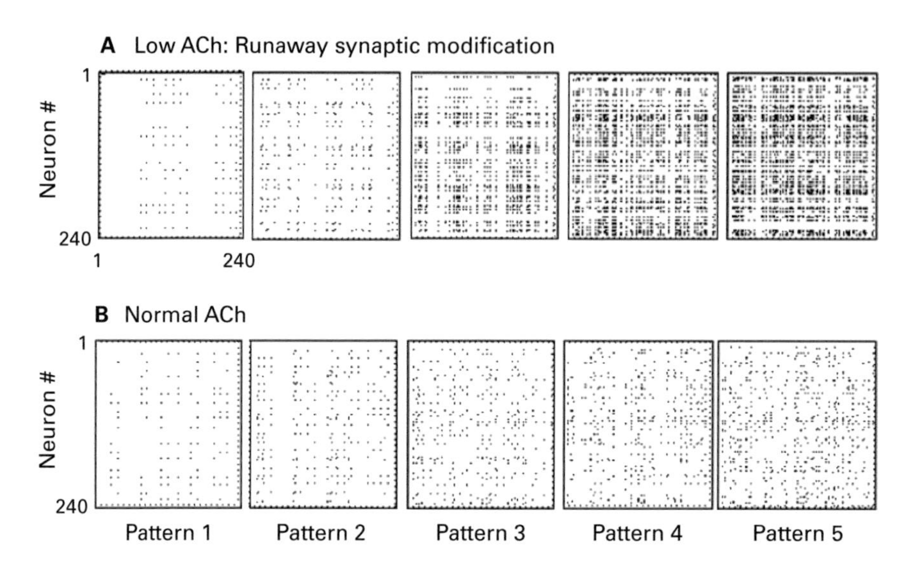

Figure 6.10

(A) Without cholinergic modulation during sequential encoding of five overlapping patterns (figure 6.9D), runaway synaptic modification builds up and strengthens a large number of undesired synapses. (B) With cholinergic presynaptic inhibition during encoding, only a small number of relevant synapses for each pattern are strengthened (adapted from Hasselmo, 1994). ACh, acetylcholine.

multiple overlapping patterns, this causes the phenomenon of runaway synaptic modification shown in figure 6.10, in which the total number of strengthened synapses increases not with the number of stored patterns but by the square of the number of stored patterns (Hasselmo, 1994). Attractor dynamics cannot prevent this problem (Greenstein-Messica and Ruppin, 1998), but the use of separate dynamics during encoding and retrieval prevents the problem.

Based on this work, the phenomenon of runaway synaptic modification was proposed to underlie the initiation and progression of cortical pathology in Alzheimer's disease (Hasselmo, 1994) in a model addressing both behavioral and molecular features of the disorder. This model indicates that the spread of pathology could be slowed by drugs that might selectively activate muscarinic presynaptic receptors mediating presynaptic inhibition. Muscarinic presynaptic inhibition has been shown to depend on activation of m4 muscarinic receptors in studies using m4 selective agonists (Shirey et al., 2008) and antagonists (Kimura and Baughman, 1997) and knockout of m4 receptors (Dasari and Gulledge, 2011). This suggests that drugs that activate muscarinic m4 receptors might be useful for slowing the spread of Alzheimer's disease. Another way of reducing runaway synaptic modification is by reducing the magnitude

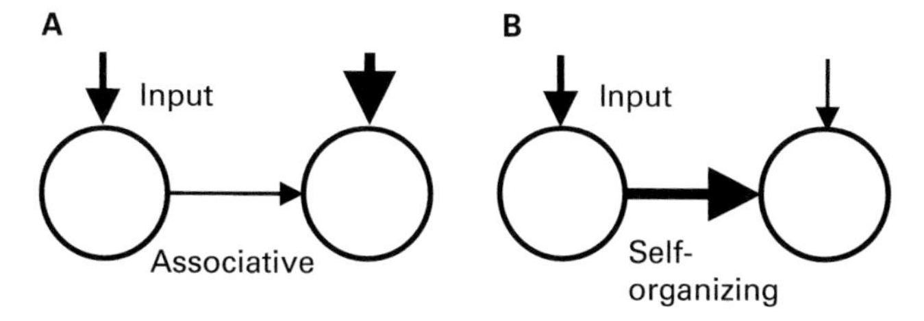

Figure 6.11
Comparison of dynamics underlying associative memory versus self-organization. (A) Modifiable synapses mediate associative memory function if they are not the dominant influence on postsynaptic activity but are instead modified on the basis of a separate dominant influence on postsynaptic activity. (B) Modifiable synapses undergo self-organization if they are the predominant influence on postsynaptic activity during learning.

of synaptic modification. This could be the mechanism for the efficacy of the NMDA antagonist memantine (Drever et al., 2007) that has been shown to reduce synaptic modification and is currently used in treatments of Alzheimer's disease. In this framework, the lack of new encoding in patient HM may have prevented the development of Alzheimer's pathology in the regions of his temporal lobes that were not surgically removed.

Differences in the dynamics of encoding versus retrieval are a defining characteristic of associative memory function. As shown in figure 6.11A, if the modifiable synapse is not the dominant influence on postsynaptic activity during application of the learning rule, then the synapse will store an association between the presynaptic activity and the separately generated postsynaptic activity. The best way to prevent the synapse from being the dominant influence on postsynaptic activity is to change the influence of the learning rule during encoding versus retrieval. In contrast, if the modifiable synapse is the dominant influence on postsynaptic activity during application of the learning rule, then the synapse will undergo self-organization of the representation as shown in figure 6.11B.

By suppressing feedback synapses, acetylcholine also allows more accurate self-organization of afferent synapses on the basis of sensory input. Modeling demonstrates that the selective cholinergic suppression of excitatory feedback would enhance self-organization of new representations for afferent input (Linster et al., 2003), which has been shown experimentally (Weinberger, 2003).

Encoding and Retrieval on Different Phases of Theta Rhythm

The behavioral and physiological effects of acetylcholine are consistent with its theorized role in setting encoding dynamics within the hippocampus and related structures. This raises the question of how the transition to retrieval dynamics is obtained. A rapid transition to retrieval dynamics would require the effects of acetylcholine to disappear very rapidly. However, the time course of acetylcholine effects in experiments in slice preparations appears to be relatively slow. A brief, pressure-pulse application of acetylcholine followed 100 milliseconds later by blockade of muscarinic receptors with pressure pulse application of the antagonist atropine still causes a slow time course of change in presynaptic inhibition. In this experiment, even though acetylcholine only temporarily activated muscarinic receptors, it caused presynaptic inhibition of potentials building up for 3–4 seconds and lasting for 10–20 seconds (Hasselmo and Fehlau, 2001). This would only allow relatively slow transitions between a cortical state focused on encoding and a state focused on internal retrieval.

This raises the question of whether other modulatory effects could more rapidly shift dynamics between encoding versus retrieval. Similar to acetylcholine, behavioral data indicate a role for theta rhythm oscillations in the encoding of new information in the hippocampal formation (Winson, 1978; Givens and Olton, 1990, 1994; Seager et al., 2002; Griffin et al., 2004). The changes in the dynamics during each cycle of the theta rhythm oscillations in the hippocampus could underlie a functional role for theta rhythm in setting encoding versus retrieval dynamics in the hippocampus (Hasselmo et al., 2002a).

The behavioral data on theta rhythm summarized in chapter 2 indicate an important role for theta rhythm in the encoding of new information, but the mechanisms for this role in memory encoding are not clear. Lesions of the medial septum or fornix that reduce theta rhythm oscillations in the hippocampus cause impairments in memory tasks. Many important studies of hippocampal function have involved lesions of the fornix to alter hippocampal function (Gaffan, 1974; Eichenbaum et al., 1990; Aggleton et al., 1995). The model presented in chapters 3 and 4 suggests a role of theta rhythm oscillations in the entorhinal cortex in providing a relative phase code for memory. The episodic memory model also required dynamics for separating the information arriving from the external world, which needs to be encoded as new, from the information being retrieved from internal circuits, which must not be confused with input from the external world. The change in dynamics during different phases of theta rhythm oscillations may provide a mechanism for separating these functional processes of encoding and retrieval (Hasselmo et al., 2002a). Both of these theoretical roles of theta rhythm provide possible explanations for the enhancement of encoding when theta rhythm is present, as well as the impairment of encoding when theta rhythm is reduced by lesions of the medial septum or the fornix.

Imagine how difficult your life would be if you could not separate external input from retrieved input. For example, imagine you are speaking with a friend on the phone, and you happen to retrieve a memory of a conversation from the previous week when you agreed to meet at a restaurant for lunch at 1 p.m. If you cannot separate retrieval from external input, then the retrieval of this conversation will be stored as if it just happened. Later that day, you arrive at the restaurant at 1 p.m. expecting to find your friend, based on your retrieval of an event from the previous week. The inability to separate retrieval from encoding would seriously impair human behavior.

Similar issues might apply to the separation of encoding real experience versus generating imaginary or future experience in a similar environment. For example, as a graduate student at Oxford, I was president of the Oxford Film Foundation and hosted visits by well-known actors, including Daniel Day-Lewis and Tim Roth. I can distinguish my episodic memories of walking through Oxford with these actors from my memories of viewing other actors in Oxford on film or television. This ability to separate encoding of reality from imagined or recalled events appears to be impaired in Korsakoff's amnesia and after aneurysms of the anterior communicating artery (ACoA), resulting in confabulation. For example, patients with ACoA aneurysms may report recent memories that appeared to arise from recall of earlier memories or even from magazine articles read in the hospital after the damage (DeLuca, 1993). Damage from the ACoA affects the basal forebrain including the medial septum whereas Korsakoff's amnesia includes damage to the mammillary bodies (Vann and Aggleton, 2004). Both structures are components of the circuits generating theta rhythm oscillations.

A number of the features of theta rhythm in the hippocampal formation support its potential role in separation of encoding and retrieval dynamics (Hasselmo et al., 2002a). The changes in physiological properties during different phases of the theta rhythm are reviewed here as background for their theoretical function.

Phasic Changes in Synaptic Current

Theta rhythm is associated with laminar segregation of rhythmic sources and sinks in region CA1 and dentate gyrus. This can be seen by moving an electrode progressively through different layers of the hippocampus and computing a current source density analysis based on the EEG recording data (Buzsaki et al., 1986; Brankack et al., 1993), as summarized in part C of figure 6.12. The current source density analysis supports possible differences in the function of the network at different phases of the theta rhythm cycle.

As shown in figure 6.12C, the strongest synaptic input from entorhinal cortex occurs at the lowest points (troughs) of EEG recorded in stratum lacunosum-moleculare, when there are prominent sinks (b) in stratum lacunosum-moleculare (Brankack et al., 1993). This is proposed as the encoding phase. Sinks usually indicate

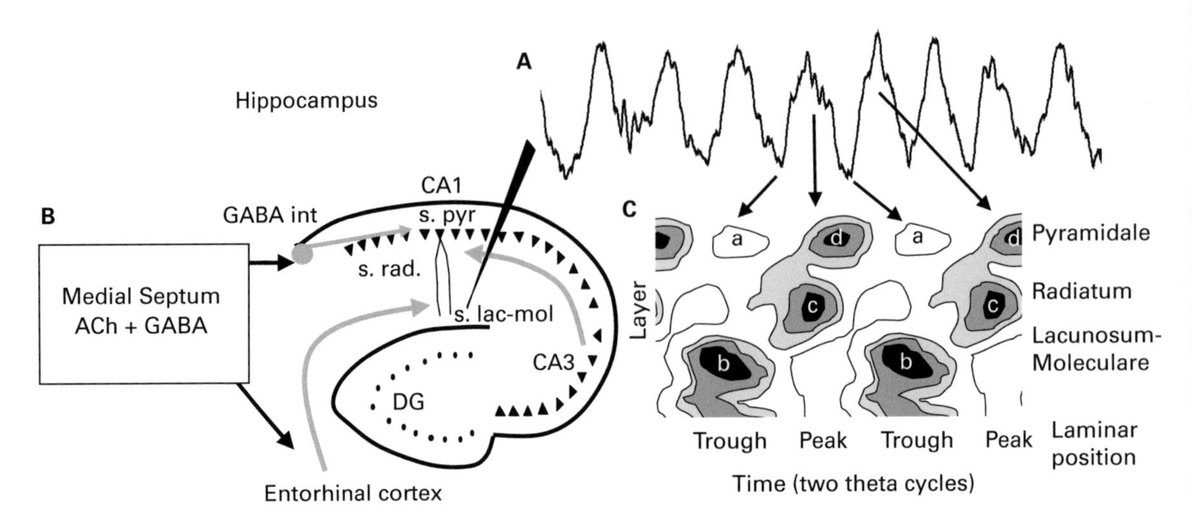

Figure 6.12
(A) Theta rhythm oscillations in the EEG from stratum lacunosum-moleculare (s. lac-mol) of hippocampal region CA1 (Wyble et al., 2000). (B) Summary of synaptic inputs from entorhinal cortex and CA3 contributing to synaptic currents in region CA1. (C) Changes in current in different layers (y axis) during two cycles of theta rhythm oscillations (x axis), showing a source (outward current) in the pyramidal layer (marked a) at the same phase as a sink (inward current) appears due to entorhinal input in lacunosum-moleculare (b). At the opposite phase (180 degrees later), a current sink occurs due to CA3 synaptic input in radiatum (c) and a sink appears due to spiking in pyramidale (d). Plot based on data from Brankack et al., 1993. s. rad., stratum radiatum; s. pyr, stratum pyramidale; DG, dentate gyrus.

excitatory (inward) synaptic currents. The strong excitatory input from entorhinal cortex to the dendrites in CA1 could allow encoding at a time when there is little retrieval spiking activity in the pyramidal cell layer, which at this phase shows prominent sources (outward currents) in stratum pyramidale (Brankack et al., 1993) due to strong inhibition at the cell body (Kamondi et al., 1998) that reduces spiking activity. These outward currents will cause the stratum pyramidale EEG to be near its peak. This is proposed as the encoding phase (Hasselmo et al., 2002a).

At a different phase about 180 degrees later, the peaks of the EEG recorded in stratum lacunosum-moleculare are associated with a sink in stratum radiatum (c) associated with CA3 synaptic input and another sink (d) in stratum pyramidale, associated with the greater firing of CA1 pyramidal cells at the peak of the EEG in stratum lacunosum-moleculare (Fox et al., 1986; Skaggs et al., 1996; Csicsvari et al., 1999). Because of the sink, the EEG recorded in stratum pyramidale will have a trough at this phase. As shown in figure 6.13, this phase is assumed to correspond to the retrieval phase of network function (Hasselmo et al., 2002a).

Phasic Changes in Presynaptic Inhibition

The phasic changes in current sinks can arise from differences in presynaptic firing rate, but could also arise from differences in presynaptic inhibition of synaptic transmission (Wyble et al., 2000). Recording of evoked synaptic potentials at different phases of theta rhythm demonstrates a change in magnitude of the rising slope of synaptic potentials (Wyble et al., 2000; Villarreal et al., 2007) as well as in the magnitude of evoked population spikes reflecting synchronous firing (Buzsaki et al., 1981; Rudell et al., 1984; Villarreal et al., 2007). The change in magnitude of synaptic transmission could be due to phasic changes in presynaptic inhibition caused by $GABA_B$ receptors (Hasselmo and Fehlau, 2001) which appears to have a sufficiently rapid time course in vivo to cause this effect (Molyneaux and Hasselmo, 2002). Activation of $GABA_B$ receptors causes presynaptic inhibition of glutamate release with a distribution similar to muscarinic presynaptic inhibition, reducing synaptic transmission at the longitudinal association fibers and Schaffer collaterals but not at entorhinal input synapses (Ault and Nadler, 1982; Colbert and Levy, 1992; Hasselmo and Fehlau, 2001) and reducing excitatory feedback but not afferent input in primary olfactory cortex (Tang and Hasselmo, 1994). As discussed below, these phasic changes could cause synaptic transmission in stratum radiatum to be weak when induction of LTP is strong during encoding and then allow strong transmission during retrieval when LTP is weak (Hasselmo et al., 2002a).

Phasic Changes in Membrane Potential

Another factor contributing to the change in magnitude of synaptic transmission and population spikes is the postsynaptic membrane potential of pyramidal cells in region CA1 and CA3. Intracellular recordings have demonstrated phasic changes in pyramidal cell depolarization during theta rhythm (Fujita and Sato, 1964; Fox, 1989; Kamondi et al., 1998). In particular, the soma membrane potential appears to be hyperpolarized at the time when dendrites are receiving the strongest depolarizing current input from entorhinal cortex (Kamondi et al., 1998). As discussed below, this hyperpolarization during a retrieval phase could prevent interference from retrieval of previously stored associations during encoding of new associations (Hasselmo et al., 2002a).

Phasic Changes in Postsynaptic Inhibition

The previous data on phasic changes in membrane potential have been refined by more detailed data on the activity of different classes of inhibitory interneurons during theta rhythm (Klausberger et al., 2003; Klausberger and Somogyi, 2008). Computational modeling based on these data indicate that axoaxonic and basket cells firing during the peak of pyramidale EEG could reduce the spiking output of neurons during encoding (Cutsuridis and Hasselmo, 2010). Conversely, the spiking activity of O-LM cells during the trough of pyramidale EEG could provide selective inhibition in

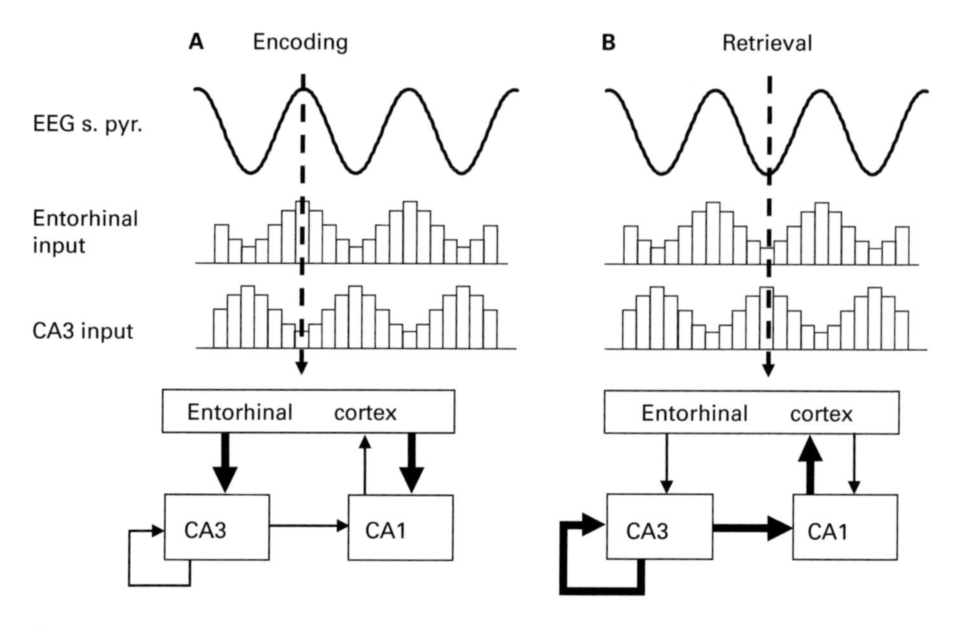

Figure 6.13

Schematic representation of the change in dynamics during hippocampal theta rhythm oscillations. (A) Encoding: Dashed line marks the encoding phase at the peak of the EEG in stratum pyramidale (s. pyr.) of CA1. At this phase, synaptic transmission arising from entorhinal cortex is strong (Entorhinal input). Synaptic transmission arising from region CA3 is weak (CA3 input), but these same synapses show a strong capacity for long-term potentiation (LTP). Bottom plot shows strong input as thick arrows, and weak transmission as thin arrows. This allows the afferent input from entorhinal cortex to set patterns to be encoded while preventing interference from previously encoded patterns on the excitatory synapses arising from region CA3. (B) Retrieval: Dashed line marks retrieval phase at the trough of the EEG in s. pyr. of CA1. At this phase, synaptic transmission arising from entorhinal cortex is relatively weak (though strong enough to provide retrieval cues). In contrast, synaptic transmission arising from region CA3 is strong, allowing effective retrieval of previously encoded associations. During this phase, synapses do not encode the retrieval activity because LTP at these synapses is weak; instead, they show long-term depression (LTD) or depotentiation. Figure adapted from Hasselmo et al. (2002a).

stratum lacunosum-moleculare that inhibits entorhinal input to the distal dendrites to allow a dominant influence of the retrieval of associations at modified synapses in stratum radiatum (Hasselmo et al., 2002a; Kunec et al., 2005).

Phasic Changes in Long-Term Potentiation

The separation of encoding and retrieval proposed here could also depend upon phasic changes in the mechanisms for changing the strength of synaptic transmission. Initially, it was shown in urethane anesthetized rats that LTP is more effectively induced in the dentate gyrus when a tetanus is delivered on positive phases of theta (Pavlides et al., 1988), and similar results have been shown in freely moving animals (Orr et al., 2001). Similar effects appear in region CA1. In slice preparations showing theta rhythm due to cholinergic agonists, stimulation on the peak of theta recorded locally in stratum radiatum causes LTP while stimulation on the trough causes long-term depression (LTD; Huerta and Lisman, 1995). In urethane anesthetized animals, stimulation delivered at the peak of the theta wave recorded locally in stratum radiatum induces LTP (Holscher et al., 1997) while stimulation delivered at negative phases of theta causes depotentiation. Note that the local peak in stratum radiatum is phase shifted from fissure EEG and would be closer to the encoding phase at the trough of fissure EEG. Induction of LTP in region CA1 was also analyzed in awake, behaving animals (Hyman et al., 2003), showing that stimulation on the peak of local theta induces LTP while stimulation on the trough induces LTD. These results suggest that induction of LTP in stratum radiatum occurs when transmission at these synapses is weak but dendrites are depolarized by entorhinal input. At this same phase, the soma is hyperpolarized to prevent spiking at the soma and interference from prior retrieval. This does not prevent LTP in the dendrites as dendritic spikes can induce LTP even when the soma is hyperpolarized (Golding et al., 2002). In this manner, even the local membrane potential at different positions along the dendrite in region CA1 neurons helps to separate encoding versus retrieval.

Summary of Theta Rhythm Model

Based on the physiological data, I developed a model (Hasselmo et al., 2002a) in which the changes in synaptic current during each cycle of the theta rhythm correspond to separate phases of encoding and retrieval that repeat during each cycle as summarized in figure 6.13. In figure 6.13, these phasic changes in physiological properties are described in reference to the EEG, but it is important to specify where the EEG is measured, as the peak of the EEG measured in stratum pyramidale of CA1 is about 180 degrees out of phase with the peak of the EEG measured in stratum lacunosum-moleculare near the fissure between CA1 and dentate gyrus.

As shown in figure 6.13A, during the encoding phase of the theta cycle, afferent excitatory input from entorhinal cortex is very strong whereas the recurrent excitatory feedback connections in region CA3, and the connections from region CA3 to region CA1, are very weak (Hasselmo et al., 2002a). This allows effective clamping of network activity to the afferent input, which is optimal for strengthening of excitatory recurrent connections to form an accurate representation of environmental features (e.g., by place cells). Note that this requires that LTP of the excitatory recurrent connections and Schaffer collaterals should be maximal at the time that synaptic transmission at these connections is the weakest. This is consistent with physiological data showing that LTP is best induced at the peak of the local EEG, which is the time when synaptic currents are the weakest (Huerta and Lisman, 1995; Holscher et al., 1997; Hyman et al., 2003). Details of the model are presented in the Appendix, pp. 284–286.

During the retrieval phase of the theta cycle (see figure 6.13B), afferent input from entorhinal cortex is at its weakest, but the excitatory recurrent collaterals in region CA3 and the Schaffer collaterals from region CA3 to region CA1 are at their strongest. This allows activity to be predominantly driven by spread of activity across previously modified synapses. Note that this retrieval will cause interference during encoding in the pattern of synaptic connectivity unless LTP is reduced at these connections during this phase (thus, LTP induction must be weakest when synaptic transmission is the strongest).

Simulations show that these mechanisms prevent interference during encoding, allowing the separation of new input and prior retrieval (Hasselmo et al., 2002a). In contrast, if retrieval is allowed during encoding, the spread of activity causes spiking to occur in a large number of neurons during the window for induction of LTP, resulting in strengthening of connections between distant, nonadjacent locations, causing a severe breakdown in network function (Hasselmo et al., 2002b). The utility of this same mechanism for separating encoding and retrieval has been shown in detailed biophysical simulations of region CA3 (Kunec et al., 2005) and region CA1 (Cutsuridis et al., 2010).

This model can account for behavioral data showing that fornix lesions (which reduce theta rhythm) cause an increase in the number of errors after reversal in a T-maze task (M'Harzi et al., 1987). Specifically, rats with fornix lesions persist in visiting an arm that was previously rewarded but is currently unrewarded. This impairment could result from loss of theta rhythm allowing the induction of LTP and synaptic transmission in stratum radiatum to be strong at the same time. After reversal, the rat makes erroneous visits to the previously rewarded location. In this case, strong synaptic transmission allows the rat to retrieve postsynaptic activity corresponding to memory of food at the now unrewarded location. This retrieval activity could cause further LTP, thus strengthening associations with the memory of food despite the fact that the location is now unrewarded. This mechanism could slow the extinction of the old association and increase the period of error generation before reversal (Hasselmo et al., 2002a). The behavioral deficits following fornix lesions might also result from loss of slow modulatory effects of acetylcholine but appear to depend on combined block of both cholinergic and GABAergic input (Pang et al., 2001). The model can also account for other studies showing impairments of memory function in other tasks after lesions that reduce theta rhythm oscillations (Winson, 1978; Givens and Olton, 1994).

The clamping of CA3 activity to afferent input could be assisted by the selective activation of dentate gyrus that appears to depend on input from the medial septum. The dentate gyrus responds more strongly to entorhinal input during specific phases of the theta rhythm oscillation (Buzsaki et al., 1981). Stimulation of the septum enhances dentate gyrus response to entorhinal input, possibly through inhibition of the dentate interneurons (Fantie and Goddard, 1982; Bilkey and Goddard, 1985), and reversible inactivation of the medial septum reduces spontaneous firing of neurons in dentate gyrus (Mizumori et al., 1989).

A number of experimental studies have supported the theorized separation of encoding and retrieval. Oscillations in human participants show reset to different phases of theta rhythm during encoding of stimuli on sample trials versus retrieval in response to stimuli on test trials (Rizzuto et al., 2006). This was found using the Sternberg task (a type of delayed match to sample task). In rats performing a delayed nonmatch to sample task, spiking occurs at significantly different phases of theta for match (old) versus nonmatch (novel) stimuli (Manns et al., 2007). Interestingly, a study of gamma frequency oscillations also supports the model (Colgin et al., 2009). That study found one phase of theta rhythm where high-frequency gamma oscillations were coherent between entorhinal cortex and region CA1 and a different phase of theta where low-frequency gamma was coherent between CA3 and CA1. This was interpreted as support for the model of separate phases involving encoding of entorhinal input versus retrieval based on CA3 input.

Because the retrieval phase is associated with long-term depression of synapses, this network has the property of retrieval-induced forgetting. This property has been used in more detail to enhance the separation of encoded memories by strengthening target memories and weakening competing memories (Norman et al., 2006; Norman et al., 2007). This model was supported by EEG data in human participants (Newman and Norman, 2010).

Low Levels of Acetylcholine Allow Consolidation

If the theta rhythm oscillations provide a rapid transition between encoding and retrieval dynamics, then what is the role of slower cholinergic modulation? Data support the hypothesis that cholinergic modulation may switch between dynamics appropriate for encoding and dynamics appropriate for consolidation of previously encoded information.

The change between encoding and consolidation dynamics could be regulated by the changes in acetylcholine levels during different stages of waking and sleep (Hasselmo, 1999b). During active waking (for example, when an animal moves around exploring an environment), acetylcholine levels are high within the hippocampus and neocortex (Marrosu et al., 1995). During quiet waking, these acetylcholine levels drop (for example, when an animal quietly grooms itself). During slow-wave sleep, acetylcholine levels in the hippocampus and neocortex drop to less than one-third of their levels during active waking, indicating a dramatic change in the modulation of cortical cells by acetylcholine (Marrosu et al., 1995). During REM sleep, acetylcholine levels return to levels similar to waking.

The concept of consolidation arises from behavioral evidence that episodic memories are more sensitive to disruption after initial encoding than after some period of time has elapsed. The change in sensitivity has been attributed to a process termed consolidation. Consolidation refers to two different phenomena: (1) the stabilization of specific cellular processes, such that functional changes (e.g., changes in synaptic strength) are rendered more permanent, and (2) the formation of additional traces such that a particular memory has a more robust representation. Here, the discussion focuses on the second definition (Eichenbaum and Cohen, 2001).

Behavioral data demonstrate that recently encoded memories are more sensitive than older memories to the effects of partial lesions of the hippocampus (Zola-Morgan and Squire, 1990; Kim and Fanselow, 1992) and electroconvulsive shock (Squire and Cohen, 1979). This has been termed temporally graded retrograde amnesia. This supports a two-stage theory of memory formation in which memories are initially encoded by rapid Hebbian synaptic modification in the hippocampus during waking (Buzsaki, 1989). In this theory, subsequent to this initial encoding, recently encoded memories are reactivated and cause gradual strengthening of additional synaptic connections within other areas including the entorhinal cortex or regions of association neocortex (Buzsaki, 1989; Wilson and McNaughton, 1994; McClelland et al., 1995). This reactivation and gradual strengthening was proposed to occur during sharp wave activity in the hippocampus, in which bursts of neural activity start in region CA3 and spread back through region CA1 to the entorhinal cortex.

The changes in acetylcholine during waking and sleep could regulate these dynamics (Hasselmo, 1999b). As shown in figure 6.4 and figure 6.14, high cholinergic levels during waking suppress excitatory feedback within region CA3 and from region CA3 to CA1, allowing dominant feedforward synaptic input appropriate for encoding, and reducing the influence of hippocampus on entorhinal cortex (Chrobak and Buzsaki, 1994). In contrast, lower acetylcholine levels during slow wave sleep remove the presynaptic inhibition, resulting in dominant feedback effects appropriate for consolidation. The strong feedback may contribute to sharp wave/ ripple activity associated with greater spiking activity in hippocampal regions CA3 and CA1 (O'Keefe and Nadel,

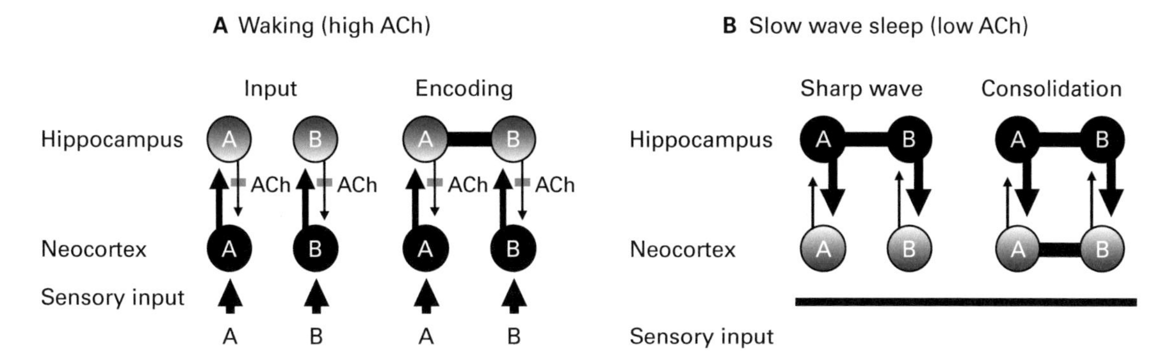

Figure 6.14
Acetylcholine (ACh) and the two-stage process of consolidation. (A) During waking there are high levels of ACh. Sensory input from two stimuli (A and B) is processed in Neocortex, but the information is not encoded there. The activity passes on to Hippocampus for initial encoding. ACh enhances the strengthening of synaptic connections between neurons responding to the stimuli (thick line between A and B in Hippocampus) and suppresses feedback mediating consolidation. (B) During quiet waking or slow wave sleep, sensory input is not present, and acetylcholine levels are low, allowing strong feedback to spread sharp wave activity within hippocampus and from hippocampus to neocortex (thick arrows). Diffuse activation of the hippocampus reactivates specific associations. The activity induced in neocortex by these reactivated associations allows slower strengthening of connections (new thick line) between the associated stimuli A and B in the neocortex (Hasselmo, 1999b).

1978; Buzsaki et al., 1992; Wilson and McNaughton, 1994) and replay of previous waking sequences (Skaggs and McNaughton, 1996; Lee and Wilson, 2002). This sharp wave activity propagates to deep layers of entorhinal cortex that receive the output from the hippocampus (Chrobak and Buzsaki, 1994), and could spread on to neocortical regions to cause storage of associations by neocortical synapses. The effects of acetylcholine in neocortical structures are consistent with this functional framework as cholinergic modulation causes presynaptic inhibition of feedback synapses from higher order somatosensory cortex while having less effect on synaptic potentials elicited in layer IV of neocortex (Hasselmo and Cekic, 1996). Similarly, acetylcholine suppresses intracortical synaptic potentials but not thalamocortical input in the somatosensory cortex (Gil et al., 1997), auditory cortex (Metherate and Hsieh, 2004), and primary visual cortex (Kimura, 2000).

The theory of the role of acetylcholine in consolidation (Hasselmo, 1999b) predicts that consolidation of memory should be impaired by experimental manipulations that increase acetylcholine levels during consolidation. This prediction is supported by data showing impairments of consolidation caused by cholinergic infusions into the medial septum after training in rats (Bunce et al., 2004) and impairments of consolidation in humans caused by injections of the acetylcholinesterase blocker physostigmine that increase acetylcholine levels during slow wave sleep (Gais and Born, 2004). The model also predicts that reducing acetylcholine levels should enhance consolidation. This was supported by a study showing that scopolamine delivered after encoding can enhance consolidation in humans (Rasch et al., 2006).

The same basic function appears in other models of consolidation, with differences in emphasis on particular features. An initial model of consolidation used binary activity patterns in two regions representing association neocortex (Alvarez and Squire, 1994). These were linked by bidirectional connections with the hippocampus, so that activation of each pattern would activate the associated neocortical patterns to strengthen connections within the neocortex. Later models using attractor dynamics in the representation of hippocampus demonstrated that it is not necessary to explicitly reactivate specific memories. Homogeneous depolarization, or random activation of the attractor network, will cause specific attractors to be activated, and the competition between attractors allows them to appear one at a time (Hasselmo et al., 1996; Shen and McNaughton, 1996). Once these attractor states have been activated, they can drive activity in the entorhinal cortex or neocortex to cause slower strengthening of associations in those structures.

Beyond the greater robustness provided by additional traces of a given episodic memory, this process of reactivation could strengthen episodic memories simply by enhancing the cross-links between components of the episodic memory. If episodic memories consist of associative networks between neurons coding multiple component features, then reactivation of the memory by hippocampal replay could result in strengthening of additional connections not formed during the initial encoding (e.g., between the memory of the cat and the memory of throwing away the melon in the kitchen). This could allow the episodic memory to be accessed by a broader range of cues.

Why should there be two stages in memory formation? Modeling suggests that these two stages are necessary to allow separate development of semantic memory representations that determine regularities in the information, independently from the associations in individual episodic memories (McClelland et al., 1995). That model proposed that semantic representations in neocortical structures would be distorted if each new piece of information were inserted without reference to other existing relations. For example, the knowledge that birds can fly would be distorted if new episodic information about penguins that are birds that swim were learned in isolation. If the information about penguins were used to train the network in isolation, the association between bird and swim would be strengthened at the expense of the association between bird and fly, such that the network could generate the output that robins are birds that swim. In contrast, if the new information is initially encoded in hippocampus and then gradually presented to the cortex interleaved with information about other birds, the old semantic information would not be distorted by the new information. The new information could be incorporated with the old in an appropriate manner. In this framework, the modulatory effects of acetylcholine could switch the network between a state of acquiring new episodic information in the hippocampus during active waking and a state of updating the semantic representations in neocortex during the occurrence of sharp waves in quiet waking and slow-wave sleep.

This leaves unanswered the role of higher levels of acetylcholine during REM sleep, which is also associated with the presence of theta rhythm oscillations (Louie and Wilson, 2001; Brandon et al., 2011a). The return to dynamics during REM sleep that resemble waking may allow exploration of alternate trajectories that could extend internal representations, or may allow exploration of trajectories associated with potential threatening events (Revonsuo, 2000). The exploration of trajectories with negative outcomes could be associated with the increase of REM sleep in depression (Hasler et al., 2004) and the negative bias of rumination in depression (Siegle and Hasselmo, 2002).

This chapter described the role of neuromodulatory agents such as acetylcholine in regulating cortical circuits between dynamics appropriate for encoding versus dynamics appropriate for consolidation. The chapter also addressed the role of theta rhythm oscillations in separating the dynamics of encoding and retrieval. These models can also address the impairment of memory caused by drugs that influence receptors for acetylcholine and GABA and alter theta rhythm oscillations. The next chapter will address how the mechanisms of encoding and retrieval of episodic memory can be used for memory-guided behavior in behavioral tasks.

7 Dynamics of Memory-Guided Behavior

How do memories guide behavior? This question becomes particularly important when studying animal behavior, as animals cannot describe the content of their memories, but most data on detailed neural activity during memory tasks has been obtained from rodents. We must understand the memory function of animals by observing how their memory guides their behavior. A number of sophisticated behavioral tasks have been designed to test episodic memory in animals. Experiments using these tasks have shown a role for the entorhinal cortex and hippocampus in memory-guided behavior. The role of these structures can be better understood by modeling the possible physiological dynamics that guide behavior. Simulation of the neural activity in rodents during performance of these tasks allows testing of the experimental validity of different models. As described here, models of the neural dynamics of memory function must account for a range of different phenomena during memory-guided behavior tasks, including the replay of neural activity, the context-dependent firing of cells in different tasks, and the theta phase precession of spiking activity.

The behavioral tasks used in rodents were reviewed at the end of chapter 1. The tasks included the Morris water maze, in which a swimming rat must use the memory of previous trials to find a platform hidden under the surface of opaque water. Another common task is the delayed spatial alternation task, in which a rat must choose what direction to go in a T-maze based on previous trials. This chapter will focus on models of the mechanisms for rodent performance in these tasks.

Behavior Guided by Trajectory Planning

The physiological dynamics of entorhinal cortex and hippocampus do not fully constrain the structure of models. Modeling shows many different possible mechanisms for memory-guided behavior, a few of which are described here. The same circuits could guide behavior by forward planning of possible trajectories, by retrieval of previously encoded trajectories, or by tracking of position along a trajectory. The guidance of behavior by forward planning will be discussed in this first section.

The Morris water maze task requires memory of the platform location, but because the start location differs on each trial and the rat follows a different trajectory on each trial, this task clearly does not depend only upon direct retracing of previously experienced spatiotemporal trajectories. In contrast to the retrieval of previously encoded trajectories, the model presented here uses grid cells to allow forward planning of trajectories to enable goal-directed behavior in a spatial memory task. The model uses current head direction to drive sequential activation of grid cells along a trajectory driven by the head direction. The sequential activation of grid cells then drives place cells along this trajectory.

Encoding for Forward Planning

Mechanisms of forward planning require some representation of the environment. The model presented here requires initial experience of the environment to sample combinations of grid cells in individual locations that activate place cells representing individual locations and cause Hebbian modification of connections between grid cells and place cells. This initial exploration of the environment can be performed by the animal over multiple trials, or as an alternative the exploration could involve internal tracing of trajectories and internal activation of place cells so that the whole environment would not need to be physically explored. Either way, the model requires the formation of associations between grid cells and place cells coding most locations in the environment.

Forward Scanning of Trajectories

When the animal encounters a particular goal location in the environment, this goal location could be held in working memory with active maintenance. Alternately, at the starting point for a new trajectory, the goal location could be retrieved from episodic memory as part of an individual spatiotemporal trajectory that was previously experienced. When the animal is placed in a new, arbitrary starting location on the perimeter of the Morris water maze, it must scan through multiple forward trajectories in order to find the correct direction to the goal. This scanning could be done by the animal's physically sampling different head directions as it turns from the wall (usually animals are placed in the tank facing the wall), or it could be done by internal virtual scanning of possible head directions.

For each head direction, the head direction activity is maintained for a period of time to drive forward scanning along a straight trajectory in the grid cell network. This could work with different types of grid cell models, but the description here focuses on the oscillatory interference model. In this model, a static head direction will cause a vector of activity in a set of head direction cells that drive different frequencies in entorhinal circuits. This causes oscillations to progress through cycles of constructive and destructive interference that are proportional to the frequency shift driven by the head direction cells and thereby the associated spacing of the grid cell firing fields. Thus, a dorsal cell will progress more rapidly through firing fields than a ventral cell. The activation of different grid cells at different times will reactivate the pattern representing different locations along the trajectory, allowing sequential activation of the place cells along the trajectory as shown in figures 7.1 and 7.2.

As shown in the figures, each head direction pattern allows forward sampling of a single trajectory through space coded by changes in grid cell firing. After one trajectory is sampled, the head direction can shift to sample a different head direction and activate a different pattern of head direction cell activity that drives a different forward trajectory. Each of the sequentially activated trajectories sequentially activates the place cells coding different positions along the trajectory. Eventually, one of the trajectories activates the place cell coding the location of the goal. When this occurs, the current head direction used to activate the trajectory can be selected as the basis for generation of movement by the rat. The rat moves forward with this head direction and goes straight to the goal.

This model works well for goal-directed navigation in an open environment without barriers to movement. However, many tasks involve barriers such as walls or edges that prevent movement directly toward the goal. Rats can learn to find the goal despite a barrier. For example, a rat can run to a reward location at the end of a complex series of hairpin turns (Derdikman et al., 2009). This requires generation of intermediate goal locations within the environment.

The generation of intermediate goal locations can be obtained within a network of place cells with Hebbian modification of connections between place cells to code transitions that have been experienced without barriers. This provides a substrate that allows propagation of activity back from the goal location through unblocked routes. This propagation of activity interacts with grid cell activity representing forward readout of potential trajectories through the environment to allow selection of pathways to a specific goal location. In particular, the network can find trajectories by selecting the forward trajectory associated with the fastest intersection of forward planning and reverse spread.

Relation to Previous Models

The mechanism of guiding navigation by reverse spread from the goal location was also developed in other models (Hasselmo et al., 2002c; Voicu and Schmajuk, 2002; Gorchetchnikov and Hasselmo, 2005; Toussaint, 2006) that do not use grid cells. In earlier models in my laboratory, the forward readout occurred in region CA3 and the reverse replay occurred with broad spatial tuning in layer III of entorhinal cortex, and the convergence was detected in region CA1 (Hasselmo et al., 2002c; Gorchetchnikov and Hasselmo, 2005).

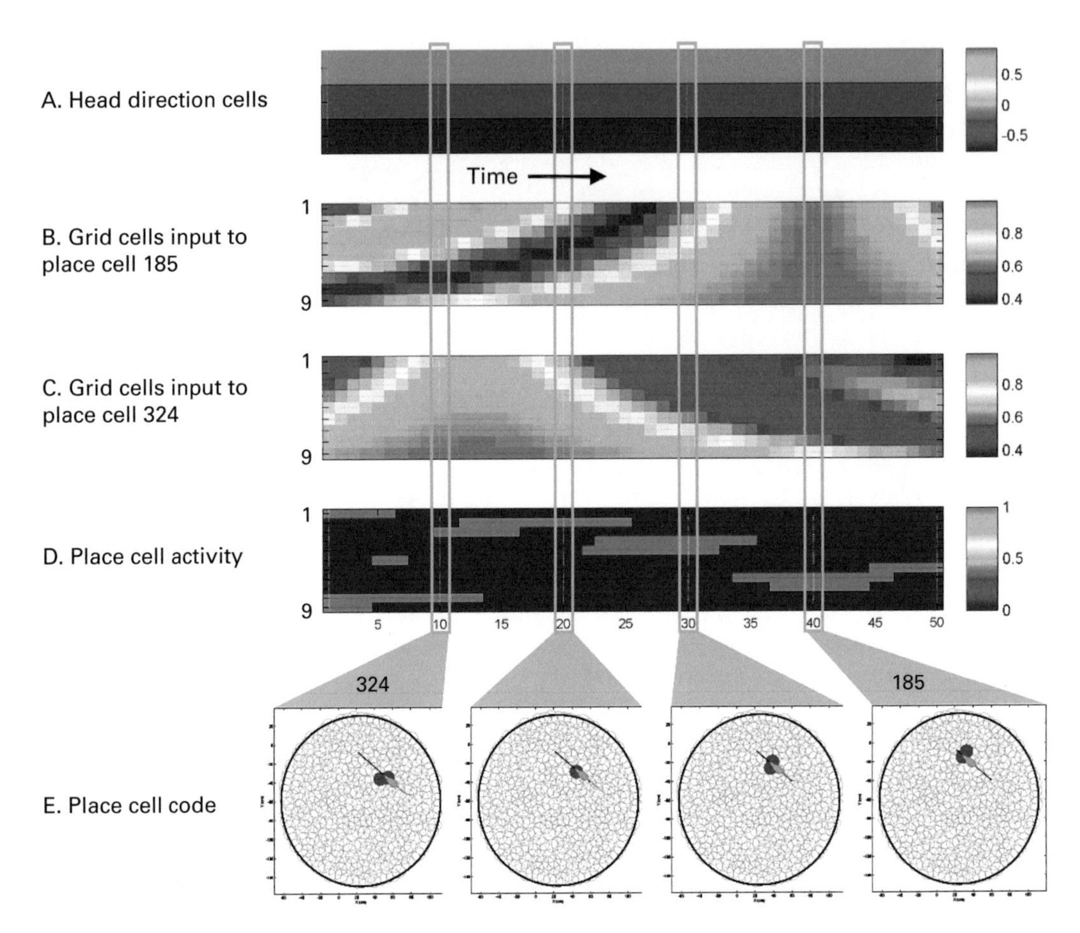

Figure 7.1 (plate 8)

Forward planning of a single trajectory. (A) The current rat head direction drives three different head direction cells with static levels of activity. The three head direction cells drive different frequencies in the grid cell model (B and C). (B) The graph shows the activity over time of one group of nine grid cells that drives place cell 185 when all grid cells are simultaneously active (red) late in the trajectory. (C) Another group of nine grid cells (C) drives place cell 324 when all grid cells are active (red) early in the trajectory. Different grid cells transition between constructive and destructive interference at different time points. (D) Input from the grid cells sequentially activates different place cells (red bars) that code sequential locations (filled circles) along a straight trajectory shown in E (figure by Murat Erdem, Erdem and Hasselmo, 2010).

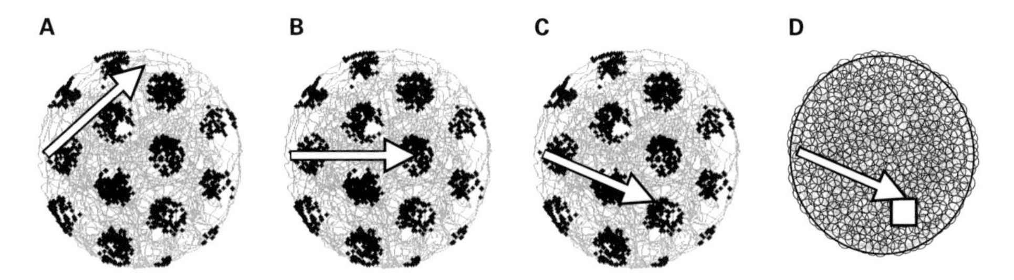

Figure 7.2
(A, B, and C). Grid cell plots showing firing fields in the environment that can be activated by sequential forward sampling of three different virtual trajectories (arrows) through the environment based on three different patterns of head direction cell activity. In each plot, the head direction cells drive forward sampling that shifts the phase of oscillations along a virtual trajectory (arrow) and sequentially activates the single grid cell due to constructive interference as it crosses the cell's firing fields (black dots). Each grid cell trajectory will cause sequential activation of associated place cells at each location. (D) Plot of place cells coding the environment. The sequential activation of place cells along the third trajectory activates the place cells coding the desired goal location (black square), allowing selection of movement based on the head direction that generated this trajectory (Erdem and Hasselmo, 2010).

A number of other previous models have addressed mechanisms of goal-directed spatial behavior using biological circuits. Many of these models do not directly address the generation of place cell responses but instead model place cells using a direct function of spatial location such as radial basis functions. These models can drive goal-directed spatial behavior based on the modified connectivity between place cells (Touretzky and Redish, 1996; Redish and Touretzky, 1998) or between place cells and units representing behavioral motor actions (Burgess et al., 1997; Arleo and Gerstner, 2000; Foster et al., 2000; Poucet et al., 2004; Hasselmo and Eichenbaum, 2005; Zilli and Hasselmo, 2008b; Sheynikhovich et al., 2009). In some cases, these models have used Hebbian modification between concurrently active place cells (Redish and Touretzky, 1998), but in other cases they have gated the modification of synapses based on the influence of a reward signal (Burgess et al., 1997; Arleo and Gerstner, 2000), for example, using the temporal difference learning rule from reinforcement learning theory (Foster et al., 2000; Hasselmo and Eichenbaum, 2005; Zilli and Hasselmo, 2008a). As an alternative, navigation can be mediated by associations between locations and the actions in prefrontal cortex that mediate transitions to other locations (Hasselmo, 2005b) and, on a more general level, between any environmental state (not just spatial locations) and the actions that transition to a different environmental state (Koene and Hasselmo, 2005).

The sampling of forward trajectories through the environment using grid cells differs from earlier models. Sequential readout of possible forward trajectories based on a progressive shift in head direction allows sampling of multiple possible paths to find the one that intersects with the goal location. This could allow a rat to select its direction based on possible pathways through the environment even if the trajectory crosses a portion of the open environment that the rat has not previously visited.

Relation to Experimental Data on Replay

The forward trajectory readout could underlie the forward spiking replay seen during rat waking behavior at choice points in a tone-cued alternation task (Johnson and Redish, 2007). Forward spiking replay refers to the sequential spiking of place cells coding locations at positions along a trajectory in front of the rat, as if the rat were imagining moving forward along a trajectory through the environment. This forward spiking replay is commonly associated with the rat performing vicarious trial and error, turning its head in different directions, as if the rat were sampling retrieval of prior episodes to choose between different possible future actions (Johnson and Redish, 2007). Neurophysiological data in other tasks suggest that rats can even do forward planning or replay of trajectories that do not start at its current location (Davidson et al., 2009; Karlsson and Frank, 2009). Data also shows that rats can preplay trajectories that have never been experienced but can be seen in the environment (Dragoi and Tonegawa, 2011). The model of forward trajectory planning presented here could account for these different types of data.

Another new feature of the model concerns the interaction of trajectory planning with barriers in the environment. The inclusion of barriers in the environment has been shown to alter the firing of hippocampal place cells (Muller and Kubie, 1987) and entorhinal grid cells (Derdikman et al., 2009). This could result from interaction of forward trajectory planning with the coding of barriers by boundary vector cells (Solstad et al., 2008; Lever et al., 2009). The framework described here shows how the differential place cell representation formed in environments with barriers can allow selection of a trajectory that reaches the goal location while avoiding the barrier locations.

Previous models require exploration of the environment for creation of place cell representations and associations between these place cells (Touretzky and Redish, 1996; Burgess et al., 1997; Redish and Touretzky, 1998; Foster et al., 2000). The current model has the advantage that it does not require association of each place cell with the direction of actions leading to other place cells or the goal location. Instead, this model can compute the direction to a goal by forward sampling of possible trajectories through the environment. The forward scanning could also provide a mechanism for greatly increasing the speed of exploration of the environment, which is an important problem for creation of maps. Exploration could be performed via the internal scanning of forward trajectories through the visible environment, and creating place cells and associations between place cells. In addition, if there is some mechanism for internal computation of the change in visual feature angle during forward scanning, then the network could form associations between the place cells and visual features even for unvisited locations.

The modeling described in chapters 3 and 4 has the property that it effectively encodes not only the individual states during an episode but also the temporal duration of transitions between these individual states, as well as the duration of individual states. Thus, it effectively maintains the timing properties of the episode. This can be considered as an element of the property of mental time travel, which has been described as an essential element of episodic memory (Tulving, 1983; Eichenbaum and Cohen, 2001; Tulving, 2002), as well as an element of future planning (Suddendorf and Corballis, 1997; Clayton et al., 2003).

Studies of patients with damage to the entorhinal cortex and hippocampus have demonstrated impairments of mental time travel into future or imaginary events, showing up as a paucity in the richness of description of both past and future events (Levine et al., 2002; Hassabis et al., 2007; Schacter et al., 2007; Kirwan et al., 2008). This suggests that the travel along future trajectories through familiar environments could utilize the same machinery as episodic retrieval. However, instead of modified synapses driving the retrieval of a previous trajectory, for imagination of future trajectories the actions along the trajectory would be determined by prefrontal input to the cells coding velocity. For example, if you wanted to imagine going forward along a familiar hallway and then turning right at a particular door, the prefrontal commands could drive head direction cells coding the initial allocentric velocity. These cells could then drive the phase code of grid cells to progressively update place cell populations representing different locations. When the desired location is reached, prefrontal input could drive a different set of head direction cells to change the imagined direction of the velocity vector to the right. In this manner, prefrontal cortex could guide traversal of the imagined trajectory.

Behavior Guided by Gated Trajectory Replay

In the fixed platform version of the Morris water maze described above, the rat must plan a novel trajectory through the environment. In fact, even the version of the water maze in which the platform is moved on each day requires planning of novel trajectories because the starting location differs on each trial. In contrast, the response made at the choice point in the delayed spatial alternation task requires less planning and can be guided based on retrieving the episodic memory of the spatiotemporal trajectory generated during the response on the previous trial.

Spatial Alternation

As noted previously, there are multiple ways of performing most memory-guided tasks. In delayed spatial alternation, the choice could be based on episodic memory alone, using retrieval of the previous spatiotemporal trajectory when encountering the choice point to allow selection of the opposite direction of movement relative to that previous trajectory. As another option, the prior response could be retrieved during the delay period and held with active maintenance until the choice point. As yet another alternative, the choice could be based on active maintenance of the previous response without any use of episodic memory. However, active maintenance would presumably be the strategy used for performance of continuous spatial alternation, which is not sensitive to the hippocampal lesions that impair delayed spatial alternation (Ainge et al., 2007). The differential sensitivity of delayed spatial alternation to hippocampal lesions suggests there is a difference in behavioral strategy between continuous spatial alternation and delayed spatial alternation, supporting the possible use of episodic retrieval of the prior response during delayed spatial alternation.

The performance of delayed spatial alternation using episodic memory alone requires selective retrieval of the most recent episodic memory without interference from other memories. This is similar to the problem you encounter when you park in the same parking garage every day. It becomes difficult at the end of each day to remember where you parked your car. You must remember where you parked it in the morning without interference from multiple other memories of parking the car in different locations on other days.

Spatial alternation was previously modeled based on episodic memory retrieval (Hasselmo and Eichenbaum, 2005) with a combined model of the hippocampus and prefrontal cortex. In this model, hippocampal circuits in the virtual rat performed encoding of the associations between adjacent locations and, at every location in the task, also performed retrieval of the episodic memory of its previous trajectory. At the choice point, the trajectory retrieval was used to select a response to the opposite arm of the maze. This could use the full circuitry described for spatiotemporal trajectories, but in the earlier paper the forward retrieval involved Hebbian associations between place cell representations of individual locations in the maze.

In the example in figure 7.3, a right turn involves a sequence of states, going from the middle of the stem to the choice point and then into the right arm. On the next trial, correct behavior requires that the recent sequence into the right arm must be retrieved separately from other sequences involving the same locations but ending in the left arm of the maze. As shown in figure 7.3A, when the rat is in the stem, forward retrieval of the trajectories based on prior associations between place cell activity alone retrieves two sequences spreading into both the left and right arm. The network requires context-dependent activity to selectively retrieve the most recent trajectory.

Context-Dependent Activity Using Gated Retrieval

In this model, the context-dependent retrieval of a single episode is obtained by gating the retrieval of forward trajectories based on temporal recency. This can be mediated

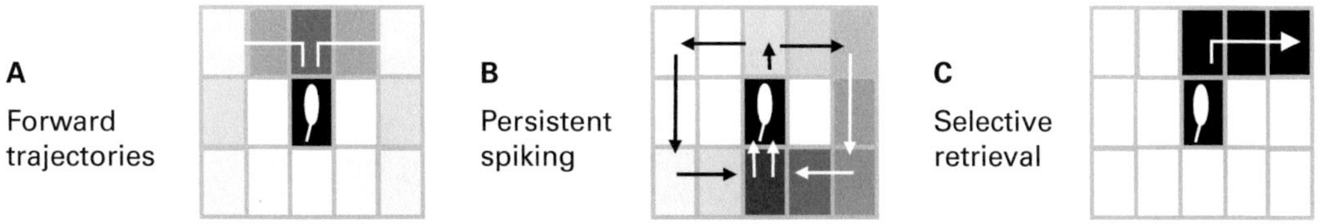

A
Forward trajectories

B
Persistent spiking

C
Selective retrieval

Figure 7.3
(A) Retrieval of possible forward trajectories from a cue representing the stem will retrieve two possible trajectories into the left and right arms of the maze. (B) Persistent spiking could provide decaying temporal context that is stronger for the more recent trajectory. (C) Interaction in region CA1 of forward replay and persistent spiking allows selective retrieval of the most recent trajectory (the right turn). Figure adapted from Hasselmo and Eichenbaum (2005).

by persistent spiking activity in layer II of the entorhinal cortex to provide a gradually decaying signal of global temporal context (Howard and Kahana, 2002; Howard et al., 2005). As shown in figure 7.3B, this results in a gradual decrease in activity for locations visited at more distant times in the past. This means that context activity is stronger for the most recent visit into the right arm and weaker for the left arm states experienced at a longer time interval. In models using temporal context (Hasselmo and Eichenbaum, 2005; Katz et al., 2007), the forward retrieval and the temporal context provide convergent input to region CA1, allowing selective retrieval of the most recent sequence (figure 7.3C) because the forward retrieval of the recent sequence interacting with context has greater activity than the other sequence. As an alternative, the context-dependent selection could be mediated by persistent spiking that drives arc length cells or time cells (as in chapter 4). Persistent spiking could drive arc length or time cells that are reset at the reward locations (Hasselmo, 2007, 2008b, 2008c), allowing robust selective activity in the stem for the most recent response.

The selective retrieval of the most recent trajectory means that some neurons respond differentially based on the previous response in the task (figure 7.3C), consistent with experimental data on hippocampal neurons showing context-dependent activity in the stem of spatial alternation tasks (Wood et al., 2000; Lee et al., 2006). In the experiments described in chapter 2 on context-dependent spiking, the rat runs at the same speed and direction through the same location in the stem, but the neuron fires differentially depending on past or future response. Figure 7.4 shows the use of the model to simulate the experimental data on context-dependent activity. When the simulated rat runs from the right arm toward the left arm, the retrieval of the previous right arm trajectory activates a neuron coding the right arm, allowing circuits in prefrontal cortex to guide selection of the left arm response. When the rat runs from the left arm to the right arm, the right arm trajectory is not retrieved, so the neuron coding the right arm is not active. Thus, the model generates firing dependent

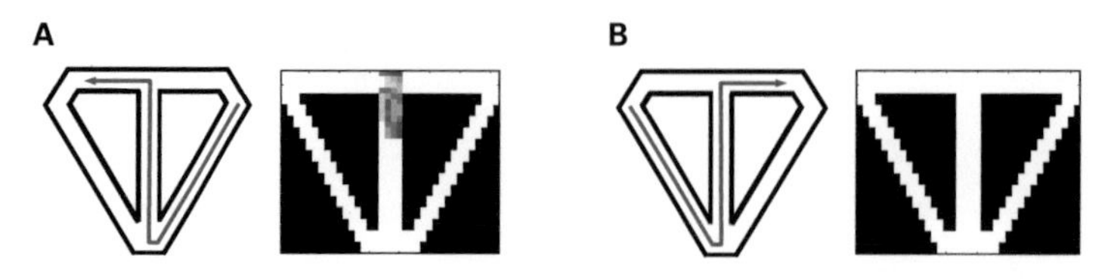

Figure 7.4
(A) When the virtual rat runs from the right return arm up the stem to the left reward arm (arrow), the stem acts as a cue for retrieval of the previous right arm trajectory, selectively activating a neuron coding the right arm (gray area in maze diagram on right). (B) When the rat runs from the left return arm up the stem to the right reward arm (arrow), the stem acts as a cue for retrieval of the left arm trajectory, and the neuron coding the right arm is not active (based on Hasselmo, 2005a).

not only on spatial location but also on the context of prior activity. This can then be used to guide correct responses in the task.

In this model, the physiological changes during different phases of theta rhythm described in chapter 6 may enhance performance of the network by allowing the selective context-dependent retrieval of individual encoded sequences during the retrieval phase without causing interference with the new sequence being encoded by synaptic modification during the encoding phase (Sohal and Hasselmo, 1998a, 1998b; Hasselmo and Eichenbaum, 2005).

The model shown in figure 7.4 was appealing because it combined generation of selective context-dependent firing with goal-directed selection of the correct behavior in the task. However, some features of the model do not match available experimental data. For example, the context-dependent responses in the model only occur when the retrieved trajectory extends into the reward arms and thus tend to appear more at the end of the stem. In addition, the model made the prediction (Hasselmo, 2005a) that context-dependent firing should occur during the retrieval phase of theta rhythm that has not yet received support from available experimental data.

Sequence Encoding Models

The sequence retrieval mechanism in the model shown in figures 7.3 and 7.4 build from models of sequence encoding in hippocampal circuits. Marr (1971) initially proposed that excitatory recurrent connections in region CA3 could provide sequential associations between individual patterns in a sequence, and this mechanism has been used in many models (McNaughton and Morris, 1987; Blum and Abbott, 1996; Jensen and Lisman, 1996b; Levy, 1996; Wallenstein and Hasselmo, 1997b; Lisman, 1999; Hasselmo and Eichenbaum, 2005). During encoding in these models, each pattern in a sequence activates a set of neurons shortly before the next pattern, and STDP (spike-timing-dependent plasticity; Levy and Steward, 1983) strengthens synapses between the sequential patterns. During retrieval, input of the first pattern will cause activity to spread across strengthened synapses to cause sequential spiking in other patterns, reading out the full sequence (Levy, 1996; Wallenstein and Hasselmo, 1997b).

This simple sequence retrieval does not have to occur in region CA3. It could occur in any network in which retrieval output can cue another retrieval step. Thus, the same mechanism could occur at recurrent synapses in the neocortex, in a loop involving dentate gyrus, region CA3, and mossy cells in the hilus (Lisman, 1999), or in a loop involving the full hippocampal circuit from entorhinal cortex back to entorhinal cortex (Hasselmo, 2008a, 2009). In addition, evidence suggests that whereas the hippocampus may encode new episodic sequences, the basal ganglia may encode highly familiar repeated sequences (White and McDonald, 2002).

Disambiguation of Overlapping Sequences

Simple sequence encoding models have difficulty encoding and selectively retrieving highly overlapping sequences (Levy, 1996), as shown in figure 7.3A. For example, after learning of both of the sequences A–B–C–D and E–B–C–F, retrieval of forward associations between each pattern would result in the cue "A" retrieving "B" and then "C," but would then result in simultaneous retrieval of both "D" and "F" as the final pattern. Selective retrieval of just one correct final pattern requires some additional mechanism for context-dependent retrieval of one but not the other overlapping sequence, through gating of retrieval output by additional synaptic activity.

In one network model of CA3, disambiguation of overlapping sequences has been obtained by having retrieval of the end of a sequence depend on synaptic input from persistent firing of additional CA3 neurons termed local context units (Levy, 1996). As an alternative, context-dependent retrieval could also be obtained by gating the output of retrieval with input from another region. For example, the retrieval of region CA3 could be gated by activity from entorhinal cortex II (ECII), or output from CA3 to CA1 could be gated by activity from entorhinal cortex III (ECIII). In the model presented above (Hasselmo and Eichenbaum, 2005), output from ECIII to CA1 was gated by context-dependent activity in the dentate gyrus and region CA3.

These mechanisms of context-dependent retrieval require a balance between the forward sequence retrieval and gating by context. Theta rhythm could provide a mechanism for sampling across different magnitudes of network variables. Once sequence retrieval activity occurs, feedback inhibition can ensure that the first, best-matching sequence is selectively retrieved. Early models of this process used phasic changes in magnitude of synaptic transmission to allow retrieval of single associations due to a global context signal representing specific environmental cues selective for one episode (Sohal and Hasselmo, 1998a, 1998b). The simulations of episodic retrieval reviewed above provided a more detailed model of the role of theta rhythm in allowing global temporal context to regulate selective retrieval of one sequence (Hasselmo and Eichenbaum, 2005). In these models, the hippocampal theta rhythm enhances the solution to this problem by allowing scanning for the first good match by phasically increasing context input due to changes in postsynaptic depolarization (Fox, 1989), or strength of excitatory synaptic transmission (Wyble et al., 2000). The activity would then be read out by a multiplicative interaction of increasing entorhinal input and decreasing CA3 input, resulting in sequential retrieval which is equivalent in magnitude for each element of a sequence, but different in magnitude for different sequences, and strongest for the sequence best matching the current context (Hasselmo and Eichenbaum, 2005).

Behavior Guided by Oscillatory Interference

The model presented in the preceding section used sequence readout to model context-dependent activity in delayed spatial alternation. As an alternative mechanism, oscillatory interference can also be used to model context-dependent activity in delayed spatial alternation based on reset of phase at reward locations. These two different types of models can also be used to simulate the features of theta phase precession via different mechanisms. Neither model effectively accounts for the full range of data, but both models are presented because some combination of these features may prove effective for simulating the full range of data. In the oscillatory interference model, the context-dependent activity does not have to depend on the retrieval of a specific sequence but can arise from reset of oscillation phase at previous locations in the task.

Context-Dependent Activity Using Phase Reset

As described in chapter 2 and in this chapter, neurophysiological data show that neurons respond to more than just spatial location. Entorhinal cells do not always fire as grid cells, and hippocampal cells do not always fire as place cells. The initial finding that hippocampal place cells respond primarily on the basis of two-dimensional spatial location (O'Keefe, 1976) seemed somewhat paradoxical in light of the role of the hippocampus in episodic memory as the responses were static over time and did not seem to code for specific events or specific times. However, even early studies showed that the response of these place cells would depend not only upon location but also upon the direction of running on the arms of an eight-arm radial maze (McNaughton et al., 1983) and could respond to nonspatial features such as individual odors or rewards in an operant task (Eichenbaum et al., 1987, 1989; Wiener et al., 1989).

Over the years, a number of studies have shown that the context of recent behavior has a strong influence on spiking activity in the hippocampus (Markus et al., 1995; Gothard et al., 1996; Wood et al., 2000; Lee et al., 2006; Griffin et al., 2007; Pastalkova et al., 2008) and entorhinal cortex (Frank et al., 2000; Lipton et al., 2007; Derdikman et al., 2009). These data potentially counter the view of grid cells and place cells as providing a spatial map as the context dependence suggests that neurons do not only respond to the two-dimensional location in the environment. These data present a broad range of phenomena that must be accounted for by models of grid cell mechanisms.

The previous section modeled context-dependent activity based on sequence readout. However, the context-dependent firing can also be modeled using the oscillatory interference model of grid cells (Burgess et al., 2007; Hasselmo et al., 2007; Burgess, 2008) or the persistent spiking model of grid cells (Hasselmo, 2008b) that were initially developed to account for regular grid cell firing in the open field.

The simulations presented in this section demonstrate how context dependence could arise from resetting the phase of persistent spiking or oscillations at specific locations (e.g., stopping locations, reward locations, or turning locations), either by resetting the relative phase of one population or by turning off the current population and turning on a new population of persistent spiking neurons. This approach to modeling context-dependent responses would be difficult to use in other types of grid cell models (Fuhs and Touretzky, 2006; Kropff and Treves, 2008).

The possibility of reset was mentioned in the original oscillatory interference model of grid cells using velocity input to code location based on Euclidean distance (Burgess et al., 2007). Phase reset can also be used in other types of interference models using different mechanism for updating frequency and phase (Hasselmo, 2007). As described previously, the frequency and phase can be influenced by time alone, or by running speed, or by velocity. In the original model described in chapter 3, a two-dimensional representation of space (based on Euclidean distance) results from frequency and phase being updated by a velocity signal coded by cells responding to head direction and speed. Without reset, this model shows spiking dependent on two-dimensional location. With reset, the model can generate separate two-dimensional maps for different phases of a task (Hasselmo, 2008b).

In the arc length model, the frequency of persistent spiking or oscillation can be shifted by running speed alone, independent of head direction (Hasselmo, 2007), resulting in firing dependent upon the one-dimensional arc length of a trajectory (Hasselmo, 2007). Phase reset results in firing dependent on arc length from the last reset location. The arc length model generates different predictions about context-dependent firing, such as the absence of mapping to a hexagonal array of firing fields, and the mirror symmetry of firing for different directions of running along a linear track with reset at both ends. In the time interval model, the phase is not shifted by any external input and the interference pattern depends on time interval alone. Reset of oscillation phase in this version results in firing dependent on the time interval since the last reset location, resulting in "time cells."

One strategy for simulating context-dependent firing involves setting phase to zero in a specific population each time reward is received in a specific location. The simulations presented here include separate populations of entorhinal neurons and hippocampal neurons. Without resetting, the entorhinal cells in the model fired as grid cells and hippocampal neurons fired as place cells. Reset shifted the spatial phase of the entorhinal population, and thereby shifted the spatial phase of the hippocampal cell firing, so that reset of entorhinal phase caused both populations to show context-dependent responses.

Spatial Alternation Using Phase Reset

As noted above, in a spatial alternation task (see figure 7.5A), some neurons show selective firing dependent upon a prior left or right turn response even when the location and direction of running is the same on the stem (Wood et al., 2000; Lee et al., 2006). The context dependence of hippocampal firing appears to correlate with context-dependent firing of entorhinal neurons in this task (Lipton et al., 2007; Lipton and Eichenbaum, 2008). This phenomenon can be simulated with reset of oscillatory interference. Without resetting, the network shows spatially consistent firing, with entorhinal cells firing according to grid cell firing fields as shown in figure 7.5B (left) and hippocampal neurons firing according to place cell firing fields as shown in figure 7.5B (right). Cells respond in the same location in the stem for both left to right and right to left trajectories.

In contrast, with reset of the phase of specific cells at reward locations, simulated entorhinal neurons show context-dependent firing on the stem that differs between

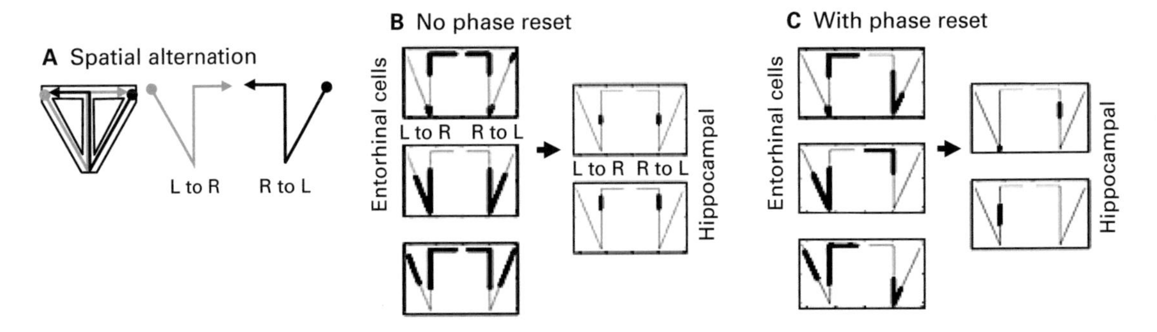

Figure 7.5
Simulation of context-dependent firing in continuous spatial alternation. (A) Behavioral task showing two trial types: L to R: Return from left arm going to right, R to L: Return from right arm going to left. (B) Each plot separately shows firing of the cell for the separate trials from L to R and from R to L with thick gray dots indicating firing locations. With no phase reset at reward location, the examples of three entorhinal grid cells (left) fire in consistent spatial locations and cause examples of two place cells (right) to fire in consistent spatial locations regardless of the trial type. (C) With phase reset at reward locations, the firing of three example entorhinal cells (left) differs for L to R versus R to L trial types. This results in differential firing of two example hippocampal cells (right), with selective firing on the stem for one cell in R to L (top) and for a different cell in L to R (bottom).

the left to right trajectory and the right to left trajectory as shown in figure 7.5C (left). This is consistent with physiological recordings showing context-dependent entorhinal activity during this task (Lipton et al., 2007; Eichenbaum and Lipton, 2008). The input from these entorhinal neurons results in context-dependent firing of hippocampal neurons on different trajectories as shown in figure 7.5C (right). This is most evident on the stem of the maze, where reset results in different firing on left to right versus right to left trajectories, consistent with previous recording of context-dependent hippocampal spiking on the stem during spatial alternation (Wood et al., 2000; Lee et al., 2006). Phase reset in the model using velocity input (Euclidean model) resembles previous simulations of context-dependent firing using phase reset in the arc length model with speed input (Hasselmo, 2007). The same techniques can also simulate the differential firing of place cells when an open field foraging task is replaced with repetitive running between different goal locations (Markus et al., 1995) if oscillations are reset at each of the different goal locations in the task.

Delayed Spatial Alternation with Running Wheel

Because these interference models update firing dependent on proprioceptive feedback of speed and head direction, they predict that the changes in firing could be decoupled from actual spatial location. Because the phase of these models integrates velocity input, the phase will continue to be updated when a rat runs on a running wheel or treadmill in a single location, as long as the perception of velocity is based on proprioceptive feedback rather than visual input. Thus, in a simulation of a rat running in a wheel or on a treadmill, the location of the rat in the environment does not change, but the model will generate spiking at specific reproducible intervals from the start of running.

This property of the model is supported by remarkable recent data on rats running in a running wheel during the delay period of a spatial alternation task (Pastalkova et al., 2008). In this experimental task, different cells fire at specific, reproducible intervals after the start of running. The structure of the task is illustrated in figure 7.6A and B. During each trial, the rat runs the left side of a spatial alternation task, then runs in the running wheel at a constant speed for a fixed time period (straight lines on the bottom left of figure 7.6B), then runs the right side of the spatial alternation task, and then runs in the running wheel at a constant speed for a fixed time period (straight line plots on bottom right side). The figure shows activity of four representative cells out of 75 modeled cells in entorhinal cortex and four representative cells out of the 400 cells in hippocampus. With phase resetting at reward location, the model shows differential firing in the running wheel after left versus right side runs. In addition, the firing appears at specific time points during wheel running that are consistent across the seven trials shown in figure 7.6C (corresponding to the perceived distance

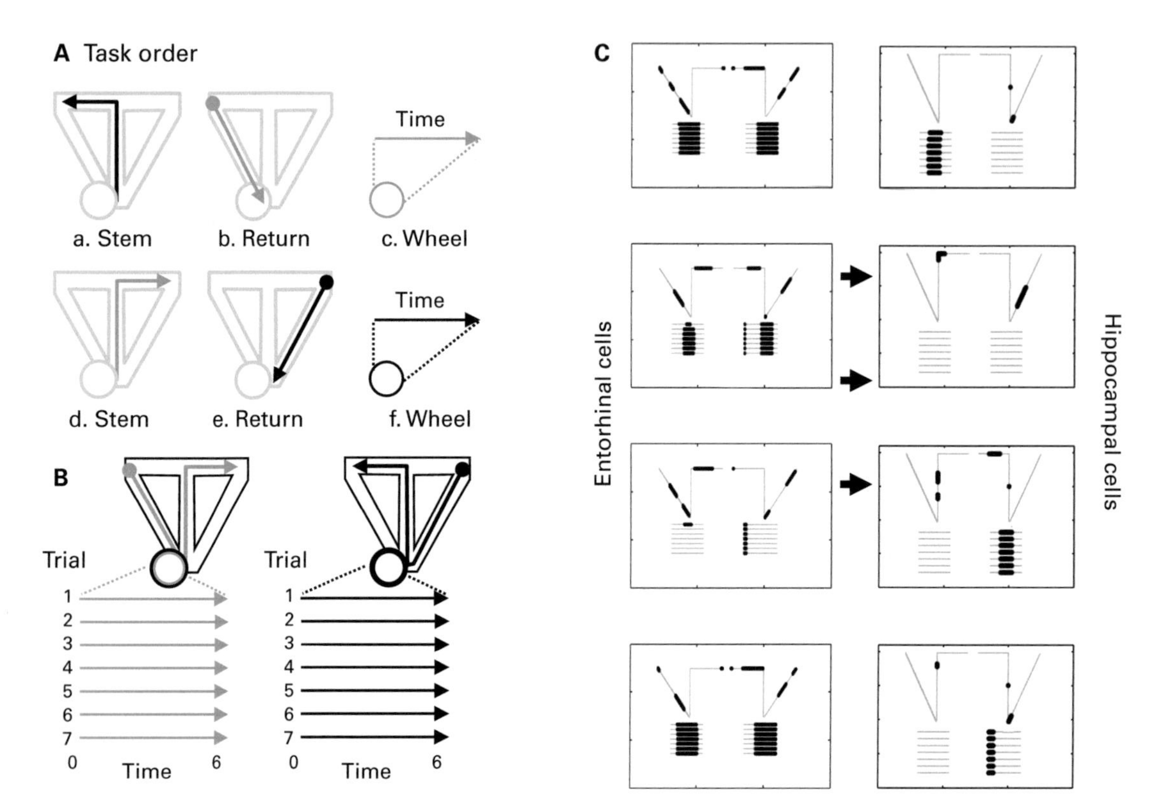

Figure 7.6
Simulated spiking activity in delayed spatial alternation task with wheel running during the delay period. (A) Order of behavior in the task. Rat runs up stem into left arm (a), then down left return arm (b), then runs in wheel during delay (c), then runs up stem into right arm (d), then down right return arm (e), and then runs in wheel during delay (f). (B) Spiking is plotted spatially for left to right trials (gray) or right to left trials (black). Spiking during running in the wheel is plotted on lines (bottom) depicting running wheel time during each trial. Lines on left depict running after left return; lines on right depict running after right return. (C) With reset at reward location, the simulation generates context-dependent spiking activity during wheel running in the delay period. Examples of four (out of 75) entorhinal neurons plotted in left column show spiking in many locations and during long time segments in the running wheel. Typical examples of four hippocampal neurons plotted in right column show spiking in discrete spatial locations and at brief restricted time points during wheel running selective for different trial types. Adapted from Hasselmo (2008b).

or arc length of the wheel run). The hippocampal neuron at the top right of figure 7.6C shows firing only during the middle period of wheel running after returning from the left side of the track. The third hippocampal neuron in figure 7.6C shows firing only during the middle period of wheel running after returning from the right side of the track.

The simulated cells in figure 7.6 match the properties of hippocampal cells in the experimental data that show firing that occurs after remarkably consistent time intervals in the running wheel on each trial (Pastalkova et al., 2008). The experimental data suggest that the hippocampal cells are explicitly integrating the time or running distance since entry into the running wheel.

This task provides an ideal opportunity for selecting between the different models described above that update frequency with different types of input during the delay period (Hasselmo, 2008b). This has been tested in experiments on rats running on a treadmill by Ben Kraus (Kraus et al., 2010). The two-dimensional Euclidean model (Burgess et al., 2007; Hasselmo et al., 2007), using frequency updated by both speed and head direction, predicts differences in firing for treadmills oriented at different angles. In contrast, the one-dimensional arc length model (Hasselmo, 2007), using frequency updated by speed only, predicts that firing in the treadmill will not depend upon the orientation of the treadmill but will depend on running speed. Finally, the time interval model (Hasselmo, 2008b) predicts firing at specific time intervals independent of orientation or running speed. These predicted differences have been tested by Ben Kraus using a treadmill that can be run at different speeds and turned to different orientations in the environment (Hasselmo, 2008b). The predictions are illustrated in figures 7.7 and 7.8.

Hairpin Task

The simulation has also been used to address data from running of a hairpin maze. This was motivated by data from cells that show grid cell firing properties in an open field but shift their pattern of firing in a hairpin maze in the same environment (Frank et al., 2000; Derdikman et al., 2009). Rather than firing in locations corresponding to the two-dimensional location of a grid cell firing field, the neurons fire in the hairpin maze at regular intervals relative to the starting point of each segment of the hairpin. This pattern can be obtained in the simulation by resetting the phase of the oscillations at each turning point, as shown in figure 7.9B, in contrast to the grid cell firing pattern that appears without reset (figure 7.9A). This model accounts for the pattern of firing of both entorhinal cells and hippocampal cells, which show firing dependent upon distance from turns in this task maze (Derdikman et al., 2009). This simulation also accounts for the pattern of firing observed in entorhinal cortex in earlier studies in a U maze and W maze (Frank et al., 2000). These data provide an important opportunity for testing reset of the one-dimensional arc length model based on running

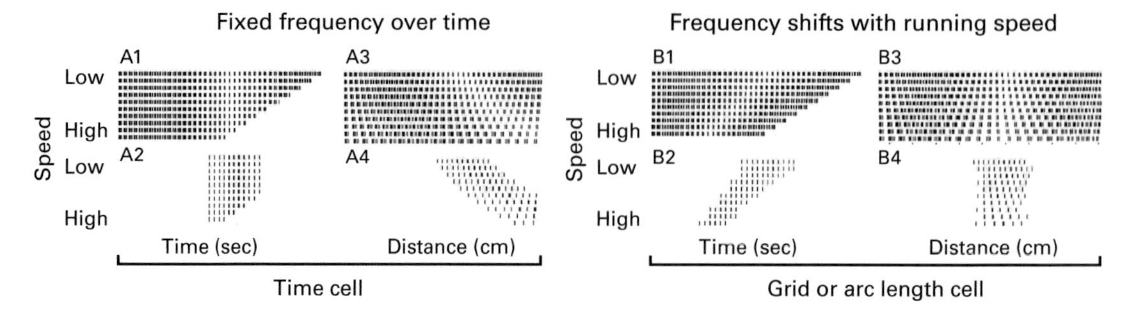

Figure 7.7

Predictions of model for treadmill task. Left: Fixed frequencies over time can generate a time cell. (A1) Two persistent spiking cells firing at different fixed frequencies transition from being out of phase to being in phase (similar to figure 3.11). (A2) When the two cells are in phase, they cause spiking in a time cell that responds at a specific time interval after the rat starts running on the treadmill. (A3) The same spiking plotted for running distance of the rat shows that with the higher treadmill speed, the rat has run a longer distance, and therefore firing appears at different distances in A4. Right: Distance sensitivity of grid or theoretical arc length cells occurs when running speed affects frequency. (B1) One persistent spiking cell changes frequency with running speed, so that high speed causes a faster shift in frequency relative to baseline spiking. (B2) The two persistent spiking cells cause firing in an arc length cell at different time intervals. (B3 and B4) The same spiking as B1 plotted for running distance shows firing at a consistent distance of running in B4.

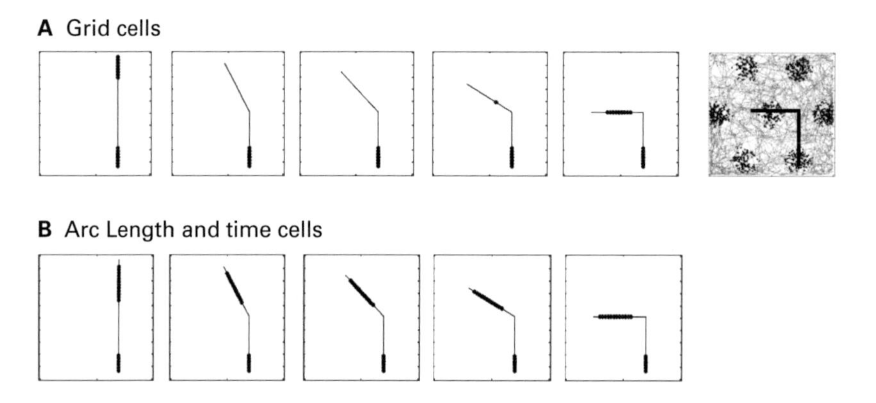

Figure 7.8

Grid cells can be distinguished from theoretical arc length or time cells by the response on a treadmill turned to different angles. (A) Grid cells respond to two-dimensional location, so their response will differ as the treadmill turns to different angles and the virtual trajectory traverses different firing fields of the overall grid cell response shown on the right. For example, the right angle trajectory traverses the lower right firing field and the central firing field. (B) Theoretical arc length and time cells will respond at the same one-dimensional distance or time interval along the track, regardless of the difference in angle that shifts the trajectory to cross different two-dimensional locations.

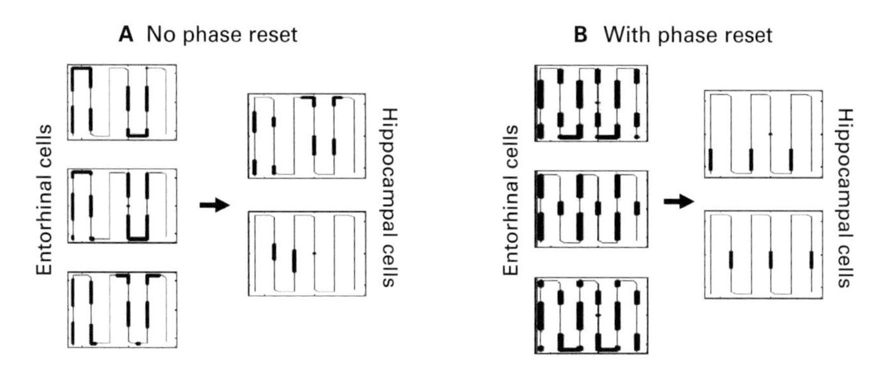

Figure 7.9
Model of context-dependent firing in the hairpin task (Hasselmo, 2008b). (A) With no phase reset, the firing of grid cells and place cells occurs in adjacent segments dependent on location relative to grid cell or place cell firing fields (three example cells out of 75 entorhinal cells are shown, and two examples out of 400 hippocampal cells). (B) With phase reset at turning locations, the firing of grid cells and place cells depends upon distance from the turns, not on two-dimensional spatial location.

speed alone versus the two-dimensional Euclidean model based on speed and head direction. As noted above, the one-dimensional model could generate mirror-image symmetry of firing for trajectories going from left to right versus right to left. The two-dimensional model would not generate the same type of mirror symmetry.

The reset at turning points that induces this pattern of firing could be due to the strong drop in running speed at the turning point, to other cues such as the increase in angular velocity at turns, or to the abrupt change in visual stimuli before and after the turn. This could also reflect the creation of intermittent goal representations for guiding sequential behavior in the maze, which could guide behavior at each point based on a goal vector created by integration of movement between each reset location (Hasselmo and Brandon, 2008). The loss of grid cell firing could result from the barriers on each segment shutting off forward retrieval (or planning) based on any action other than along the one-dimensional trajectory of the task. In contrast, the grid firing in open fields might result from retrieval (or planning) of multiple alternate pathways of movement. The data in these tasks appear to support the Euclidean distance model using velocity input in contrast to the model responding to the arc length of the trajectory or the time intervals. Even though it does not depend on location alone, the context-dependent firing in the data appears to consistently start in the same location even after a slightly longer or shorter running trajectory.

In summary, the use of phase reset in interference models provides an effective means of simulating context-dependent firing in a number of behavioral tasks shown above. Thus, this constitutes an important alternative to the sequence retrieval models presented in the previous section. The ability to account for context-dependent firing also gives the interference models an advantage over attractor dynamics models of grid cells that cannot so easily account for context-dependent changes in firing. The simplicity of this mechanism and its capacity for accounting for context-dependent firing supports its potential role in memory-guided behavior.

Comparing Models of Theta Phase Precession

The sequence retrieval model and the oscillatory interference models can both be used to model context-dependent activity in spatial memory tasks. These two types of models have also been used to simulate the phenomenon of theta phase precession. Ultimately, a complete model of physiological mechanisms in the hippocampus and entorhinal cortex must account for the mechanisms of theta phase precession. To give insight to the strengths and weaknesses of current models, some properties of these and other models of theta phase precession will be described here.

Detection of theta phase precession involves simultaneous recording of the timing of spikes generated by an individual place cell or grid cell (unit data), along with recording of the local field potential to determine the relative phase of the theta rhythm oscillation (O'Keefe and Recce, 1993; Skaggs et al., 1996; Huxter et al., 2003). As a rat runs along a continuous track, individual place cells will fire as the rat traverses the place field of that cell. In the phenomenon of theta phase precession, the spiking of the place cell shifts in a systematic manner relative to the theta rhythm. As the rat enters the place field associated with a particular place cell, the firing of that cell will occur late in the theta cycle. As the rat crosses and leaves the place field, the firing of the place cell will occur earlier and earlier, reaching its earliest phase near the end of the place field. This phenomenon primarily appears when the rat runs in a single direction along a linear track but has also been shown during foraging in an open field (Skaggs et al., 1996).

Oscillatory Interference Model

As shown in figure 2.11, the initial paper presenting data showing theta phase precession was accompanied by a model of how theta phase precession could arise from interference of oscillations with different frequencies (O'Keefe and Recce, 1993). This oscillatory interference model (sometimes known as the dual oscillator model) was simulated in more detail by Lengyel, Szatmary, and Erdi (Lengyel et al., 2003), who showed that nearly 360 degrees of precession could be obtained with a higher amplitude of the higher frequency oscillation. This model formed the basis for the oscillatory interference model of grid cells (O'Keefe and Burgess, 2005; Burgess et al., 2007). The basic features of this model of precession are summarized in figure 7.10.

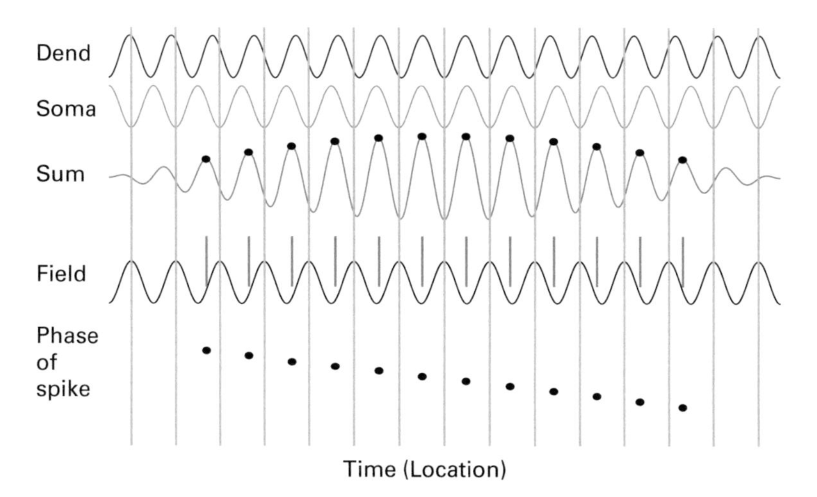

Figure 7.10
Oscillatory interference model of theta phase precession. During running at a constant velocity, the oscillation in the dendrite (Dend) has a constant higher frequency relative to the baseline oscillation caused by network inhibition on the soma (Soma). The sum of the two oscillations (Sum) shows a peak that gradually shifts in phase relative to the soma oscillation. To replicate the local field potential in the stratum pyramidale, the soma membrane potential is inverted (Field). Plotting the time of peak of the Sum as short vertical lines shows how spikes generated at the peak (short lines) would shift in phase relative to the peaks of the local field potential in stratum pyramidale (long vertical lines). Plotting the phase of the short vertical lines relative to the long vertical lines shows the progressive shift in phase (Phase of spikes). The oscillations and phase of spikes are plotted against time. For constant running speed, time is proportional to location. For variable running speed, precession will more directly depend on spatial location if the dendritic oscillation is regulated by running velocity (Burgess et al., 2007).

After the original O'Keefe and Recce paper, a paper from the McNaughton laboratory replicating the phenomenon of theta phase precession (Skaggs et al., 1996) was rapidly followed by several publications about a different type of model of theta phase precession based on sequence retrieval (Jensen and Lisman, 1996b; Tsodyks et al., 1996; Wallenstein and Hasselmo, 1997b). These sequence retrieval models focus on the idea that theta phase precession could result from a stored sequence of place cells, which is read out sequentially as the rat traverses the environment. In particular, for an individual place cell, the responses late in the theta rhythm could reflect activation of the place cell at the end of a sequence being read out in the hippocampus. As the rat crosses the place field, the cell becomes an earlier component of the sequence (and fires at earlier phases) until, near the end of the place field, it is driven directly by sensory input and evokes retrieval of a sequence of place cells encoding future locations to be traversed.

Buffer for Sequence Encoding

In contrast to the oscillatory interference model, sequence encoding models depend on the retrieval of a sequence of place cells that must first be encoded by synaptic modification. As a rat repeatedly runs in the same direction along a linear track, it will cause sequential spiking of place cells that are here designated as place cells 1, 2, 3, 4, and 5. Behavioral transitions between place cell firing fields in the environment occur over a time course of several hundred milliseconds or even seconds. The time course of behavior is much slower than the time intervals important for STDP in the hippocampus, which results in strengthened synapses only when a presynaptic spike precedes a postsynaptic spike by less than about 40 milliseconds (Levy and Steward, 1983; Bi and Poo, 1998).

A rat does not move very far in less than 40 milliseconds, so the input from the environment does not change very much, raising the question of how a rat could form associations between neurons spiking at much slower intervals during sequential visits over longer periods in a task. If there is sufficient overlap between the firing fields, this could allow STDP between sequentially activated neurons. However, overlapping place fields would not necessarily have the correct timing for activation of STDP. In particular, spikes might come in the wrong relative order, or there may be long gaps due to the rat's slowing down in the field or due to intervals between the phases of theta rhythm cycles when cells show their predominant spiking.

How can the short interval of STDP be used to store associations between locations encountered at longer time intervals? The spike timing requirements for STDP in the hippocampus could be obtained by a mechanism that involves persistent spiking in entorhinal cortex. As described above, the persistent spiking could hold information about prior location over hundreds of milliseconds or longer. A model initially proposed by Lisman (Lisman and Idiart, 1995) shows how spiking induced by intrinsic afterdepolarization in neurons, interacting with theta rhythm oscillations, allows cells to spike sequentially within the 40-millisecond time window necessary for STDP (Lisman and Idiart, 1995; Jensen and Lisman, 1996b; Koene et al., 2003; Koene and Hasselmo, 2007).

Simulations of detailed models with spiking neurons have shown how buffering could result from input eliciting sustained spiking activity in the entorhinal cortex (Fransén et al., 2002; Hasselmo et al., 2002b; Koene et al., 2003). Cholinergic modulation activates the CAN current (Klink and Alonso, 1997a), described in chapter 2 and the Appendix, which allows rhythmic reactivation of the elements of the input sequence. In these models, theta rhythm in entorhinal cortex serves to time the reactivation and updating of the working memory buffer. This buffer maintains grid cell firing field representations for a period of time sufficient to drive spiking activity of place cells in region CA3 so that STDP can modify synapses between CA3 place cells activated by adjacent locations, or between each CA3 place cell and the subsequently activated CA1 place cell. These modified synapses can form the basis for encoding of specific sequences traversed through the environment. The buffer is updated by each new spiking pattern causing first-in, first-out replacement to knock out the start of the sequence when a new pattern is added to the end of the buffered sequence (Koene and Hasselmo, 2007).

As an alternative, the persistent spiking model of grid cells presented in chapter 3 (Hasselmo, 2008b) could also provide the necessary spike timing for sequence encoding. Instead of a new spiking pattern causing first-in, first-out replacement, the persistent spiking model of grid cells would update the phase of relative firing of grid cells that could drive the relative phase of firing of place cells. The shift in frequency with velocity would then update the phase in a systematic manner that would only update the sequence in proportion to movement through the environment.

Slow Sequence Retrieval Model

Once the sequence has been encoded in these models, then each place cell can cue retrieval of a sequence of subsequent place cells in CA3, and CA3 place cells could also drive a sequence of place cells in CA1. For example, entry to the location coded by place cell 1 would cause retrieval of a sequence in the form of sequential spiking in place cells 2, 3, 4, and 5. As shown in figure 7.11A, if one observes the response of a single cell (coding for location 5), the response of cell 5 will initially occur late in theta when the rat is at location 1 and cell 5 is activated at the end of the readout sequence. Then, when the rat moves to location 2, the place cell activated by location 2 will drive retrieval of place cells 3, 4, 5, and 6. The same process will occur when the cue is place cell 3, 4, or 5. Recordings of the single cell coding location 5 will show spiking that moves to earlier and earlier phases as the cue location gets closer to location 5. For example, when the rat is at location 3, the retrieval of place cells 4, 5, 6, and 7 causes spiking of cell 5 in the middle of the theta cycle. When the rat is at location 5, then cell 5 is being directly activated at the start of the sequence (early phase). Thus, place cell 5 will shift to earlier phases until it is driven by sensory input at the start of the cycle. When the rat moves past this location, then the cell will stop firing.

Models of this type (Jensen and Lisman, 1996b; Tsodyks et al., 1996) require that input for the start of the sequence occur only at one phase of the theta cycle. This implies that sensory input drives firing only at the end of the place field of a given place cell. The start of the place field is due to retrieval of a previously encoded sequence. These models also require relatively slow readout of sequences across the full cycle of the theta rhythm (which has a period of about 125 to 200 milliseconds). In the model by Tsodyks, Skaggs, McNaughton, and Sejnowksi (1996) this slow readout is obtained with weak excitatory connections and very long sequences. In a model by Jensen and Lisman (1996), this slow readout is obtained with the slow dynamics of

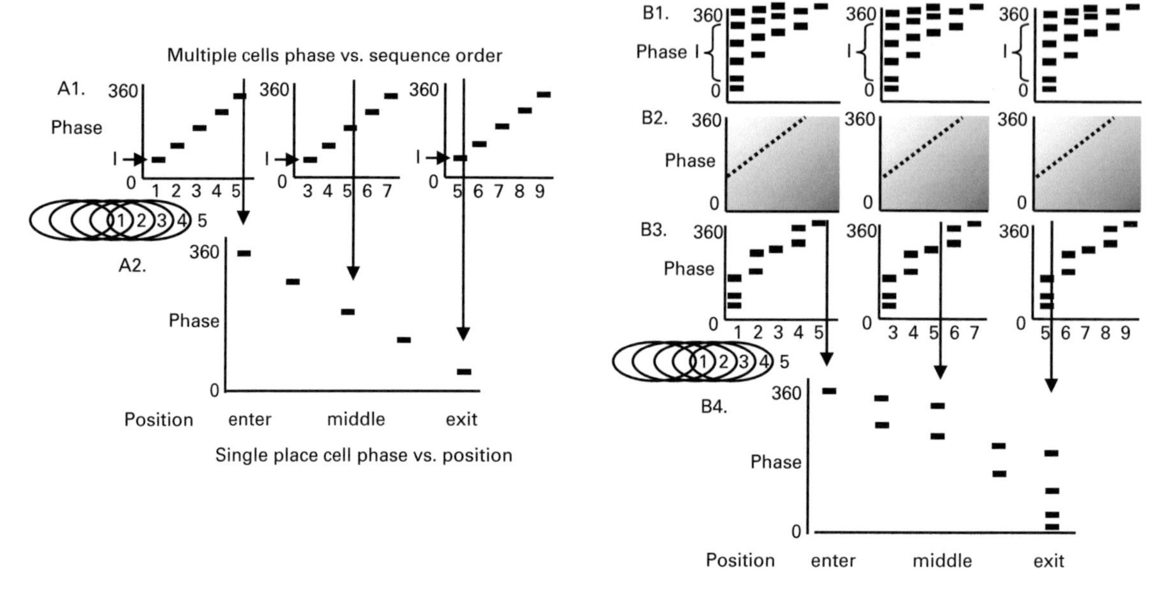

Figure 7.11
Left: Model of phase precession based on sequence retrieval. (A1) Phase of spike versus place cell number. When the rat is at location 1, input I activates place cell 1. Place cell 1 activates 2, place cell 2 activates 3 and so on for cells 3, 4, and 5 across the theta cycle. At location 3, place cell 4, 5, 6, and 7 spike sequentially. At location 5, cells 5, 6, 7, 8 and 9 spike. Input I is present only at early phase. (A2) Plot of the spiking phase of just place cell 5 (from top graphs) as the rat moves through the locations. In location 1, place cell 5 fires at late phase. In location 3, place cell 5 spikes at middle phase. In location 5, place cell 5 fires at early phase. Right: Model of theta phase precession with fast, context-dependent retrieval of sequences. (B1) Forward retrieval in entorhinal cortex that spreads to more neurons as theta phase increases. At early phase only cell 1 fires. At late phase cells 1,2,3,4 and 5 fire rapidly in sequence. Input is present at all phases of theta. (B2) Temporal context input from CA3 gates CA1 activity. Temporal context is stronger (darker) for more recent locations (right). For locations to the left of the dotted line, temporal context is too weak and CA1 activity falls below threshold. This input decreases over phases within each cycle, causing activity to fall below threshold for more locations. (B3) CA1 activity depends upon the multiplicative gating of entorhinal cortex input by temporal context from CA3. Input cue is present during the full cycle, but region CA1 spikes only when entorhinal cortex input converges with strong temporal context. (B4) Firing phase of a single place cell (cell 5) as the rat moves through its firing field. Place cell 5 activity appears as a wide, slightly scalloped distribution of spiking that shifts from late to early phases. Figure based on Hasselmo (2005a).

the NMDA receptor. Figure 7.11A illustrates the proposed mechanism for theta phase precession dependent upon slow sequence retrieval.

Fast Sequence Retrieval Model

In contrast, more rapid readout with AMPA receptor kinetics is used in a separate model (Wallenstein and Hasselmo, 1997b; Hasselmo and Eichenbaum, 2005). In this model, the theta phase precession is obtained by readout of sequences to different lengths during different phases of the theta cycle, due to phasic changes in postsynaptic depolarization or phasic changes in regulation of synaptic strength by activation of $GABA_B$ receptors. The mechanism for theta phase precession dependent upon rapid readout of sequences is summarized in figure 7.11B. This model of theta phase precession can have the input cue present during the full theta cycle (Hasselmo and Eichenbaum, 2005) as shown in figure 7.11B.

The model of fast readout of sequences for theta phase precession has also been adapted to include context-dependent retrieval of the sequence (Hasselmo and Eichenbaum, 2005). In this model, the phase of firing of region CA1 neurons in the model depends on the relative strength of forward sequence retrieval and the backward temporal context retrieval at different phases of theta (as described above). Due to the change in dynamics, the forward retrieval in entorhinal cortex primarily occurs during half of the theta cycle. During the earliest phases of theta, entorhinal cortex responds primarily to afferent input. This results in a firing response for current location that spreads across a broad range of phases similar to experimental data, accounting for the increase in phase variance that appears in later portions of the place field.

As shown in figure 7.12, the context-dependent retrieval mechanism effectively addresses a previously puzzling property of theta phase precession, which other models do not address. In experimental data, theta phase precession is initially weak on a unidirectional one-dimensional track and becomes more prominent as the rat runs subsequent loops around the track on each day (Mehta et al., 2002). This is puzzling as it was initially expected that once a sequence is learned it would be retrieved in the same manner on subsequent days. The authors of that study proposed that synapses between sequential place fields would have to be weakened overnight.

In contrast, in our model (Hasselmo and Eichenbaum, 2005), temporal context is weak on the first run around the track on each day due to the lack of previous exposure to the task on that day. The absence of the temporal context input from region CA3 to region CA1 neurons leaves only the input from forward sequence retrieval in entorhinal cortex. This entorhinal cortex input is strongest for the current input, but requires convergent input of temporal context from region CA3 to activate region CA1. Because temporal context is not present on the first loop around the track on a given day, the context input is not present, and activity in region CA1 primarily reflects current input, resulting in no sequence retrieval and therefore no precession.

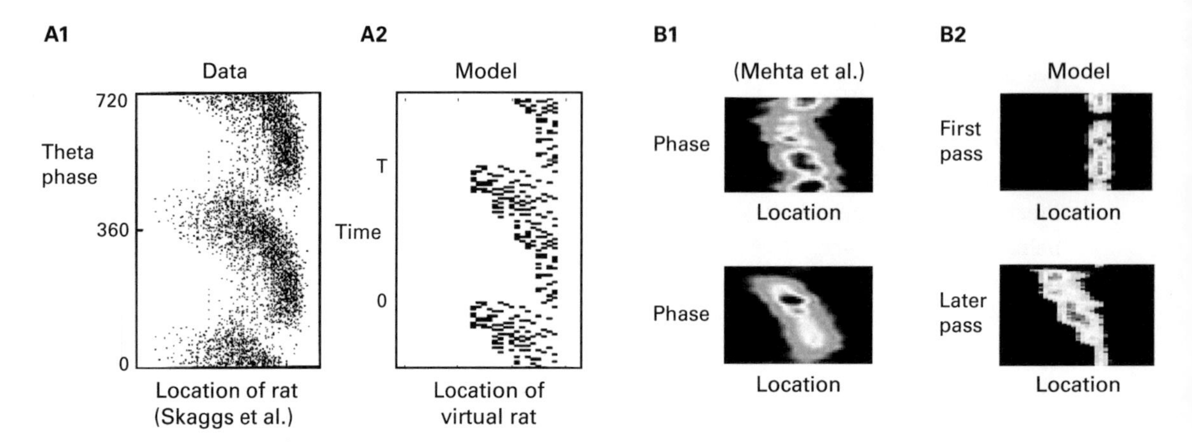

Figure 7.12
Theta phase precession of hippocampal place cells. (A1) Experimental data (Skaggs et al., 1996) show that the theta phase of firing (y axis) moves to earlier phases as the rat moves from left to right (x axis). (A2) The simulation shows the same pattern for activation of a single place cell in different locations as the virtual rat moves in one direction around a rectangular track. (B1) Experimental data (Mehta et al., 2002) show that theta phase precession is not strong on the first pass through a location on a given day (top) but is stronger on later passes through the same location (bottom). (B2) The simulation shows that the absence of temporal context results in an absence of phase precession on the first loop (top). On later loops (bottom), temporal context results in increased theta phase precession (Hasselmo and Eichenbaum, 2005).

As the rat completes the unidirectional loop used in this task, it encounters previously visited locations that provide strong temporal context, resulting in stronger temporal context input to region CA1 neurons. This temporal context input allows retrieval of the sequence of place cells in CA1 which causes theta phase precession (Hasselmo and Eichenbaum, 2005). This same context mechanism could also underlie the backward expansion of place fields that appears on each day of recording (Mehta et al., 1997, 2002).

Other Models of Precession

In addition to the oscillatory interference model and the sequence retrieval models described above, other models of theta phase precession have been presented. One of these models is the depolarizing ramp model, in which a depolarizing input at the dendrite interacts with rhythmic inhibition at the soma arising from network theta rhythm oscillations (Mehta et al., 2002). The depolarizing ramp causes the cell to cross firing threshold at earlier and earlier phases relative to network theta. Another theory of theta phase precession proposed an interaction of dendritic and somatic oscillations at different phases based on experimental data rather than simulations (Magee, 2001;

Gasparini and Magee, 2006). In this theory, a progressive increase in the amplitude of the dendritic oscillation causes a partial shift in the phase of the summed oscillation (Magee, 2001; Losonczy et al., 2010). This is the somatodendritic model.

Intracellular Recording of Phase Precession in Vivo

Mechanisms of phase precession were tested by intracellular recordings from awake, behaving mice (Harvey et al., 2009) that were kept in fixed head position while running on a spherical treadmill as a virtual world was projected in proportion to their running movement. Despite the mouse running in a single location, place fields were recorded in this task in response to the virtual world, and theta phase precession was observed during intracellular recording. Intracellular recordings show a shift in the phase and amplitude of oscillations as well as an overall depolarizing shift. The clear precession of intracellularly recorded subthreshold membrane potential oscillations relative to the network theta rhythm (Harvey et al., 2009) over almost 360 degrees supports the oscillatory interference model more than other models. The sequence retrieval models do not predict a phase shift in subthreshold oscillations within a single neuron (Jensen and Lisman, 1996b; Tsodyks et al., 1996; Wallenstein and Hasselmo, 1997b). The somatodendritic interference model cannot account for precession over more than 180 degrees, and the input amplitude change in this model causes a strong link between firing rate and firing phase (Gasparini and Magee, 2006) that is inconsistent with the dissociation of firing phase and firing rate in the experimental data on phase precession (Huxter et al., 2003; Harvey et al., 2009). The intracellular data show spiking at the peak of each intracellular oscillation that does not support the depolarizing ramp model (Mehta et al., 2002) because earlier crossing of threshold should cause spiking to shift in phase relative to the peak of membrane potential oscillations. Thus, the oscillatory interference model receives the most support from the intracellular data, though it must incorporate the ramp depolarization found in intracellular data that could contribute to a shift in oscillation frequency and phase.

Interaction of Memory Systems

The various behavioral tasks addressed here could be solved with different mechanisms of memory. In some of the examples above, I have focused on solving the tasks using episodic encoding and retrieval of trajectories. However, I have also mentioned in some cases that there are other mechanisms that could be used to solve the tasks, such as active maintenance of prior activity. In addition, even if the episodic encoding and retrieval of trajectories is used to solve a task, this still requires mechanisms of active maintenance for maintaining representations of spatial location and time during the process of encoding.

Previous work has shown how specific behavioral tasks can be solved with different types of memory strategies (Zilli and Hasselmo, 2008a, 2008c) or the interaction of memory systems (Zilli and Hasselmo, 2008b). Human imaging data show that active maintenance of activity in the absence of a stimulus is correlated with the subsequent memory for that stimulus at a later time (Schon et al., 2004; Schon et al., 2005). The new modeling framework presented here provides potential mechanisms for simultaneously modeling the interaction of memory systems such as working memory and episodic memory (Hasselmo and Stern, 2006). For example, the active maintenance of a phase representation in multiple different neurons in the grid cell model can be considered to be a working memory for current spatial location. However, once this phase code of working memory for a state causes activity to spread through previously modified synapses to hippocampal place cells and head direction cells to alter the current phase, then the working memory has cued retrieval of an episodic memory. The retrieval of the episodic memory shifts the phase code of working memory, so the new retrieved state can be held in working memory. Thus, this framework provides a shared physiological mechanism to allow the interaction of working memory with episodic memory for solution of the task. This mechanistic description is consistent with more abstract mathematical analysis showing how different types of memory can disambiguate individual states in a behavioral task (Zilli and Hasselmo, 2008b, 2008c).

Stoichiometry of Memory Tasks

The analysis of the role of different memory mechanisms in solving a task provides a useful perspective on memory-guided behavior. In the past, there has been a tendency to treat behavioral tasks as if they test a single, monolithic form of memory function. For example, people might refer to a "working memory task" or an "episodic memory task." However, these terms should not be interpreted as excluding other memory mechanisms. Certain tasks might put a strong demand on one memory system, but they almost always put demands on other memory systems as well.

A more realistic approach to the analysis of behavioral tasks may be to consider a stoichiometry of memory tasks. This draws an analogy with chemistry, in which individual compounds are made up of measurable components of particular elements. For example, two molecules of diatomic hydrogen gas (H_2) combine with one molecule of diatomic oxygen gas (O_2) to produce two molecules of water (H_2O). Similarly, the contributions of different memory systems could be analyzed at different states within a task. For example, a unit of episodic retrieval may occur at the choice point in spatial alternation. Or, a unit of episodic retrieval may occur at the base of the stem, and a unit of working memory at each state along the stem might hold this information until it is utilized at the choice point.

This suggests that the memory demands of a task should be analyzed with reference to individual behavioral states within the task. The analysis of different memory mechanisms showed that it is most fruitful to analyze memory contributions for disambiguation of individual states in a task (Zilli and Hasselmo, 2008b, 2008c).

The process of analyzing task demands in terms of individual behavioral states has been used in reinforcement learning theory (Sutton and Barto, 1998). However, most initial models in reinforcement learning theory do not focus on episodic memory or working memory mechanisms, instead focusing on what could be considered procedural memory or semantic memory. In reinforcement learning theory, the set of individual decision points in a task usually make up a Markov decision process. A Markov decision process refers to a process in which all of the necessary information for a decision at a particular state on a given trial is available at that state. For example, a visual discrimination task has all the information available for a decision at each point in the task. For example, a fixation cue tells the monkey where to fixate, and a pair of visual stimuli then dictate what type of response the monkey should generate. The response can depend on learning of the reward contingencies of a given state on previous trials, so it could depend upon semantic memory or procedural memory. However, the response in a Markov decision process does not depend upon episodic memory or working memory.

The tasks that depend upon episodic memory or working memory can be described as partially observable Markov decision processes. This is an abstruse and confusing term that sounds mysterious and impressive but unfortunately tends to impede the influence of reinforcement learning theory on neuroscience. The term refers to the fact that all the necessary information for making a decision is not available at each state in the task. Therefore, the agent performing the task needs to maintain information from previous states in order to reduce ambiguity at an individual state in order to make a decision.

The manner of maintaining the information from previous states can differ depending on the type of memory mechanism. For example, working memory is a limited capacity buffer that holds information from prior states. Episodic memory is a larger capacity system that holds information about sequences of prior states (Zilli and Hasselmo, 2008a, 2008c). The difference between working memory and episodic memory may primarily reflect their difference in capacity based on the use of active neurons versus modified synaptic connections. For example, working memory can hold sequences and may underlie the human capacity for immediate recall of sequential verbal information using mechanisms that may depend upon phase codes or rate codes for temporal order (Henson, 1998; Burgess and Hitch, 2005). However, working memory can only hold this information about a small number of recently experienced sequences. In contrast, episodic memory can hold a lifetime's worth of complex spatiotemporal trajectories for access at remote times. This is due to the larger capacity of synaptic connectivity matrices that encode sequential associations between items and location dependent upon a specific context. These can code the various permutations of single items on trajectories that occur in different episodes.

Working memory for multiple items based on persistent activity could be used to directly solve behavioral tasks (Zilli and Hasselmo, 2008a, 2008c) or to provide input or output for an episodic store based on synaptic modification (Zilli and Hasselmo, 2008a, 2008b). In support of this, human imaging data show that persistent activity in the absence of a stimulus is correlated with the subsequent memory for that stimulus at a later time (Schon et al., 2004; Schon et al., 2005) and shows load effects dependent on number of items held during a delay (Schon et al., 2009). The mechanisms of persistent spiking might play an important role in the neural activity present in the hippocampus and parahippocampal cortices during working memory for novel stimuli (Stern et al., 2001; Hasselmo and Stern, 2006) and during the encoding of novel information into long-term memory (Stern et al., 1996; Wagner et al., 1998; Kirchhoff et al., 2000).

Selection of Memory Actions

Another important question concerns the mechanism for controlling the processes of memory function. For example, the previous work on memory-guided behavior in a spatial alternation task (Hasselmo and Eichenbaum, 2005) used episodic retrieval of sequences at every single location in the maze. Thus, even though retrieval was only necessary for guiding the choice at the end of the stem, the model performed retrieval at every single position in the maze, including the reward arms and return arms that are traversed after making the choice. This was because there was no mechanism in this model for selection of the states where encoding and retrieval were necessary.

A more recent model was able to selectively perform memory actions such as "encode" and "retrieve" only when necessary for task performance (Zilli and Hasselmo, 2008a). This model used reinforcement learning mechanisms to select necessary actions at specific states in the environment. In addition to physical actions such as "go East" or "go West" the model also could select memory actions such as "encode" or "retrieve." This allowed selection of memory actions at the appropriate time in the maze. For example, the "retrieve" action was selected more frequently on the stem near the choice point, with less selection at other times in the task.

Similar mechanisms could be used to mentally project novel trajectories through state and action space to allow mental time travel through imaginary or future locations when these processes are necessary to perform a task. Instead of temporal intervals due to interference or a recurrent loop driving the retrieval of a previous trajectory, the actions along the trajectory could be determined by prefrontal input to the cells coding velocity. For example, to imagine arbitrary movement through a familiar house, semantic memory could activate memory of the front hallway, and prefrontal cortex could generate a representation of action (going forward). These cells could then drive the phase code of grid cells to progressively update place cell populations representing different locations and thereby activate associations with items at particular locations. At the end of the imagined hallway, the prefrontal cortex could generate an action to go left or go right. This new action would then update the phase code of grid cells to drive place cells representing a location in a different room, and associations with items in that room. Thus, the same cellular mechanisms described here may underlie the role of parahippocampal and hippocampal structures in encoding and mental time travel during retrieval of episodic memory (Steinvorth et al., 2006) as well as the mental time travel involving imagination of future experiences (Schacter et al., 2007). The framework presented here resembles the overall framework used in previous simulations of interactions of prefrontal cortex with medial temporal and parietal cortices (Byrne et al., 2007), but the model presented here focuses on understanding the role of specific cellular intrinsic properties mediating coding of both time and space in the hippocampus and entorhinal cortex.

The theoretical mechanisms for coding of space and time along spatiotemporal trajectories presented here provide a useful framework for addressing topics of memory-guided behavior. In particular, the framework provides shared mechanisms for representing space and time within episodic memory and working memory, thereby simplifying the analysis of the interaction of these memory systems. The framework also provides shared mechanisms for retrieval of past spatiotemporal trajectories and the planning of future spatiotemporal trajectories that simplify the study of internal actions that control memory and planning. As described here, the coding of space and time has been inspired by cellular data on grid cell firing properties in entorhinal cortex and place cell firing properties in the hippocampus, as well as by cellular intrinsic properties within the entorhinal cortex. This link to the experimental data provides an important means for analyzing the variety of different modeling frameworks to determine which one is most likely to reflect the true brain mechanisms of episodic memory function.

Appendix: Mathematical Models of Memory

Elements of the Episodic Memory Model

The model of episodic memory uses a number of different computational components described in this Appendix. This appendix covers the following topics: (1) representation of the spatiotemporal trajectory, (2) a simple representation of membrane potential and spiking activity in populations of neurons, (3) a simple representation of oscillations in single neurons and networks, and the coding of space and time by a phase shift in oscillations, (4) the oscillatory interference model of grid cells, as well as theoretical arc length cells and time cells, (5) encoding of associations by associative memory models, (6) the full model for encoding of episodic memories, (7) a mathematical example of the separation of encoding and retrieval by cholinergic modulation, (8) a mathematical example of the separation of encoding and retrieval by theta rhythm oscillations, and (9) a description of biophysical simulations of membrane potential dynamics.

Episodic Memory as a Spatiotemporal Trajectory

As described in the main text, one aspect of an episodic memory consists of the spatiotemporal trajectory traversed by the agent experiencing the episode, sometimes referred to here as an episodic trajectory. The individual items or events encountered during the episode can be defined in relationship to this spatiotemporal trajectory. Other agents perceived in the episode might follow their own separate trajectories.

The description of a spatiotemporal trajectory can use the standard representation of movement used in physics. The location of the agent experiencing the episode can be described by three-dimensional spatial coordinates x_1, x_2 and x_3 combined together in an array or vector $\vec{x} = [x_1, x_2, x_3]$. The arrow over the x indicates it is a vector. In most cases in this book, we will focus on only two spatial dimensions.

The time of an event or position on a trajectory will be described with the variable t. The spatial position of an agent in time can be described with the evolution of the spatial coordinates over time, $\vec{x}(t) = [x_1(t), x_2(t), x_3(t)]$. This mathematical description of the spatiotemporal trajectory can indicate a continuous change of position in time or a discrete set of locations at discrete time steps. Note that if the agent is stationary with no change in spatial coordinates, the trajectory still evolves over time.

The velocity of the agent at a particular point in time is the derivative of the continuous trajectory of spatial positions: $\vec{v}(t) = d\vec{x}(t)/dt$. The velocity can be approximated by discrete changes in each spatial coordinate $\Delta x(t) = x(t) - x(t - \Delta t)$ over each discrete period of time Δt, so $\vec{v}(t) \approx [\Delta x_1(t), \Delta x_2(t)]$. Remember that the physical definition of velocity is not just speed in one dimension but a vector that represents the change in all spatial dimensions. The velocity vector combines both the speed and direction of movement of the agent.

The speed of the agent can be estimated from the discrete change in spatial location as follows: $s(t) = \sqrt{\Delta x_1(t)^2 + \Delta x_2(t)^2}$. Imagine this as the hypotenuse of a right triangle with movement in each dimension as the sides of the triangle. Based on this same triangle, the angle $h(t)$ of the direction of movement can be estimated from the arctangent of the dimension opposite from the angle divided by the dimension adjacent to the angle $h(t) = arctan2(\Delta x_1(t)/\Delta x_2(t))$. Sometimes this estimate of movement direction is used as an estimate of the head direction of the agent, but this makes the assumption that the head is pointed in the direction of movement, a convenient approximation that data shows is not always true during real behavior.

Neuron Population Activity Represented by Vectors

The spatiotemporal trajectory and the items and events encountered on the trajectory cause patterns of activity in the populations of neurons within individual brain regions. The activity of populations of neurons in this appendix will be represented with activity vectors $\vec{a}(t)$, as shown in figure A.1. In a neural activity vector, individual neurons are assigned an index i out of a total number n of neurons. If there is more than one population, then the index for neurons in one population might be different integer values of i and for neurons in another population might be different values of j. Different patterns may be represented by a superscript. Activity in different regions may be represented with different letters. In some cases here the activity of a population of grid cells is designated as $\vec{g}(t)$ and place cells as $\vec{p}(t)$.

In the activity vector, the membrane potential of each individual neuron j over time is represented by an individual variable, $a_j(t)$, with the subscript index indicating the individual neuron. When vector notation is used, the whole vector can be represented by a single lowercase letter, sometimes using boldface or a line above the letter, $\vec{a}(t)$. When the full vector is used in linear algebra, it is commonly written as a column vector

$$\vec{a}(t) = \begin{bmatrix} a_1(t) \\ \dots \\ a_n(t) \end{bmatrix}.$$

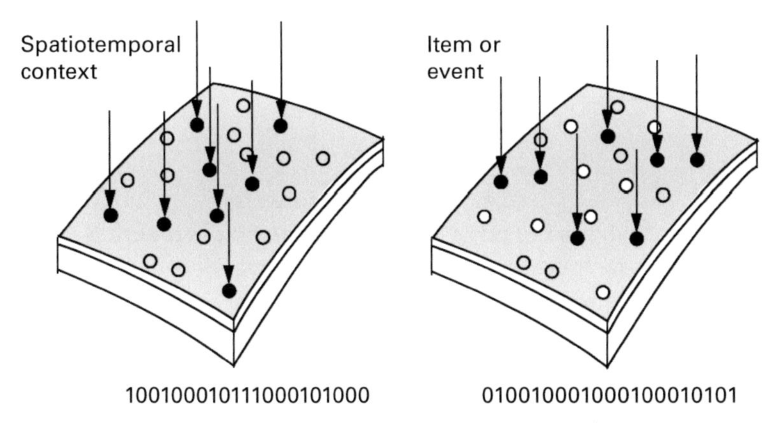

Figure A.1
Examples of two different binary row vectors representing two arbitrary different patterns of spiking activity within a population of neurons. Patterns like this can be used to represent position on a spatiotemporal trajectory or the pattern of features in individual items or events.

The column vector notation is very awkward for formatting in text but is the more standard notation in linear algebra. If vectors are defined as column vectors, a superscript T indicates they have been transposed into a row vector $\vec{a}^T(t) = [a_1(t), \ldots a_n(t)]$. As described below, in the notation of linear algebra the mathematical operations to be computed with vectors is defined by the order of interactions of column and row vectors.

In biophysically detailed models described later, the membrane potential is directly simulated. If the membrane potential stays constant in the absence of input, this is called the resting potential and is often simulated to be around –70 millivolts. This resting potential arises from a difference in concentration of ions on either side of the membrane, with high concentrations of Na^+ and Cl^- outside the cell, and high concentrations of K^+ inside the cell. In the abstract models described first, a neuron at resting potential has activation $a(t) = 0$. Positive values of activation represent depolarization, and negative values represent hyperpolarization. Depolarization and hyperpolarization can be caused by direct current injection from an input vector $A_{in}(t)$ given to the model or by synaptic input from other neurons in the model that activate synaptic receptors that change ionic conductances.

When the membrane potential of a neuron crosses a firing threshold, the activation of voltage-sensitive channels in the neuron generates action potentials (spikes) that spread along axons to cause synaptic transmission at synapses on other neurons. In many models, the relationship between membrane potential and spiking rate is represented by an input–output function $g(a_i(t))$ that takes the neural activation $a_i(t)$ as input. The input–output function represents the increased firing rate with depolarization of the neuron above the firing threshold θ. Biologically, the firing threshold is defined as the membrane potential where a neuron has a 50% probability of generating an action potential.

Input–Output Functions

The input–output function is based on experimental data obtained during intracellular recording from a neuron that shows the firing frequency f of the neuron during injection of different levels of depolarizing current I, giving an *f-I* curve. As shown in figure A2D, the *f-I* curve in most neurons starts out at zero frequency when the current injection does not push the cell above threshold. Once the current injection depolarizes the neuron above firing threshold, then the frequency increases with greater current injection. At high levels of current injection, the neuron levels off at a maximum rate (often several hundred spikes per second). However, most neurons in behaving animals never fire at a rate close to the maximum rate found with current injection, indicating that network interactions such as feedback inhibition usually limit their firing rate below the maximum. Many neurons in the cortex show adaptation, in which the *f-I* curve flattens over the time period of current injection. Detailed biophysical models represent the dynamics of membrane potential over time. Integrate-and-fire models represent the generation of a single binary spike each time the activation crosses threshold. Continuous firing rate models do not model individual spikes but represent a continuous change in firing rate over time.

Input–output functions can be modeled with a number of mathematical forms. The output function can be a step function or Heaviside function as shown in figure A.2A. $o_i = g(a_i) = [a_i]_H$. The Heaviside function generates an output of 0 when input $a(t)$ is below threshold θ and generates an output of 1 when the input is above threshold. Another common function is the logistic function that has a sigmoid shape:

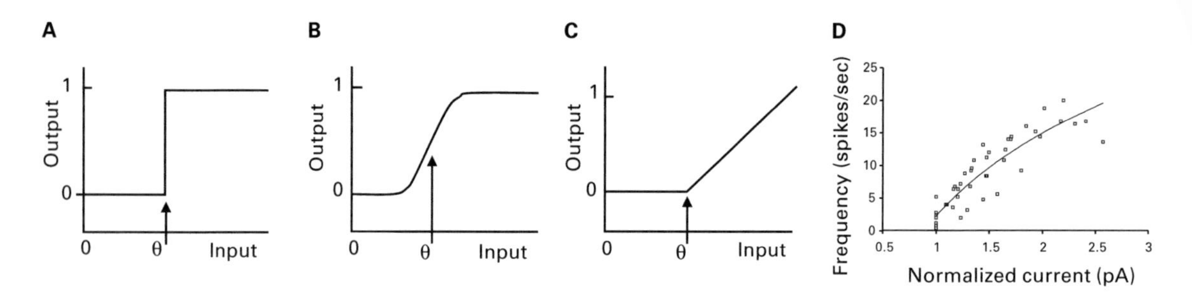

Figure A.2
Input–output functions. (A) Step function (Heaviside). (B) Sigmoid function. (C) Threshold linear function. (D) Frequency-current (f-I) relationship from intracellular recording data (Giocomo and Hasselmo, 2008a).

$o_i = g(a_i) = (1 + e^{\beta(a_i - \theta)})^{-1}$. The threshold θ in this function does not match the threshold in physiological studies but is instead half the maximum output of the neuron.

A more realistic input–output function is the threshold linear function $o_i = g(a_i) = [a_i - \theta]_t$, in which the output stays at zero when activation is below threshold θ (as indicated by the plus sign) and then increases linearly in proportion to how much activation exceeds the threshold. The grid cell model presented later could also be considered a type of complex input–output function as it transforms a pattern of input into a pattern of output.

Sometimes models include direct input to the neurons that is presented via current injection. In this case, the postsynaptic neurons i receive direct current injection represented by I_i. In many cases it is necessary to represent sensory input without representing the many preceding stages of synaptic input. In this case, the activation vector is directly updated by an input pattern representing synaptic input from other regions. Because this replaces a matrix of synapses, the notation presented here uses a capital letter A for afferent input. For example, the activation $a(t)$ at time step 1 can be updated by an afferent input vector $a_i(t = 1) = A_i^{(k)}$ representing the pattern of activation induced in a region induced by input A from a particular sensory stimulus labeled with index k.

Current injection into a neuron causes a gradual increase to a final depolarization level, and when current injection ceases, the membrane potential decays back to resting potential at a slower rate than the change in current injection. These effects arise because of the neuron properties of membrane capacitance (commonly referred to as C_m) and membrane resistance (commonly referred to as R_m). This can be represented by a model based on an equivalent circuit representation of the membrane potential. In this circuit, the input current A_i equals two currents: the current that charges the membrane capacitance $I_C = C_m da_i/dt$ and the current that crosses the resistance $I_R = a_i/R_m$. These currents also regulate decay of membrane potential back toward resting potential. Combining $A_i = I_C + I_R$ and rearranging algebraically gives the change in membrane potential for a given input as the differential equation

$$\frac{da_i(t)}{dt} = \frac{A_i}{C_m} - \frac{a_i(t)}{R_m C_m}.$$

This has the solution $a_i(t) = A_i R_m (1 - e^{-t/R_m C_m})$ that rises asymptotically to the steady state of $a_i(t) = A_i R_m$ with a time course described by the membrane time constant $\tau = R_m C_m$. The time constant describes the time to reach a value proportional to the steady state multiplied by (1-1/e), which is about 63%. In the absence of A_i, the membrane potential decays from starting value $a(0)$ back to resting potential with time course $a_i(t) = a(0)e^{-t/R_m C_m}$. This decay is also described by the same membrane time constant $\tau = R_m C_m$ that describes the time it takes for the potential to reach 1/e (about

37 percent) of the potential at the start of decay. Time constants are useful for characterizing the temporal dynamics of neural activity.

Synaptic Connections

When a neuron generates an action potential, the action potential propagates down the axon, which usually splits into many collaterals and terminates at multiple presynaptic terminals making synapses on many other neurons. At the presynaptic terminal, the action potential causes calcium influx that triggers release of neurotransmitter that causes a transient activation of postsynaptic receptors that briefly change the conductance of the postsynaptic neuron. This causes a depolarizing (usually excitatory) or hyperpolarizing (inhibitory) synaptic potential. The rapid excitatory transmission of information is usually mediated by release of the neurotransmitter glutamate that activates AMPA and NMDA receptors. These receptors contain channels that increase the conductance of the membrane to the cations Na^+ and K^+, causing depolarization of the membrane potential toward a reversal potential of zero millivolts. (The reversal potential is the membrane potential where current passing through a conductance changes direction). Rapid feedforward or feedback inhibition is mediated by the neurotransmitter GABA that activates $GABA_A$ or $GABA_B$ receptors. Activation of $GABA_A$ receptors increases the conductance of the membrane to chloride ions (Cl^-) and pushes the membrane potential down toward a reversal potential near resting potential. This moves the membrane potential away from firing threshold and thereby inhibits the spiking of the cell. $GABA_B$ receptors increase the conductance to K^+ and push the membrane potential toward a reversal potential of –90 millivolts.

Models of memory function commonly change the "weight" or "strength" of excitatory glutamatergic synapses between neurons. The synaptic strength of an excitatory synaptic connection corresponds to the amount of postsynaptic depolarization induced by a given presynaptic firing frequency. The pattern of strength of synaptic connectivity between one population of neurons and another population of neurons is commonly represented by the "weight" matrix W_{ij}. This can also be used to represent excitatory feedback connections on the same population as in figure A.3. In the figure, the presynaptic index is j and the postsynaptic index is i. In the synaptic connectivity matrix, the row of each synapse is coded by the index i and the column is coded by the index j. Activity flows from presynaptic columns j to postsynaptic rows i. (The meaning of the indices can differ in other models depending on whether the standard vector is a row vector or a column vector.)

Weight matrices will often appear in vector notation without the individual indices. In this notation, capital letters usually indicate matrices and lowercase represent vectors. When multiple different matrices connect multiple different regions, then it is convenient to designate the individual connectivity matrices with letters indicating

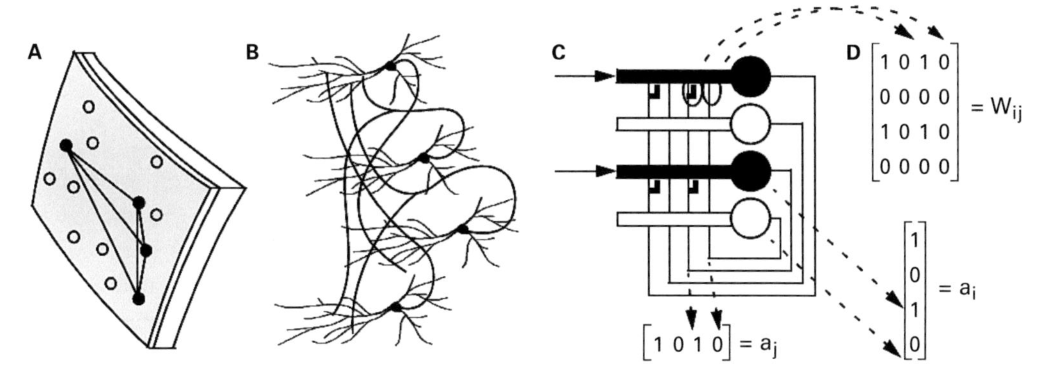

Figure A.3
(A) Illustration of a population of neurons with four active neurons marked by filled circles. Lines between active neurons indicate synaptic connections strengthened by Hebbian synaptic modification. (B) Illustration of the group of four active neurons with dendritic trees and axonal branches making excitatory recurrent connections. (C) Schematic representation of a small network showing connections in a more orderly manner. Two neurons are active (black), and two are inactive (white). (D) Mathematical representation of this network. Top: The pattern of excitatory recurrent connections in C shown as a synaptic connectivity matrix (W_{ij}). Bottom: The pattern of postsynaptic activity shown as a vector (a_i) that spreads along excitatory recurrent connections to also generate the presynaptic activity vector (a_j).

the presynaptic and postsynaptic population. For example, the synaptic connections from a population of grid cells $g(t)$ to a population of place cells $p(t)$ might be designated as W_{PG}. Note the reversed order of the letters consistent with the use of column vectors designating that region p gets input from region g.

In neurobiology, a single neuron usually releases the same transmitter or transmitters from all of its presynaptic terminals. This is referred to as Dale's law, after the codiscoverer of acetylcholine who first discussed this principle. The rapid effects of glutamate are mediated by AMPA or NMDA receptors that cause excitatory depolarization and cannot cause inhibitory effects. This means that the strength of synapses in a matrix representing excitatory synapses should all be positive and cannot change sign to become negative, as this would require a large change in ionic concentrations or a presynaptic change from glutamate release to GABA release that does not occur. Dale's law is violated by some early neural network models such as the early attractor networks.

In cortical regions, the connections between regions are usually mediated by populations of excitatory neurons. Inhibitory connections usually arise locally from neurons referred to as interneurons because they do not project out of the region. If these interneurons are driven by excitatory connections from the same region, they mediate feedback inhibition. If they are driven by excitatory connections from a different region, they mediate feedforward inhibition.

Oscillations and Interference

Oscillations are a common phenomenon of neural circuits, both at a single neuron level and at a network level. At a network level, oscillations appear in EEG recordings of field potentials in the hippocampus and entorhinal cortex. These field potential oscillations result from the net synaptic currents across a large population of neurons. Network oscillations commonly arise from interactions of populations of pyramidal cells and interneurons (Buzsaki, 2002).

Oscillations also appear in the membrane potential of single neurons within the entorhinal cortex when synapses are blocked. The membrane potential oscillations arise from interactions of voltage-dependent membrane currents within single neurons (Alonso and Llinas, 1989; Fransén et al., 2004) as summarized in a later section of the appendix. Single neurons can also show rhythmic persistent spiking activity with a stable firing frequency regulated by calcium-sensitive membrane currents (Klink and Alonso, 1997; Fransén et al., 2002) even without synaptic input.

Oscillations Modeled as Trigonometric Functions

The function of single-cell and network oscillations can be modeled on the simplest level by using trigonometric functions, such as the sine and cosine function. The mathematical representation of such simple models will be described in this section.

The sine and cosine functions can be used to model simple oscillations in potential as a function of time. Most of us are familiar with using the sine and cosine function to compute the relative length of the sides of a right triangle based upon the angle θ of one corner. For example, as shown figure A.4, the sine function (sin) describes how the angle θ determines the length of the side (y) opposite from the angle relative to the hypotenuse (h) of the triangle, $\sin(\theta) = y/h$. Similarly, the cosine function (cos) describes how the angle θ determines the length of the side (x) adjacent to the angle relative to the hypotenuse (h) as follows: $\cos(\theta) = x/h$. Usually, h is assumed to be of magnitude 1.

The angle can be described either with degrees or radians. When using degrees, a full circle contains 360 degrees, bringing the angle back to zero, and a right angle is 90 degrees. An alternative is to describe the angle with radians, which for a given angle describes the distance along the perimeter of the circle relative to the radius. The full distance around the perimeter of a circle (360 degrees) is 2π times the radius (2π radians), and a right angle (90 degrees) is one quarter of this distance, or $\pi/2$ radians.

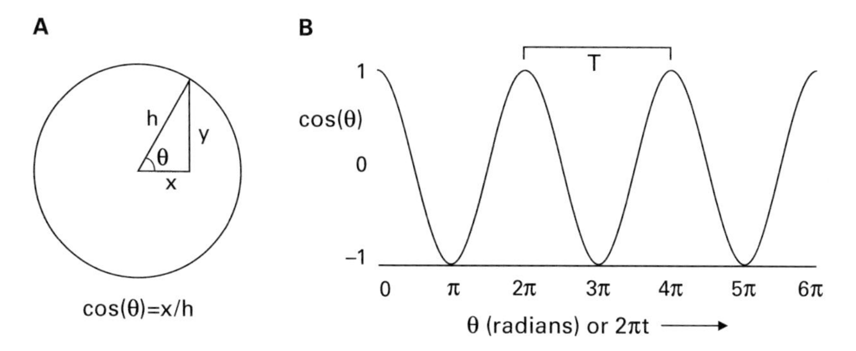

Figure A.4
(A) The functions sine and cosine describe the size of the adjacent side (x) and opposite side (y) of a right triangle relative to the hypotenuse (h) as the angle θ of the triangle progresses through the full circle of angles. For this circle, the radius $h = 1$. (B) Plot of $x = \cos(\theta)$ for different values of θ (in radians) results in an oscillation with complete cycles each time the angle θ passes integer multiples of 2π. For the function $\cos(2\pi t)$, the peaks are reached at $t = 1, 2, 3$, and so on.

The description of oscillations over time involves having the angle θ change with time t and then plotting the resulting value of x at different times (assuming $h = 1$). Thus, oscillations are described by the equation

$$x(t) = \cos(2\pi ft).$$

The number of cycles of the oscillation per unit time is determined by multiplying time by the frequency f. Because a full cycle of the oscillation will be complete at 2π radians, time is multiplied by 2π to scale time to the number of cycles. As shown in figure A.4, if the frequency $f = 1$, then between $t = 0$ and $t = 1$ the values of x will go through a full cycle that includes values of $x(t = 0) = 1$, $x(t = 0.25) = 0$, $x(t = 0.5) = -1$, $x(t = 0.75) = 0$, and $x(t = 1) = 1$. If the unit of time is seconds, and the frequency $f = 7$, the function will go through 7 cycles in one second. The unit of frequency is Hertz (abbreviated as Hz), which is the number of cycles per second. The time duration of one cycle is the period $T = 1/f$.

Note that the oscillations can also be described by their phase angle (usually referred to as phase). The phase corresponds to the angle being used in the cosine function as follows:

$$\varphi = 2\pi ft.$$

Thus, for frequency $f = 1$ and time $t = 0.25$, the phase angle is $\varphi = \pi/2$. The new phase angle at each new time step $t + \Delta t$ can be computed by adding the old phase angle at time t to the change in phase angle for each time interval Δt as follows: $\varphi(t + \Delta t) = \varphi(t) + 2\pi f t \Delta t$. This is equivalent to $2\pi ft$ if the frequency does not change. If the frequency changes as a function of time $f(t)$, then the phase angle needs to be updated for each time step.

Memory as Phase Angle

Neural oscillations in single neurons or networks could hold memory for prior inputs in the form of the relative phase angle of the oscillation. In the simplest example, the memory being encoded could consist of a depolarization of the neuron that shifts the frequency of oscillation for the period of time that the depolarization is present. This change in frequency will shift the phase angle relative to other reference oscillations. Thus, the stored memory takes the form of a difference in the phase angle of a neural oscillation relative to a reference phase angle. The shift in phase angle will be proportional to the magnitude and duration of the encoded input.

Consider a neuron that can hold memory for the velocity and duration of a previous movement by an animal. The shift in phase angle can be induced by changing the frequency of the oscillation for a period of time according to $f(t) = f + h(t)$, where f is a baseline frequency and $h(t)$ is sensory input influencing frequency. For example, imagine there is no initial animal movement until time $t = 2$, then movement occurs with speed of 0.25 until time $t = 3$, at which time movement stops. As shown in figure A.5, the input about movement speed causes a shift in the frequency of the oscillation that results in a shift in the phase of the oscillation as follows:

$$\varphi(t+\Delta t) = \varphi(t) + 2\pi(f + h(t))\Delta t.$$

Updating the phase over a period of time steps starting with time zero would require summing the values at each step between zero and t (as shown in the middle of the equation below). Summing over infinitesimally small time steps corresponds to taking the integral over a period of time, as shown on the right side of the equation below:

$$\varphi(t) = \sum_{0}^{t} 2\pi(f + h(\tau))\Delta\tau \approx \int_{0}^{t} 2\pi(f + h(\tau))d\tau.$$

This shows integration up to a point in time t. The end point t appears over the integral sign, and the Greek letter tau (τ) designates the variable of integration and the infinitesimal time steps $d\tau$ used for the integration. The integral of the constant baseline frequency f is just ft, so this part of the integral can be solved separately as shown in the following equation. If we put the computation of phase into the cosine function, then we get the oscillation

$$\cos(\varphi(t)) = \cos(2\pi(ft + \int_{0}^{t} h(\tau)d\tau)).$$

Thus, because the phase changes accumulate the effect of frequency over time the phase of the oscillation integrates the frequency change caused by an input. The

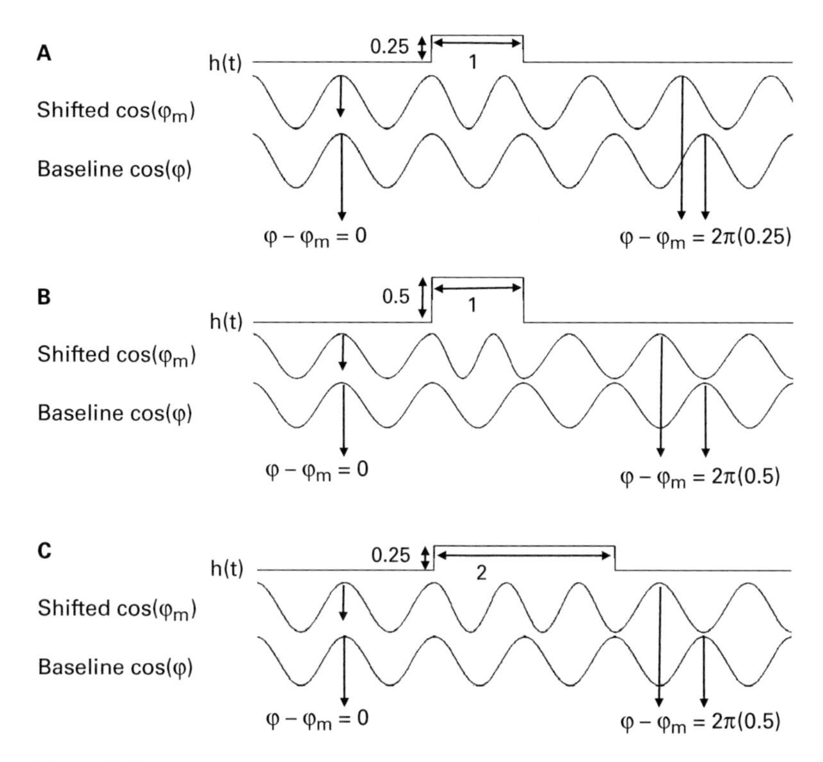

Figure A.5
Examples showing how the input $h(t)$ can cause a shift in phase angle that integrates the magnitude and duration of the input. (A) Input of magnitude 0.25 for duration 1 second gives phase shift of 0.25. (B) Input of magnitude 0.5 for 1 second gives phase shift of 0.5 (180 degrees). (C) Input of 0.25 for 2 seconds gives the same magnitude of shift as B.

integration of input by oscillation phase can be seen in figure A.5 above when comparing the phase φ_m of the oscillation holding the memory to the phase φ of the baseline oscillation. Before the movement input $h(t)$, the phase difference between the two oscillations is zero. During the movement input, the phase difference increases by $2\pi h(t)\Delta t$, so that at the end of the input in part A of the figure the phase difference is $2\pi(0.25)(1)$. The total phase shift is proportional to the integral of $h(t)$ over the interval of input. In the absence of further input, this phase shift persists in the network over all of the subsequent cycles of oscillation. Thus, the figure illustrates how memory of a previous input can be maintained in the form of a shift in the phase of one oscillation relative to a baseline oscillation.

The phase shift can integrate any function $h(t)$ over any interval. Two more examples are shown in the figure. In part B, $h(t)$ increases from zero to a speed of 0.5 for a period of 1 second, resulting in a phase shift of $2\pi(0.5)(1) = 2\pi(0.5)$. This is visible as a shift in the peak of φ_m to an earlier phase that precedes the peak of the baseline oscillation by half a cycle. Part C shows that the same phase shift is obtained with half the magnitude of speed (0.25) in the same direction for twice the duration (2 seconds), giving $2\pi(0.25)(2) = 2\pi(0.5)$. The examples in parts B and C were given to show how integration of different velocities for different durations gives position in a one-dimensional example. Movement at half the speed for twice the duration results in the same total shift in position and the same shift in oscillation phase. In the physical world, if you move in one direction at a specific speed for a specific duration and note your location, and if you then return to the starting point and move in the same direction at half the speed for twice the duration, you will end up in the same location. Thus, the network can keep track of position by integrating velocity.

By integrating the magnitude and duration of previous input, the phase shifts described here encode a continuous representation of previous input and maintain this memory by holding the phase (in the absence of further input). However, this leaves open the problem of reading out the phase angle difference that codes the memory.

Oscillatory Interference Provides Memory Readout

To be useful, the memory encoded by phase differences in the previous section must be accessible to readout. A potential mechanism for reading out the difference in phase is through the interference of oscillations. In the example in figure A.6, the input causes a difference between the phase of the oscillation holding the memory and the phase of the baseline oscillation. This can be read out in the form of interference between these two oscillations that appears in the sum v(t) of the two oscillations:

$$v(t) = \cos(2\pi ft) + \cos(2\pi(ft + \int_0^t h(\tau)d\tau))$$

As shown in figure A.6, the sum of two oscillations that are π radians out of phase with each other will undergo destructive interference so that the summed oscillation is a flat line. However, if one of the oscillations is shifted by input of magnitude 0.2 for 2 seconds, the integral of the shift is 0.4. Thus, the phase has shifted by about half a cycle $(0.4(2\pi) = 0.8\pi)$. This brings the two oscillations close in phase with each other. In this case, the sum of the oscillations now shows constructive interference, resulting in a large amplitude oscillation. If we consider a neuron that spikes whenever the oscillation crosses a threshold of 1.4, this results in regular spiking that indicates that a previous input of a particular magnitude was presented.

This readout specifically indicates the integral of the prior input, giving the same response for an input of 0.8 for 0.5 time steps or 0.1 for 4 time steps. In addition, it codes magnitude in a repeating manner, responding the same for 0.4 and 1.4. However, the readout can be made more specific by utilizing oscillations with different

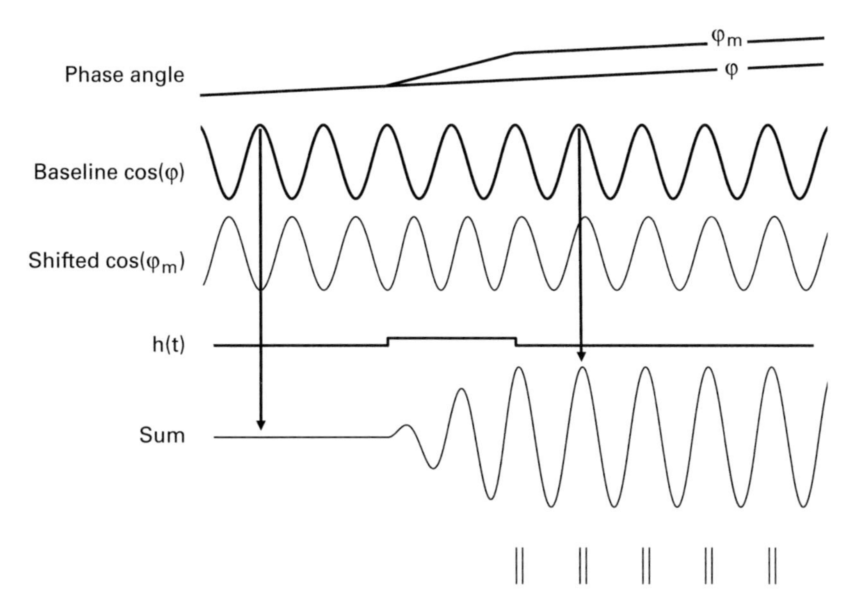

Figure A.6
Sum of two oscillations provides readout. When the oscillations are out of phase (left), the sum shows destructive interference (flat line) and does not generate spiking activity. Input $h(t)$ causes a shift in phase of one oscillation. When the oscillations are close in phase (right), the sum shows constructive interference and crosses threshold to generate spiking activity (bottom lines).

sensitivity to input that shifts them by different amounts. For example, consider a second pair of oscillations in which the input is scaled by 2/7. This second pair of units will integrate 0.4 and 1.4 times 2/7, giving phase shifts of 0.11 and 0.4, thereby distinguishing the two states.

In summary, this section described how a shift in phase of oscillations can provide a memory for prior input and shows how interference between shifted phases can provide readout of the prior memory.

Theta Phase Precession

The interference of oscillations described above has experimental support as it provides an effective model of the features of theta phase precession in the hippocampus (O'Keefe and Recce, 1993; Skaggs et al., 1996) and entorhinal cortex (Hafting et al., 2008) as shown in figure A.7. Consider a rat running at a constant speed along a linear track. In the oscillatory interference model of theta phase precession (O'Keefe and Recce, 1993; Lengyel et al., 2003), this will provide a constant velocity input that causes a constant difference between one theta rhythm frequency (here labeled Dend)

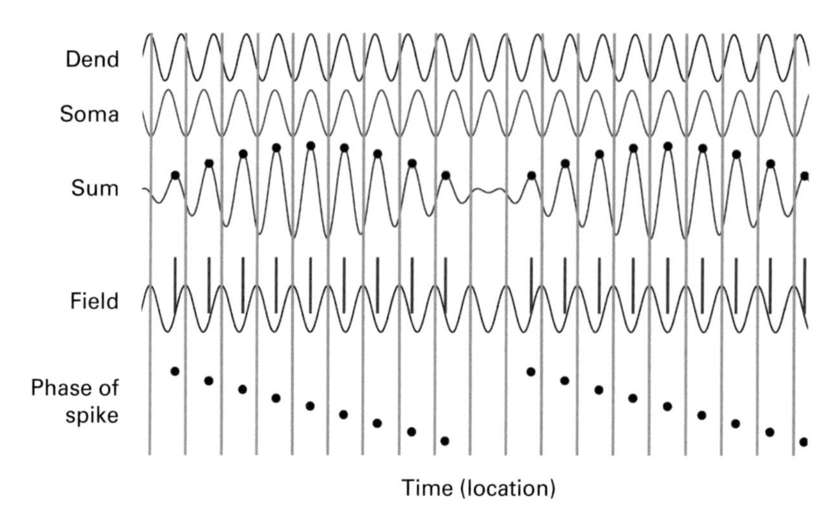

Figure A.7
Simulation of the oscillatory interference model of theta phase precession. Movement at a constant velocity could cause a constant increase in the intrinsic theta rhythm frequency in the dendrite (Dend) relative to a lower stable baseline theta rhythm oscillation frequency caused by network inhibition on the soma (Soma). The sum of the two oscillations shows a peak that gradually shifts in phase relative to the soma oscillation (Sum). To replicate the local field potential in the stratum pyramidale (by the soma), the soma membrane potential is inverted (Field). Plotting the time of peak of the sum as short vertical lines shows how spikes generated at the peak (short vertical lines) would shift in phase relative to the peaks of the local field potential in stratum pyramidale (long vertical lines). Phase of spikes is plotted based on the full cycle between peaks. For constant running speed, time is proportional to location, so x axis is labeled as Time (Location). This mechanism was initially proposed by O'Keefe and Recce (1993) and further simulated by Lengyel et al. (2003) and applied to grid cells by Burgess et al. (2007).

relative to a stable baseline theta rhythm frequency (labeled Soma). The interaction of these two oscillations produces theta phase precession. The simulation of theta phase precession uses an interaction of theta frequency oscillations that differ slightly in frequency.

As shown in figure A.7, during movement the intrinsic properties of the neuron may generate subthreshold membrane potential oscillations in the dendrite that are slightly higher in frequency than the network theta rhythm oscillation regulated by the medial septum. At the same time, the rhythmic inhibition arising from the medial septum causes rhythmic disinhibition in the hippocampus and entorhinal cortex (Toth et al., 1997; Buzsaki, 2002). This results in stable oscillations of lower frequency in the membrane potential at the soma (Soma). In the model, spiking activity is assumed to occur at the peak of the sum (Sum) of dendritic and soma membrane potential (though it could simply occur at the peak of the dendritic oscillation). In figure A.7, the time of the spikes at the peak of the sum are plotted as short vertical lines. These spikes are plotted next to a representation of the local field potential in stratum pyramidale that would be recorded by the same electrode that records the spikes. The same rhythmic inhibition that causes negative deflections in the soma membrane potential will cause positive deflections in the local field potential, so the field potential is the inverse of the soma membrane potential. The phase of spikes is measured based on the position of the spike between the two peaks of the field potential oscillation. For example, if it is halfway in between the two peaks, it is at 180 degrees (π radians). The phase of the spike measured relative to the field potential peaks is plotted on the y axis versus time or location on the x axis (because of the constant speed, location is proportional to time in this example). The phase of spiking shifts with location as the rat runs. If the dendritic oscillation frequency is driven by running speed, then the shift in phase will depend more on location than on time.

The model shown in the figure uses the following equation:

$$v(t) = \cos(2\pi ft) + \cos(2\pi(f + h)t)$$

where $v(t)$ is the sum of the soma oscillation with a constant frequency ($f = 6$) and of the dendritic oscillation with a constant frequency determined by running speed h ($f + h = 6.4$) which is constant in this example.

Grid Cell Model

The oscillatory interference model of theta phase precession essentially predicted the existence of grid cells. Right from the start, the model automatically generated multiple firing fields (O'Keefe and Recce, 1993) as shown in figure 2.11D and in figure A.7. These repeated firing fields were seen as a problem with the oscillatory interference model until the discovery of grid cells that suggested the repeating firing fields were a prediction rather than a problem. However, the original version of the oscillatory interference model for theta phase precession needed to be extended in a number of ways to become the oscillatory interference model of grid cells (Burgess et al., 2007). The model needed to be extended to two dimensions and needed to include the influence of running velocity on oscillation frequency.

The modulation of frequency by running velocity is essential to the grid cell model. Without this velocity input, frequencies will remain fixed in time. This will result in patterns of constructive interference that occur at regular time intervals but are not consistent relative to space. As shown in figure A.8, part A, if constructive interference occurs at regular intervals as a rat moves in some arbitrary trajectory through space (e.g., a triangle), the spiking occurs at regular temporal intervals, but this spiking occurs at arbitrary spatial locations.

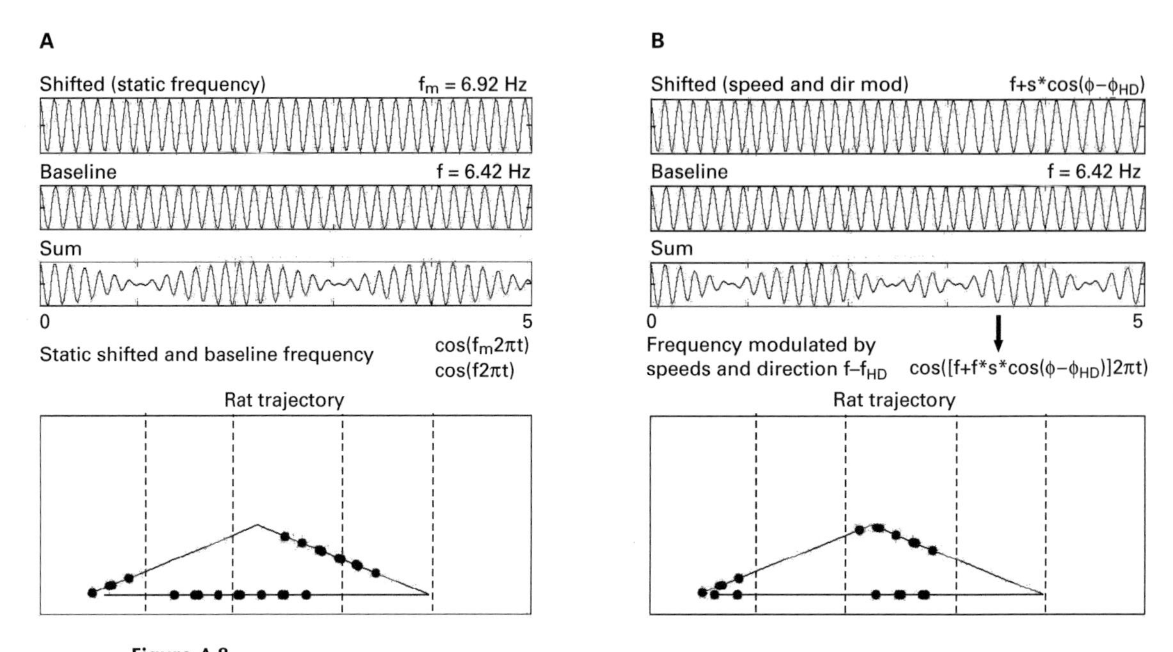

Figure A.8
(A) With two oscillations at different fixed frequencies, the constructive interference (Sum) occurs at regular time intervals. However, when plotted on the trajectory of a rat moving through space (Rat trajectory), the firing at regular time intervals does not show spatial regularity. (B) In contrast, when the top (shifted) oscillation frequency is regulated by speed and head direction (speed and dir mod), the constructive interference (Sum) occurs at irregular intervals in time, but the resulting spiking occurs in regular spatial bands on the Rat trajectory.

In contrast, part B of the figure shows how constructive interference can occur at regular spatial locations. This was obtained in a simple manner by driving oscillations with input representing the velocity of the animal moving through the environment. This velocity input can come in the form of cells coding head direction and running speed.

Neurons Coding Running Speed

Neurons coding running speed have been shown in the postsubiculum (Taube et al., 1990a; Sharp, 1996), and the mammillary nuclei (Sharp and Turner-Williams, 2005; Sharp et al., 2006). Figure 3.12 showed a neuron changing firing rate with running speed in the entorhinal cortex (Brandon et al., 2011c). Sensitivity to running speed has also been shown in recordings of axons running through the hippocampus (O'Keefe et al., 1998). Neurons coding running speed could shift the frequency of oscillations so that a shift in relative phase of oscillations accurately codes the location

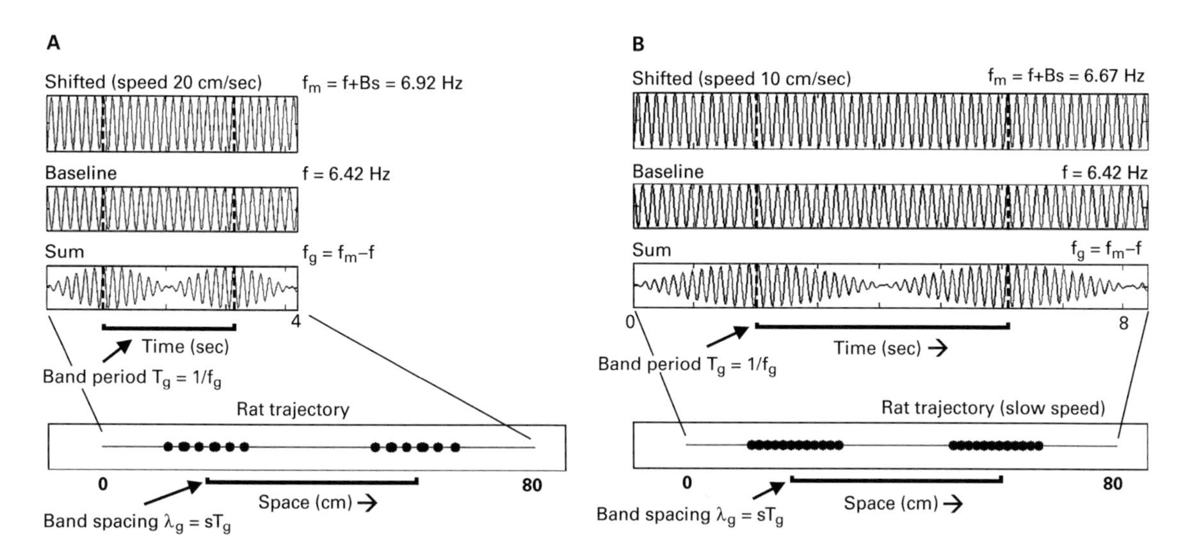

Figure A.9

(A) Oscillations plotted over 4 seconds in a rat running at 20 cm/sec, resulting in a shifted frequency of 6.91 Hz relative to a baseline of 6.42 Hz. Two peaks of constructive interference (Sum) occur with a period between them of 2 seconds. Plotted on the rat trajectory, the constructive interference occurs at a spacing of 40 cm. (B) Oscillations plotted over 8 seconds in a rat running at 10 cm/sec. The lower speed causes a smaller difference in the shifted frequency (6.67 Hz) relative to the baseline of 6.42 Hz. The sum of oscillations (Sum) shows two peaks of constructive interference with a period between them of 4 seconds (note that the full plot is 8 seconds long, not 4 seconds long). Because the frequency was scaled with running speed, the constructive interference occurs at the same spatial locations as in part A with a spacing of 40 cm. Adapted from Hasselmo et al. (2007).

of the animal. This could involve synaptic connections from neurons coding running speed to neurons undergoing oscillations. The synaptic input could drive changes in the oscillation frequency of individual neurons or the oscillation frequency of populations of neurons.

The left side of figure A.9 (A) shows oscillations plotted for a period of 4 seconds as the rat runs at 20 centimeters per second. One oscillation frequency (Shifted) is shifted in proportion to the running speed of the rat. (The running speed in centimeters per second must be scaled to frequency by a scaling factor B. Only a single input to a single neuron is being considered here, so a weight matrix is not modeled.) A speed of 20 centimeters per second raises the frequency to 6.92 Hz, and the baseline frequency is 6.42 Hz (Baseline). This causes two waves of constructive interference during a single 4-second interval (Sum). During the 4-second period of running at 20 centimeters per second, the rat runs a distance of 80 centimeters. The constructive interference generates two periods of spiking that appear as two firing fields spread along the rat trajectory. The intervals of constructive interference can be determined by finding the frequency of the envelope of interference (the increase and decrease in the sum), which is equal to the difference in frequency of the two oscillations $f_g = f_m - f$. This corresponds to the "beat" frequency of the interference. The difference in frequency 6.92 – 6.42 is approximately 0.5 Hz. The period of this envelope is $T_g = 1/f_g$, or about 2 seconds per cycle. The spacing between the peaks of constructive interference can be found by multiplying speed $s = 20$ centimeters per second times the envelope period of about $T_g = 2$ seconds/cycle, giving a spacing of about $sT_g = 20*2 = 40$ centimeters per cycle. Note that we call this running speed rather than velocity in this example because the direction of the rat does not change.

The right side of the figure (A.9B) shows oscillations plotted for 8 seconds as the rat runs at half the speed (10 centimeters per second). The slower speed of 10 centimeters per second raises the shifted frequency by a smaller amount to 6.67 Hz relative to the baseline frequency of 6.42 Hz. (Note that plots are twice as long because the rat takes twice as long to cover the same running distance). Because the frequency difference is smaller, the two oscillations shift in relative phase at a slower rate. The frequency difference 6.67 – 6.42 is 0.25 Hz. This means that the peaks of constructive interference in the sum come at longer intervals in time. The period of the envelope is $T_g = 1/f_g = 4$ seconds. However, the rat is running more slowly at 10 centimeters per second. The spacing between peaks of constructive interference can be found by multiplying speed $s = 10$ centimeters per second times the envelope period of 4 seconds, giving a spacing of about $\lambda_g = sT_g = 10*4 = 40$ centimeters per cycle. Therefore, the firing fields appear in the same locations on the rat trajectory, even though they occur at different points in time. Note the different timescales in A and B. Part A plots from 0 to 4 seconds, and part B plots from 0 to 8 seconds. Both figures plot over the same distance of 80 centimeters.

In this example, the shift in frequency proportional to speed requires the use of a scaling factor B to translate the speed in centimeters per second into the frequency in cycles per second. Thus, $f_m = f + Bs$. In this example, B is 0.025 cycles per centimeter. The frequency of the envelope of interference will be $f_g = f + Bs - f = Bs$, and the period T_g between peaks of constructive interference will be 1/Bs. The interference bands will occur at a spacing $\lambda_g = sT_g = s/Bs = 1/B$. In this example, 1/B is 40 centimeters per cycle. The spacing between the interference bands will decrease as the scaling factor B increases. In the figure, the running speed s in each section is constant over time, so the following equation describes the sum of oscillations relative to running speed: $v(t) = \cos(2\pi ft) + \cos(2\pi(f + Bs)t)$. The persistent spiking version of this model simply adds a Heaviside input output function with a threshold near the peak so that the oscillation produces one or more spikes on each cycle. If the running speed changes over the time of the simulation, this will change the frequency over time. In this case, the simulation must update the oscillation phase according to the integral of the running speed

$$v(t) = \cos(2\pi ft) + \cos(2\pi(ft + B\int_0^t s(\tau)d\tau)).$$

Neurons Coding Head Direction

Even if the rat runs at the same speed, constructive interference will occur at irregular locations when the rat runs in different directions in the environment as shown on the left side of figure A.10 (part A). However, if the change in frequency is regulated by the cosine of head direction angle relative to a specific preferred head direction, this will cause firing to occur at regular spatial intervals in one dimension. Extensive

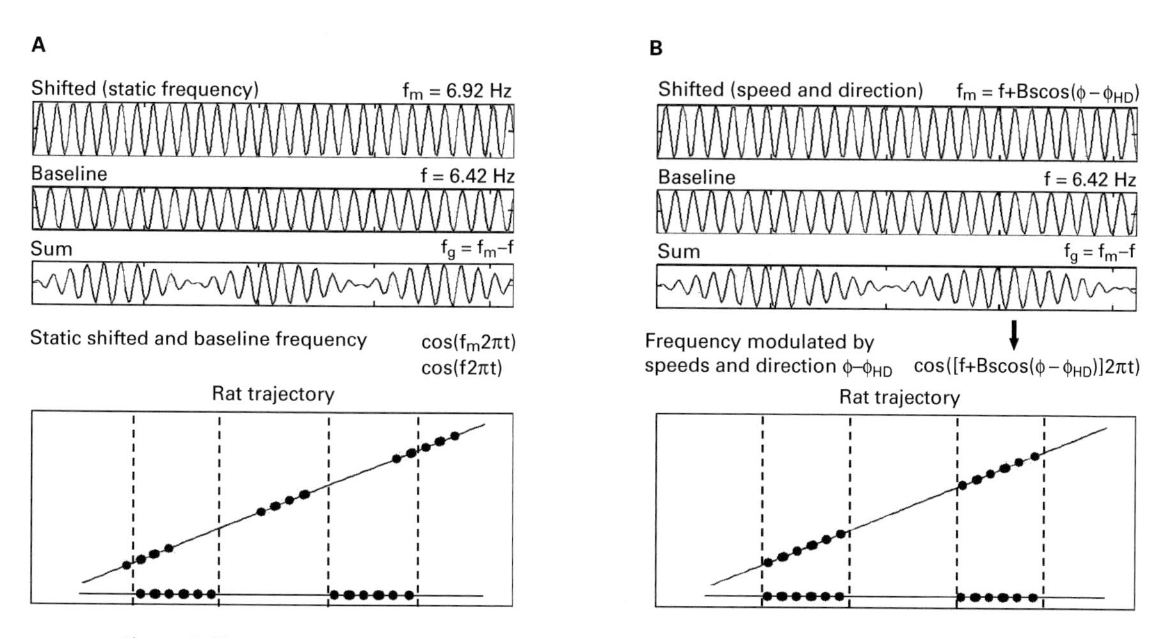

Figure A.10

(A) Without modulation by head direction, a fixed difference in oscillation frequencies causes constructive interference at inconsistent spatial locations when the rat trajectory runs in different directions through the environment. (B) The constructive interference occurs at regular intervals in one spatial dimension when the oscillation frequency is regulated by a cosine-tuned head direction cell. In this example, the head direction cell has a preferred head direction of $\phi_{HD} = 0$ and responds based on the cosine of the angle of rat head direction relative to this preferred head direction. The constructive interference on different trajectories occurs at intervals that line up in bands in one spatial dimension (west to east) within the environment. Adapted from Hasselmo et al. (2007).

experimental data describe cells that respond to the head direction of the rat that are found in areas including the postsubiculum (dorsal presubiculum), the anterior thalamus (Taube et al., 1990b; Taube and Burton, 1995; Sharp, 1996; Taube et al., 1996; Sharp et al., 2001; Taube and Bassett, 2003), and in deep layers of the entorhinal cortex (Sargolini et al., 2006; Boccara et al., 2010; Brandon et al., 2011b). These neurons show selective firing when the rat is facing in a specific range of directions. In the grid cell model, head direction cell firing is modeled with a cosine function of head direction angle ϕ relative to the preferred angle of a given head direction cell ϕ_{HD}. Thus, the output of a cosine tuned head direction cell as a function of the head direction angle is $h(\phi) = \cos(\phi - \phi_{HD})$. Do not confuse this cosine function with the cosine function of time used to model oscillations. This cosine function is not an oscillation in time but a change in response based on spatial head direction angle. The first examples below will assume constant head direction angle on each individual trajectory.

The right side of figure A.10 (part B) shows modulation of frequency by a head direction cell with orientation preference of zero (to the east). When the rat runs directly in the zero direction (east), the head direction cell responds at maximal strength (cos(0) = 1). The head direction cell drives the oscillation frequency to a high level and the oscillation transitions quickly between bands of constructive interference. When the rat runs at an angle of 30 degrees ($\pi/6$ radians) to the northeast, the head direction cell responds at a lower level of cos(30 degrees) = 0.866. This causes a smaller difference in frequency that causes the periods of constructive interference to occur at longer intervals in time (1.15) that give the same spacing in one dimension along the preferred direction of the head direction cell (0 degrees). Overall, the modulation by cosine results in the periods of constructive interference lining up in spatial bands on the rat trajectory that are perpendicular to the orientation preference of the head direction cell regulating frequency. In this example, because the preferred direction of the head direction cell is zero, the bands of maximal constructive interference are vertical (90 degrees). A similar example is shown in figure A.11 with more directions shown. The equation for the sum of oscillations with constant head direction and speed is

$$v(t) = \cos(2\pi ft) + \cos(2\pi(f + Bs\cos(\phi - \phi_{HD}))t)$$

The examples above used a fixed, stable head direction on each individual trajectory in a different direction. However, when a rat forages in an open field environment, this will cause continuous changes in speed and head direction, so that functions of time must be used to express both speed $s(t)$ and head direction $\phi(t)$. In this case, the phase must continuously integrate the speed and head direction as follows:

$$v(t) = \cos(2\pi ft) + \cos(2\pi(ft + B\int_0^t s(\tau)\cos(\phi(\tau) - \phi_{HD})d\tau)).$$

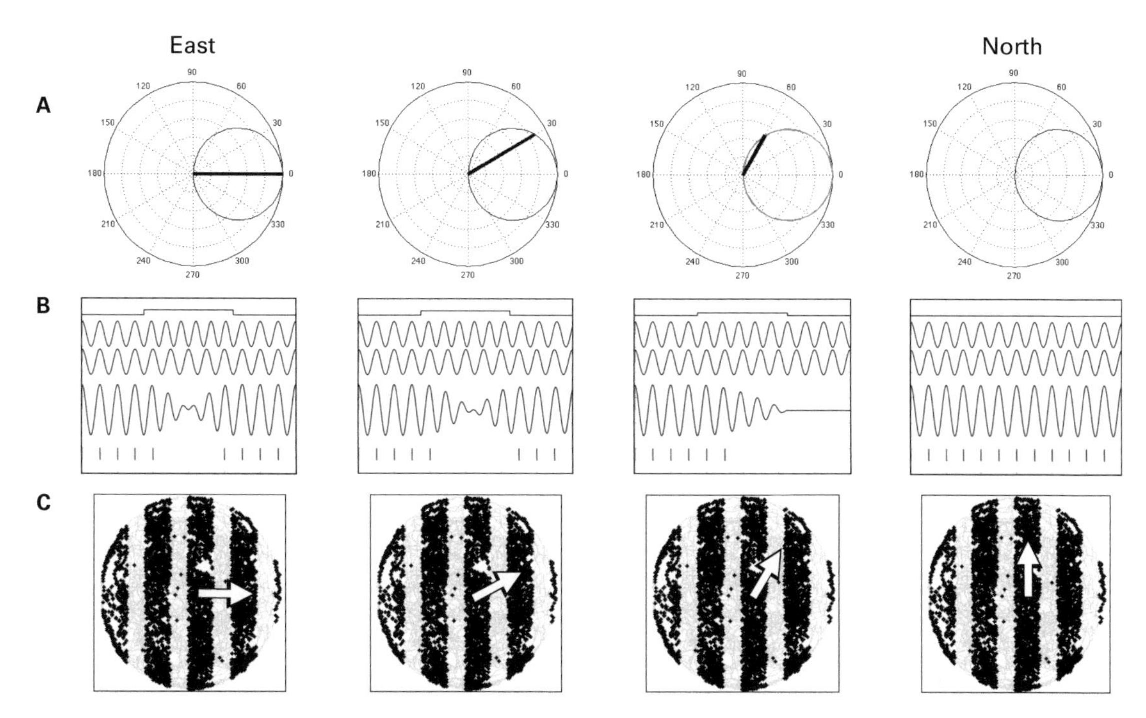

Figure A.11

(A) The top row shows how different directions result in different firing rates for a single head direction cell with preference angle of zero. The small circle is a cosine tuned head direction cell plotted in polar coordinates. The length of each thick line represents the firing rate of the head direction cell for that angle of movement relative to zero. (B) The middle row shows how directions further from the preferred direction result in a smaller frequency difference and slower periods between peaks of constructive interference in the summed output. (C) The bottom row shows the transition between constructive interference bands for different directions. Left: For movement in the preferred direction (east) the frequency difference is large and causes short periods between peaks of constructive interference. Right: For movement to the north, the head direction cell does not fire at all (cos(90 degrees) = 0) and the movement causes no change in the level of constructive interference.

Equation for Grid Cell Model

The modulation of frequency by speed and head direction will generate patterns of constructive interference that fall in bands (figure A.11) that are oriented perpendicular to the head direction preference of the head direction cell providing input. How do these bands of constructive interference combine to generate a grid cell firing pattern? The interaction of two or more bands of constructive interference at intervals that are multiples of 60 degrees will generate the characteristic hexagonal firing pattern of grid cells as shown in figure A.12. The correct combination of orientations may arise from self-organization of the connections to the medial entorhinal cortex from head direction cells in postsubiculum or deep layers of entorhinal cortex (Burgess et al., 2007). The generation of the grid cell firing pattern from the interaction of

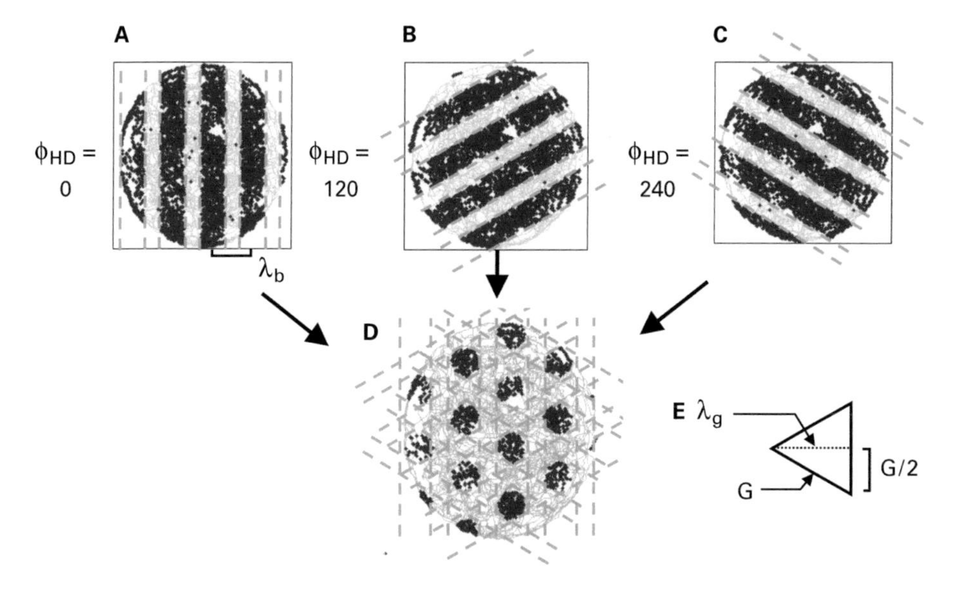

Figure A.12
Combination of bands of constructive interference from different cells or populations of cells acting as velocity controlled oscillators results in the grid cell firing pattern. Each head direction cell causes interference with consistent spatial wavelength in bands running perpendicular to the selective direction preference of the individual head direction cell. (A–C) Interference bands due to modulation of frequency by speed-modulated head direction cells with specific direction preferences angles at zero (A), 120 degrees (B), and 240 degrees (C). (D) Summation of interference bands due to different oscillations causes the simulated grid cell to cross firing threshold in a hexagonal pattern matching grid cell firing fields. (E) Equilateral triangle illustrating the relationship between the spatial wavelength λ_g of band interference shown in previous figures (height of triangle) and the grid cell spacing G (length of one side). Adapted from Hasselmo et al. (2007).

different bands of constructive interference is shown in the figure. Note that the interference bands in the figures above have a spacing λ_g that relates to the grid cell spacing according to the equilateral triangle shown in the figure. Thus, the grid cell firing field spacing G = λ_g/cos(30). Both the band spacing λ_g and the spacing G of grid cell firing fields will decrease as B increases.

In the equation describing the oscillatory interference model of grid cells (Burgess et al., 2007), the interference patterns resulting from the summed oscillations for each head direction are then summed together. Thus, the final equation sums across the three different head directions as shown by the summation sign Σ with index *i*. This sum then goes through a Heaviside function to produce the spiking pattern of a grid cell shown in the figure:

$$g(t) = [\sum_{i=1}^{3} \{\cos(2\pi ft) + \cos(2\pi(ft + B\int_{0}^{t} s(\tau)\cos(\phi(\tau) - {}_{i}\phi_{HD})d\tau))\}]_H .$$

This single equation sums together the pairs of summed oscillations for the three head direction cells (with preferred angles 0, 120, and 240) to generate the firing pattern of the single grid cell shown in figure A.12 above. (The persistent spiking version of the model can be obtained by moving the Heaviside function inside the summation sign.)

Different Frequency Response Causes Different Field Spacing

In the model, the spacing between firing fields of a grid cell will differ if the magnitude of the change in frequency differs for a given speed and head direction. This magnitude of the change in frequency is determined by the scaling factor B that determines how much speed and head direction influence frequency. The scaling factor B was included in the equations above. This multiplier can be seen as the slope of the relationship between frequency and running velocity. Different values of B will generate different spacing. As described in chapter 3, intracellular recording from stellate cells in medial entorhinal cortex shows different frequencies of membrane potential oscillations and resonance in response to depolarizing current injection, with higher frequencies in dorsal neurons and lower frequencies in ventral neurons (Giocomo et al., 2007; Giocomo and Hasselmo, 2008a, 2008b, 2009). These differences in frequency with depolarization could correspond to different values of B in the model and could underlie the difference in spacing between grid cell firing fields shown in unit recording from awake, behaving animals (Hafting et al., 2005; Moser and Moser, 2008).

Figures 3.12 and 3.13 in chapter 3 show the difference in spacing for two cells with different B generating a different frequency response to running at the same speed for the same duration. Both examples start with the animal motionless in a single firing field of the grid cell and with the two oscillations in phase with each other. The animal then starts running at a constant speed for an interval of 2 seconds and then stops. In figure 3.12, the cell has a large value of B and shows a large frequency change with running speed. This causes the phase to rapidly shift by 360 degrees (2π radians), shifting from constructive to destructive to constructive interference. This results in much narrower spacing between the firing fields of the cell. In figure 3.13, the rat runs at the same speed, but the cell has a smaller value of B and the speed causes a small shift in frequency. This results in a slower shift in phase in the figure by 180 degrees (π radians), shifting from constructive interference and spiking to destructive interference and a loss of spiking. This results in wide spacing between the firing fields of the cell. This requires linear frequency shifts with velocity. Depolarization causes linear shifts in cell resonance but not single cell oscillations (Yoshida et al., 2010).

Theoretical Arc Length and Time Cells

The representation of grid cells is effective for linking specific spatial locations with specific items or events. However, episodic memories that involve trajectories that overlap in the same spatial location require other mechanisms to disambiguate the overlapping segments of the trajectory. One possible mechanism described in chapters 3 and 4 is to update the oscillation frequency based on different types of input. The property of the simulations will differ depending on the nature of the input influencing oscillation frequency. If the shifted oscillation frequency and phase are updated by both speed and head direction $s(t)\cos(\phi - \phi_{HD})$ (i.e., by velocity), then oscillatory interference will generate grid cells that code two-dimensional location. If the shifted oscillation frequency and phase are updated by speed alone $s(t)$, then oscillatory interference will generate cells that code the arc length of the trajectory as described in chapter 4, responding at different arc lengths along the trajectory. If the shifted oscillation frequency and phase has a constant frequency difference from baseline (not responding to speed or direction), then the oscillatory interference will generate cells that code time intervals (time cells) as shown in figure 3.11. This is a very effective mechanism for coding time of events that was used in the simulations shown in figure 1.1 and in figure 7.7. The timing could be initiated by stimuli associated with a specific item or event that initiates the shift in frequency that then persists to allow time cells to code the interval from the initial appearance of that item or event. Another possible mechanism described in chapter 7 is the reset of oscillation phase at specific locations.

Associative Memory Function

Associative memory representations are used in a number of ways in the model of episodic memory. For example, associative memory mechanisms are used to associate individual positions along the spatiotemporal trajectory with individual items or events. In addition, individual positions on the trajectory are associated with the actions performed at those positions.

Most associative memory models utilize the mathematical notation of linear algebra to describe associative networks. These models represent the activity of populations of neurons with one-dimensional arrays (vectors) containing elements describing the activity level of each neuron (see figure A.1). The pattern of synaptic connectivity between the various neurons in the population, and between one population and another, is represented with two-dimensional matrices, in which each element represents the strength of one synapse. Most network models utilize two major descriptive equations: (1) a learning rule and (2) an activation rule.

Learning Rule

Experimental data from the hippocampus and other cortical regions indicate that the strength of synapses can be altered by the pattern of presynaptic and postsynaptic activity. Changes in the synaptic potential's size induced by patterned input were initially referred to as long-term potentiation (LTP; Bliss and Lomo, 1973; Bliss and Collingridge, 1993). Changes in synaptic potential size induced by direct manipulation of spike times are referred to as spike-timing-dependent synaptic plasticity (Levy and Steward, 1983; Bi and Poo, 1998). Experimental data performed on synapses arising from region CA3 cells and terminating in CA3 or in CA1 show that changes in their synaptic strength depend upon the conjunction between presynaptic and postsynaptic activity (Wigstrom et al., 1986; Kelso et al., 1986; Zalutsky and Nicoll, 1990). Data also indicate that the postsynaptic contribution does not have to depend on postsynaptic spiking output at the soma but can also depend on postsynaptic depolarization that causes dendritic spikes (Golding et al., 2002).

The dependence of synaptic modification on correlated presynaptic and postsynaptic activity is referred to as the Hebb rule. Mathematically, Hebbian synaptic modification can be written as a change in the synaptic strength matrix W based on the outer product combining presynaptic output and postsynaptic depolarization as follows: $\Delta W_{ij}(t) = \mu a_i(t) g(a_j(t))$. This equation shows how the individual synapse W_{ij} from neuron j to neuron i changes in proportion to output based on the activity (a_j) of presynaptic neuron j and the activity (a_i) of postsynaptic neuron i. The parameter μ determines the speed of learning. Thus, the learning rule describes how synaptic modification in response to different input patterns determines the pattern of synaptic connectivity. The change in strength of a synapse is written as ΔW_{ij}. The static value of a synapse at a point in time is W_{ij}. The amount of change can be regulated by a synaptic plasticity constant μ, which indicates the amount of change in synaptic strength per unit time for a given level of correlated activity. Figure A.13 illustrates the application of this learning rule to the matrix of connections in a neural model.

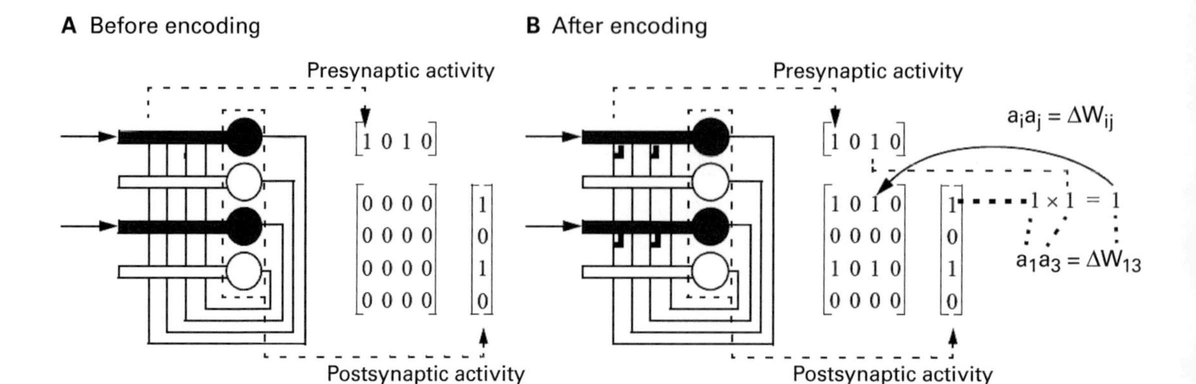

Figure A.13

Mathematical representations used for describing the encoding of a new pattern. (A) Before encoding: In abstract associative memory networks, the initial connectivity is initially assumed to be zero (as shown by the matrix of zeros). Afferent input activates two neurons in the network, resulting in the patterns of pre- and postsynaptic activity shown as vectors. (B) After encoding: When Hebbian synaptic modification is applied within the network, connections are strengthened at synapses where both pre- and postsynaptic activity is present as shown by the new matrix. The change in strength is described by the equation $\Delta W_{ij} = a_i a_j$, where the row of each synapse is coded by the index i and the column is coded by the index j. On the right, an example is shown for the synaptic connection from the axon in column $j = 3$ to the neuron in row $i = 1$.

Activation Rule

The activation rule describes how the activity of neurons in the network at one point in time depends upon the activity at previous points in time. Thus, the activation rule shows how the presynaptic activity (a_j) crossing synapses of strength W_{ij} influences the postsynaptic activity (a_i). The synaptic input updating the membrane potential $a_i(t+1)$of postsynaptic neuron i can be modeled by multiplying the output $g(a_j(t))$ of a presynaptic neuron j by the synaptic weight between those neurons W_{ij} as follows: $a_i(t+1) = W_{ij}g(a_j(t))$. Usually, a given postsynaptic neuron i will receive input that sums up from a large number of presynaptic neurons j. This can be written as a sum over the synaptic input from n different neurons with index j as follows:

$$a_i(t+1) = \sum_{j=1}^{n} W_{ij}g(a_j(t)).$$

Figure A.14 illustrates the use of this activation rule in a neural model.

Linear Algebra Notation

The notation of linear algebra provides simpler ways of writing these operations without using the indices or the summation sign. In the simpler notation, the learning

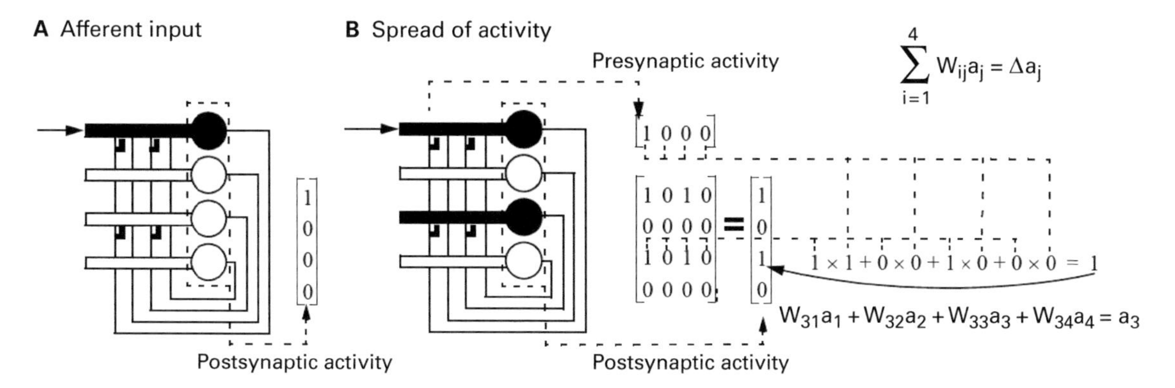

Figure A.14
Spread of activation in an associative memory. This network contains the synaptic connectivity matrix induced by the encoding in the preceding figure. (A) Afferent input (arrow on left) initially induces activity in one of the four neurons (black), represented as the vector of postsynaptic activity which is transmitted along axons (lines) to become the presynaptic activity vector in B. (B) The presynaptic activity in the axon (column 1) induces synaptic transmission. Mathematically, this synaptic transmission is represented by matrix multiplication

$$\begin{bmatrix} 1 & 0 & 1 & 0 \\ 0 & 0 & 0 & 0 \\ 1 & 0 & 1 & 0 \\ 0 & 0 & 0 & 0 \end{bmatrix} \begin{bmatrix} 1 \\ 0 \\ 0 \\ 0 \end{bmatrix} = \begin{bmatrix} 1 \\ 0 \\ 1 \\ 0 \end{bmatrix},$$

in which the vector representing the pattern of presynaptic activity is multiplied by each row of the matrix. The product of each element of presynaptic activity with each element of the matrix is summed up across each row and used to update the postsynaptic activity for the postsynaptic neuron represented by that row of the matrix. As an example, the figure illustrates the computation of postsynaptic activity just for neuron 3, showing summation of the product of each element of presynaptic activity with each element of the weight matrix for row 3. Summation across various elements of a row is written with the symbol Σ.

rule forming a matrix W from two column vectors p and g can be represented by the notation: $W = \vec{p}\vec{g}^T$, where the "T" indicates the transpose operation that changes the column vector into a row vector. In this simpler notation, the activation rule equation $\vec{p}(t) = W\vec{g}(t)$ indicates the spread of activity from population g to population p. The order of row and column vectors implicitly indicates the operation being computed. If a column vector precedes a row vector, the multiplication produces a matrix, and the operation is known as an outer product. If the row vector precedes the column vector, the multiplication produces a single number (a scalar) and the operation is known as an inner product or dot product. If a matrix precedes a vector, this represents matrix multiplication that computes the inner product of each row of the matrix with the following vector.

The example of a learning rule below shows how the synaptic matrix W encodes an association between two vectors by computing the outer product of two vectors p and g in the equation $W = \vec{p}\vec{g}^T$:

$$\begin{bmatrix} p_1g_1 & p_1g_2 & p_1g_3 & p_1g_4 \\ p_2g_1 & p_2g_2 & p_2g_3 & p_2g_4 \\ p_3g_1 & p_3g_2 & p_3g_3 & p_3g_4 \\ p_4g_1 & p_4g_2 & p_4g_3 & p_4g_4 \end{bmatrix} = \begin{bmatrix} p_1 \\ p_2 \\ p_3 \\ p_4 \end{bmatrix} [g_1 \quad g_2 \quad g_3 \quad g_4] \qquad \begin{bmatrix} 0 & 0 & 0 & 0 \\ 0 & 1 & 0 & 1 \\ 0 & 1 & 0 & 1 \\ 0 & 0 & 0 & 0 \end{bmatrix} = \begin{bmatrix} 0 \\ 1 \\ 1 \\ 0 \end{bmatrix} [0 \quad 1 \quad 0 \quad 1].$$

As an example, the vectors [0 1 1 0] and [0 1 0 1] result in the matrix shown above on the right.

The example of an activation rule below uses matrix multiplication to retrieve the association stored on the weights W by the spread of activity from the full population of presynaptic neurons across the pattern of synaptic connections to the full population of postsynaptic neurons. The example below shows the computations for the equation $\vec{p} = W_{cue}\vec{g}$:

$$\begin{bmatrix} p_1g_1g_1 + p_1g_2g_2 + p_1g_3g_3 + p_1g_4g_4 \\ p_2g_1g_1 + p_2g_2g_2 + p_2g_3g_3 + p_2g_4g_4 \\ p_3g_1g_1 + p_3g_2g_2 + p_3g_3g_3 + p_3g_4g_4 \\ p_4g_1g_1 + p_4g_2g_2 + p_4g_3g_3 + p_4g_4g_4 \end{bmatrix} = \begin{bmatrix} p_1g_1 & p_1g_2 & p_1g_3 & p_1g_4 \\ p_2g_1 & p_2g_2 & p_2g_3 & p_2g_4 \\ p_3g_1 & p_3g_2 & p_3g_3 & p_3g_4 \\ p_4g_1 & p_4g_2 & p_4g_3 & p_4g_4 \end{bmatrix} \begin{bmatrix} g_1 \\ g_2 \\ g_3 \\ g_4 \end{bmatrix}$$

$$\begin{bmatrix} 0 \\ 1 \\ 1 \\ 0 \end{bmatrix} = \begin{bmatrix} 0 & 0 & 0 & 0 \\ 0 & 1 & 0 & 1 \\ 0 & 1 & 0 & 1 \\ 0 & 0 & 0 & 0 \end{bmatrix} \begin{bmatrix} 0 \\ 0 \\ 0 \\ 1 \end{bmatrix}.$$

In the example shown underneath, the cue input vector containing one element of g can retrieve the full pattern of p.

Each row of the postsynaptic vector represents the activity of a single postsynaptic neuron influenced by the synaptic connections from the full population of presynaptic neurons. Notice that if the presynaptic activity matches elements of the previously associated presynaptic activity, then postsynaptic activity resembling the previously associated postsynaptic activity will be induced.

Each row of postsynaptic activity can be referred to as an inner product or dot product of the presynaptic activity vector multiplied by the weights to the single postsynaptic neuron. The inner product or dot product can also be used to compute a value indicating the amount of similarity between two vectors. For example, the inner product c of vectors a and b is

$$c = \vec{a}^T\vec{b} \qquad [a_1b_1 + a_2b_2 + a_3b_3 + a_4b_4] = [a_1 \quad a_2 \quad a_3 \quad a_4]\begin{bmatrix} b_1 \\ b_2 \\ b_3 \\ b_4 \end{bmatrix}.$$

It is important to notice for different models in different papers whether vectors are initially defined as row vectors or column vectors and to notice the order of the elements. An outer product will result from multiplication of a column vector by a row vector, and an inner product will result from multiplication of a row vector followed by a column vector.

Implementation of Episodic Memory Model

The previous sections describe the modeling of grid cell firing patterns, and the formation of associations between patterns of neural activity. This section will describe the use of grid cells and associative memory function in the full implementation of the episodic memory model (Hasselmo, 2009) that generated the figures discussed in chapter 4 and figures 1.1 and 1.2.

In the episodic memory model (Hasselmo, 2009), the vector $\vec{h}(t)$ is mathematically equivalent to the representation of speed and head direction angle used in the equations of the grid cell model presented above. The use of the vector $\vec{h}(t)$ is based on a standard formula for the inner product (dot product): $\|\vec{u}\|\|\vec{v}\|\cos(\phi - \phi_{HD}) = \vec{u}^T\vec{v}$. Rearranging this equation gives $\|\vec{v}\|\cos(\phi - \phi_{HD}) = \vec{u}^T\vec{v}/\|\vec{u}\|$, where the left side matches the way that speed and head direction were shown in previous equations, and the right side shows how to obtain the vector $\vec{h}(t)$. This equation shows that the speed ($\|\vec{v}(t)\| = s(t)$) and the angle ϕ between the velocity vector and a preferred head direction ϕ_{HD} is equivalent to the right side of the equation which shows the inner product of the velocity vector v with a unit vector $\vec{u}/\|\vec{u}\|$ defined by each head direction cell preference angle $\vec{u}/\|\vec{u}\| = [\sin\phi_{HD} \ \cos\phi_{HD}]$. The activity of a group of speed-modulated head direction cells $\vec{h}(t)$ with different preference angles can be obtained by the matrix multiplication $\vec{h}(t) = H\vec{v}(t)$. This computes the inner product of the velocity vector $\vec{v}(t)$ with each row of the head direction transformation matrix:

$$H = \begin{bmatrix} \sin\phi_1 & \cos\phi_1 \\ \sin\phi_2 & \cos\phi_2 \\ \sin\phi_3 & \cos\phi_3 \end{bmatrix}$$

In this matrix, ϕ_i represents the selectivity of individual head direction cells with index i. The multiplication maps velocity into each element of the speed modulated head direction vector $h_i(t) = s(t)\cos(\phi(t) - \phi_i)$ which code velocity in the form of speed in a given direction. As the velocity vector changes over time $\vec{v}(t)$, this changes the speed modulated head direction vector over time $\vec{h}(t)$.

Integration for Spatial Location

The integration of the speed modulated head direction over time corresponds to position transformed by the matrix H:

$$\int_{\tau=0}^{t} \vec{h}(\tau)d\tau = H \int_{\tau=0}^{t} \vec{v}(\tau)d\tau = H(\vec{x}(t) - \vec{x}_0).$$

In the figures in chapter 4, for the purpose of demonstrating how the entorhinal grid cells represent location, the integral of speed modulated head direction was transformed back into Cartesian spatial coordinates (Hasselmo, 2009). This transforms from the dimensions of the head direction vector at angles of 120 degrees into the 90-degree angle between the x and y dimensions of Cartesian coordinates. This is done by multiplying the integral of head direction activity by the inverse of the head direction transformation matrix:

$$H^{-1}\int_{0}^{t} \vec{h}(\tau)d\tau = \vec{x}(t) - \vec{x}_0.$$

The location coded by grid cell firing phase can be obtained from two components of the integrated head direction separated by 120 degrees, allowing use of a two-by-two inverse matrix (Strang, 1988) transforming the two dimensional phase code back into two dimensional Cartesian coordinates as follows:

$$H^{-1} = \begin{bmatrix} \cos\phi_2 & -\sin\phi_2 \\ -\cos\phi_1 & \sin\phi_1 \end{bmatrix} / (\cos\phi_2 \sin\phi_1 - \cos\phi_1 \sin\phi_2).$$

Grid Cell Model

How does the model perform integration of the velocity signal? In the model, the integration necessary to keep track of spatial location is performed by shifts in the relative phase of oscillations in a simulation of entorhinal grid cells. Baseline oscillations with fixed frequency can be described by the instantaneous phase at each time point $\phi = 2\pi ft$ (this assumes initial phase zero at time zero). As described above, the relative phase used in this model is the difference between the instantaneous phase of one oscillation and the instantaneous phase of a baseline oscillation at that same time point.

The integration of velocity in the model arises due to input from head direction cells causing a shift in frequency in oscillations in entorhinal cortex. There are excitatory projections to entorhinal cortex stellate cells from the postsubiculum (Caballero-Bleda and Witter, 1993) and deep layers of entorhinal cortex (Sargolini et al., 2006), both of which contain head direction cells. Thus, active head direction cells could cause changes in the spiking frequency of single entorhinal stellate cells or in the oscillation frequency of entorhinal networks.

If we consider a pure velocity input, then the instantaneous phase of shifted oscillations depends upon the baseline frequency f and the shift in frequency due to velocity as follows:

$$\varphi_s(t) = 2\pi(ft + \int_{\tau=0}^{t} \vec{v}(\tau)d\tau).$$

The relative difference in instantaneous phase between the shifted oscillation (with index s) and an oscillation with the baseline frequency (index b) will be

$$\varphi_s(t) - \varphi_b(t) = 2\pi(\int_{\tau=0}^{t} \vec{v}(\tau)d\tau) = 2\pi(\vec{x}(t) - \vec{x}_0).$$

The oscillatory interference model (Burgess et al., 2007) in this formulation (Hasselmo, 2008c, 2009) uses the speed modulated head direction $\vec{h}(t)$ to regulate different oscillation frequencies as summarized by the following equation:

$$_{jkm}g(t) = [\prod_{i=1}^{3}\{\cos(2\pi ft) + \cos((2\pi(ft + {}_jB\int_0^t \vec{h}(\tau)_i d\tau) + {}_{km}\vec{\varphi}(0))\}]_H$$

where $_{jkm}g(t)$ represents the firing over time of an array of grid cells (see figure A.15) with different groups of cells with different spacing j, and within each group different cells with spatial phases in different directions k and m. Π_i represents the product of the different oscillations receiving input from different components of the head direction vector with index i, and spiking is computed by the Heaviside step function output (the model has output 1 for any value above a threshold). The intrinsic oscillations change in frequency in proportion to a constant $_jB$, where j represents different scales of frequency to velocity at different dorsal to ventral positions in medial entorhinal cortex. The oscillations have a two-dimensional range of different spatial phases $_{km}\varphi(0)$ giving an array of spatial phases covering the environment along two dimensions of spatial phase offset with indices 'k' and 'm'.

In simulations presented here, this grid cell model equation was replaced with an equation that computes the relative phase difference more rapidly and generates denser grid cell firing and more effective place cell formation. Rather than computing the full sum of multiple oscillations, this equation directly computes the relative phase difference caused by head direction input as follows:

$$_{jkm}g(t) = [\prod_{i}\cos(2\pi\, {}_jB\int_0^t h(\tau)_i d\tau + {}_{km}\vec{\varphi}(0))]_H$$

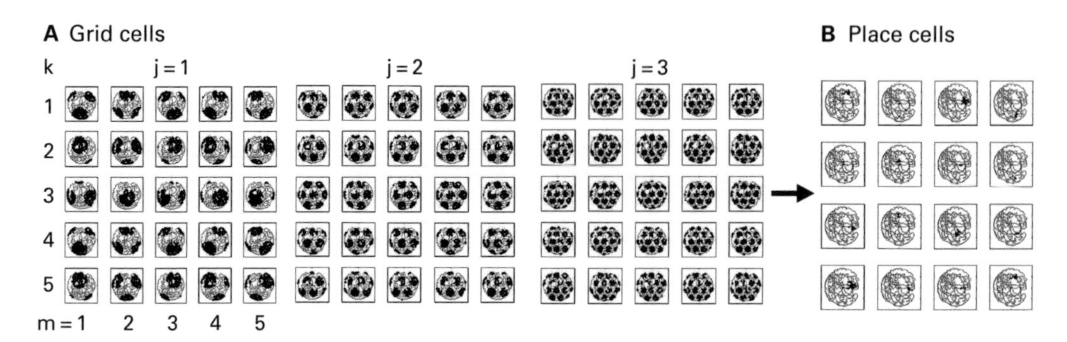

Figure A.15
(A) The model used an array of 75 grid cells with three different spatial frequencies j, and a 5-by-5 array of spatial phases indexed by 'k' and 'm.' Subsets of grid cells were combined to generate the activity of hippocampal place cells. (B) Examples of 16 out of 400 place cells generated in the simulation.

Figure A.15 shows simulations from the model of different grid cells with differences in size and spacing of fields (due to different values of B_j) and different initial phases 'k' and 'm.' These differences in spacing and phase of grid cells are essential to the coding of spatial location and generation of place cell responses in the model. The grid cell spacings shown here prove effective for coding location. Even though the grid cells have repeating fields, the combination of input from different grid cells can result in place cells that are selective over large distances. This mechanism can function on a larger scale as well as some cells receiving output from these grid cells will respond on the basis of the least common integer multiple of the spacing of a single grid cell (Fiete et al., 2008; Gorchetchnikov and Grossberg, 2007), so that spacings of 30, 40, and 50 centimeters can drive selective coding at a scale of 450 centimeters. The scale of spatial specificity can be further expanded by including grid cells with larger spacing intervals up to 10 meters as shown in more ventral regions of entorhinal cortex (Brun et al., 2008).

The difference in size and spacing of grid cell firing fields in the model depends upon differences in the effect of depolarization on intrinsic frequency that have been shown along the dorsal to ventral axis of medial entorhinal cortex (Giocomo et al., 2007; Hasselmo et al., 2007; Giocomo and Hasselmo, 2008b). The original Burgess model proposes that grid cells arise from interference, but it is open to different mechanisms for the oscillations undergoing interference. The equation probably does not correspond to the interaction of membrane potential oscillations within a single neuron (Giocomo et al., 2007; Hasselmo et al., 2007) but could represent the interaction of phasic synaptic input from neurons firing rhythmically due to persistent spiking (Hasselmo, 2008) or due to oscillatory network dynamics (Zilli and Hasselmo,

2010). The different size and spacing of grid cell firing fields proves essential to coding of location in this model (Hasselmo, 2009). The coding of different scales by neurons with different resonance properties could prove to be a more general phenomenon that appears in other areas such as medial prefrontal cortex and piriform cortex.

Place Cell Model

The phase of grid cell oscillations effectively represents location, but how can this be accessed by other neural structures? The readout of spatial location can be provided by hippocampal place cells. Thus, the next stage of the model involves grid cell effects on place cells. Examples of place cells that can be generated by the model are shown in figure A.15.

The influence of grid cells on place cells has been modeled in different ways (Fuhs and Touretzky, 2006; McNaughton et al., 2006; Rolls et al., 2006; Solstad et al., 2006; Gorchetchnikov and Grossberg, 2007; Savelli and Knierim, 2010), but research has not converged on a single model. In general, the place cell representation $p(t)$ arising from the grid cell representation $g(t)$ in entorhinal cortex would depend upon the synaptic connectivity W_{PG} from grid cells to place cells.

In the model, place cells are created by selection of random subsets of three grid cells' inputs and computation of the overlap in firing for these three grid cells. The place cell is assumed to fire anywhere that these three grid cells fire simultaneously. This generates a spatial firing pattern for each place cell that is evaluated by taking the standard deviation of spiking location in the x and y dimension. If the standard deviation of spiking location is smaller than a previously set parameter in both dimensions, then this place cell is selected for inclusion in trajectory encoding. The input synapses from the randomly selected subset of three grid cells to this place cell are then strengthened according to the outer product equation:

$$W_{PG} = \sum \vec{p}(t)\vec{g}(t)^T$$

where $\vec{g}(t)^T$ is a vector of presynaptic grid cell activity including the three currently selected grid cells, and $\vec{p}(t)$ is a vector of place cell activity with only the currently selected place cell active. This ensures that the same place cells are reliably activated in the same location dependent on the pattern of grid cell spiking induced by the phase of grid cells. Examples of place cell firing generated by the model are shown in figure A.15B.

During both encoding and retrieval of an episode, the place cell vector $\vec{p}(t)$ results from the pattern of activity in the grid cell vector $\vec{g}(t)$ dependent upon multiplication by the previously modified synaptic connectivity W_{PG} between these regions as follows:

$$\vec{p}(t) = W_{PG}\vec{g}(t).$$

Encoding of Trajectory

How does the model encode a trajectory? One mechanism depends upon sequentially encoding associations between place cells and velocity and later sequentially retrieving these associations. Alternate mechanisms involve encoding associations between arc length or time cells and the associated velocity.

The third stage of the model plays the primary role in encoding of a new trajectory. This encoding is performed by modification of synaptic connections between active hippocampal place cells and active head direction cells. This synaptic modification allows formation of associations between individual states (spatial locations) and the actions (movements) performed in those states.

During initial encoding of a trajectory, the activity of all cells is determined by the actual velocity of the animal during movement. The sensory input about actual speed and head direction drives the head direction cell activity vector $h(t)$. The head direction input updates the grid cells according to the grid cell model equation, and the grid cells activate individual place cells according to spread of activity across synapses W_{PG}.

Encoding of a trajectory involves strengthening a set of synapses W_{HP} to associate locations coded by place cells $p(t)$ with the associated action performed at that location as coded by the speed modulated head direction cell vector $h(t)$. Thus, a trajectory is stored by formation of associations via synaptic modification of the synaptic connectivity matrix W_{HP} between the population of place cells $\vec{p}(t)$ and the population of head direction cells $\vec{h}(t)$ as follows:

$$W_{HP}(t+1) = (W_{HP}(t) + \vec{h}(t)\vec{p}(t)^T)/2.$$

This synaptic modification during encoding of the trajectory completes the circuit that forms the basis for later retrieval of the trajectory as described in chapter 4 and figure 4.1.

The synaptic connections between place cells and head direction cells could correspond to direct projections from region CA1 of the hippocampus to the postsubiculum (van Groen and Wyss, 1990), or to polysynaptic effects of projections from the hippocampus to subiculum (Swanson et al., 1978; Amaral and Witter, 1989), combined with anatomical projections from the subiculum to the postsubiculum and medial entorhinal cortex (Naber and Witter, 1998).

Retrieval of Trajectory

The initial cue for retrieval can take a number of forms. The cue can be the actual current location, coded by place cells updated by grid cells. The cue could also be environmental stimuli that have been associated with a particular pattern of place cells that can reactivate the place cell pattern. Sensory stimuli are used to cue retrieval in the examples of episodic retrieval, using the mathematical representation of associations between sensory stimuli representing items or events and places as described below. Finally, a cue that does not correspond to the actual current stimulus or location in the environment could be activated by internal mechanisms.

Whatever the source, the place cell cue can initiate retrieval of a trajectory by activating the associated action. Thus, activity $h(t)$ can be retrieved from the current place cell activity $p(t)$ as follows:

$$\vec{h}(t) = W_{HP}\vec{p}(t)/\sum \vec{p}(t).$$

This forms one component of a functional loop for retrieval in the model. This loop retrieves trajectories via three stages: (1) place cells activate associated head direction activity via matrix W_{HP}, (2) head direction cells update the phase of velocity-controlled oscillators driving entorhinal grid cells, and (3) grid cells $g(t)$ update place cells via matrix W_{PG}. The new place cell activity then activates the associated head direction pattern, and the loop continues driving retrieval of the trajectory. This loop is summarized in the main text (see figure 4.1).

Retrieval activity in the model can be summarized with the following equation:

$$\vec{p}(t+\Delta t) = W_{PG}g\left(\int_0^t W_{HP}\vec{p}(\tau)d\tau\right).$$

The previous place cell activity spreads across the matrix W_{HP} to drive activity in the head direction cells. The grid cell function $g(t)$ integrates this activity to update grid cell phase. Then the grid cell activity is transformed through a matrix W_{PG} representing the synaptic drive of grid cells on hippocampal place cells. Thus, the pattern of activity in the two synaptic matrices encodes the episodic memory for the trajectory in the model.

Alternately, the retrieval equation could be written as

$${}_{jkm}g(t) = \left[\prod_{i=1}^{3}\{\cos(2\pi ft) + \cos((2\pi(ft + {}_{j}B\int_0^t W_{HP}W_{PG}g(\tau)_i d\tau) + {}_{km}\vec{\varphi}(0))\}\right]_H$$

where the head direction activity driving the grid cells at each time point is replaced by the spread of activity across matrices of previous associations between grid cells and place cells and between place cells and head direction cells. These matrices could be combined into a single matrix W representing prior learning of the environment or causal relationships in any environment. The effect of head direction activity on grid cell activity during retrieval requires that the head direction vector (coding a velocity vector corresponding to rat action) needs to maintain activity for the period Δt until a new place cell representation causes retrieval of the next head direction response. This is justified by tonic levels of persistent firing that have been shown in intracellular recording from neurons in deep layers of entorhinal cortex slice preparations (Egorov et al., 2002; Fransén et al., 2006) and intracellular recording in the postsubiculum (Yoshida and Hasselmo, 2009). Head direction cells in the deep layers of entorhinal cortex (Sargolini et al., 2006) may be involved in encoding and retrieval of trajectories.

Note that retrieval has different dynamics from encoding as described in chapter 4 and figure 4.1. During encoding, the sensory input of velocity determines activity of head direction cells, which then drives grid cells and place cells. During retrieval, sensory input does not influence the system, and head direction activity is determined internally. The agent's awareness of the difference in dynamical state of the network allows differentiation of retrieved events from current sensory input. The difference in dynamics could be determined by modulatory or attentional influences on the postsubiculum or deep entorhinal cortex determining the relative influence of different synaptic inputs. As described in chapter 6, encoding and retrieval of single associations or items have been modeled as occurring on different phases of each theta cycle (Hasselmo et al., 2002), but the sequential retrieval of a full trajectory requires updating of phase over multiple cycles and might not be feasible to interleave with encoding on each cycle. A slower transition between encoding and retrieval could involve modulatory regulation of afferent versus feedback transmission by muscarinic acetylcholine receptors (Hasselmo, 2006) as described in chapter 6 and described mathematically later in the appendix.

Enhancement of Retrieval by Theoretical Arc Length Cells or Time Cells

Trajectory retrieval can be greatly enhanced by combining the mechanism based on place cells $p(t)$ with cells that code time intervals alone, or by cells that code arc length of the trajectory (Hasselmo, 2007). Cells with fixed differences in frequency will cause interference at a fixed time interval that could generate spiking in cells coding specific time intervals. In contrast, arc length cells will activate at a specific arc length (travel distance) along the trajectory from a given reference location. Both types of cells can be obtained in simulations by assuming that a specific item or location activates persistent spiking neurons in the entorhinal cortex (Klink and Alonso, 1997; Fransén et al., 2002). Persistent spiking neurons with fixed frequency will interfere with a baseline frequency at a specific time interval from their onset. Theoretical arc length cells could fire at a rhythmic rate that is modulated by speed but not by head direction, and theoretical time cells could have a constant difference in spiking frequency initiated by a specific event. In both cases, firing frequencies of some cells differ from the baseline frequency of other persistent spiking cells. The neurons receiving convergent input from these persistent spiking cells will fire dependent upon when the shift in frequency causes constructive interference with the baseline cells.

In mathematical terms, the mechanism of determining the activity a of arc length cells uses an interaction of two oscillations as follows:

$$a_i(t) = [\cos(2\pi ft) + \cos(2\pi(ft + B\int_0^t s(\tau)d\tau) + {}_i\varphi(0))]_H.$$

This resembles the model of grid cells, except that the interaction involves only two oscillations and the modulation of frequency depends on speed $s(t)$ only rather than speed-modulated head direction. Note that each arc length cell with index 'i' will fire at an arc length dependent upon its initial phase ${}_i\varphi(0)$. The threshold and frequency of arc length cells was set so that they would fire once during the trajectories.

Simulations of arc length cells (Hasselmo, 2007) are able to replicate a number of features of the physiological data from the hippocampal formation, including the phenomenon of context-dependent "splitter" cells that fire selectively for right or left turn trials in continuous spatial alternation (Wood et al., 2000; Lee et al., 2006), the forward shifting of these splitter cells toward goal locations (Lee et al., 2006), and the context-dependent firing of neurons in a delayed nonmatch to position task (Griffin et al., 2007) or plus maze tasks (Ferbinteanu and Shapiro, 2003; Smith and Mizumori, 2006a, b).

The coding of pure temporal intervals by time cells can be obtained if the interference does not depend upon velocity or speed but arises from a cue stimulus inducing a difference in frequency that remains fixed and stable over the subsequent period of time (Hasselmo et al., 2007). This could take advantage of the tendency of persistent spiking neurons to fall into the same persistent spiking frequency (Yoshida and Hasselmo, 2009). The equation for generating time cell activity is

$$a_i(t) = [\cos(f2\pi t) + \cos(2\pi(f + B)t + {}_i\varphi(0))]_H$$

where the difference in frequency between f and $f + B$ drives the interference. Because the frequency does not change over time, the integral can be solved as $(f + B)t$. The simulation of time cells was used to encode and retrieve the spatiotemporal trajectories in chapter 1. This allowed the network to accurately encode and retrieve the time of events when the agent was stationary (e.g., I could remember the separate events of meeting with Chantal and with Mark even though I was at a single location at my desk) as shown in figure 1.2.

For simulations in chapter 4, a circuit composed of arc length cells, speed modulated cells, and cells with oscillation frequencies modulated by speed can model sequential activation of arc length cells. Analogous to the connection between place cells and head direction cells (W_{HP}), a connectivity matrix W_{SA} links the vector $a(t)^T$ representing arc length cell activity with the associated speed $s(t)$ at each position along the trajectory:

$$W_{SA} = \sum_t s(t)\vec{a}(t)^T.$$

This process does not depend upon retrieval causing activation of an accurate vector of head direction activity at each step. Instead, it only requires that retrieval activate a representation of speed. Thus, retrieval drives phase changes in one dimension along the trajectory instead of two dimensions. The activity a of arc length cells drives the activity of speed cells as follows:

$$s(t) = W_{SA}\vec{a}(t).$$

Note that the arc length cell could be associated with the mean speed

$$S = \sum_{t_a}^{t_b} s(t)/(t_b - t_a)$$

on a segment of the trajectory rather than the instantaneous speed.

In addition to the associations with speed, theoretical arc length and time cells are concatenated with the place cell activity and represented together by the vector of place cell activity $p(t)$ so they form associations with speed modulated head direction cells in the matrix W_{PH} created. This forms associations between arc length and time cells and the action at the corresponding position on the trajectory to drive retrieval of the spatiotemporal trajectory coded by the grid cells. These cells and their phases are also associated with stimuli representing items or events as described in the next section.

During encoding, arc length cells and time cells are created at regular intervals of trajectory when the magnitude of interference crosses a threshold. In these simulations, the threshold and frequency shift of arc length cells were set so they would be active at a single contiguous segment on a trajectory. At the start of retrieval, the arc length phases are reset to the appropriate values and subsequent activity is computed based on retrieved speed rather than external speed. This causes sequential activation of arc length cells that can drive speed-modulated head direction and thereby drive the correct update of grid cell phase and place cell firing over time.

Associations with Items or Events

The simulations included formation of associations between the place cell vector $p(t)$ (which included the vector of arc length cells and time cells) and a vector $o(t)$ representing the stimulus features of individual objects or events experienced at specific positions on a trajectory. For simulations in chapter 3, individual objects were assigned locations near the trajectory and a unique object vector, and the object vector was active when the agent was within a specific distance of the object location. For the simulation at the start of chapter 1, events were assigned locations and time intervals, so they were only active when the agent was within a specific distance during a specific time interval. During encoding, Hebbian modification of synapses W_{OP} formed associations between the concatenated place cell vector $p(t)$ (that included the vector of arc length and time cells) and the stimulus feature vector $o(t)$ (potentially arriving from the lateral entorhinal cortex) as follows:

$$W_{OP}(t+1) = (W_{OP}(t) + \vec{o}(t)\vec{p}(t)^T)/2.$$

The retrieval of items shown in some figures in chapter 4 resulted from the place cell vector $p(t)$ (that included arc length cells) activating the stimulus feature vector $o(t)$ according to

$$\vec{o}(t) = W_{OP}\vec{p}(t)/\sum \vec{p}(t).$$

Because the place cell vector $p(t)$ includes arc length cell and time cell activity, the matrix W_{OP} also allows arc length and time cells activated at a specific position along a trajectory to cue the retrieval of stimulus features associated with objects or events.

In some examples, retrieval was cued by a specific stimulus feature vector. This was done by using bidirectional Hebbian modification so that the inverse of W_{OP} would code associations between the object stimulus feature vector and the place cell vector (including arc length cells). In addition, a matrix coded associations between individual arc length activity and the vector of oscillation phases in the entorhinal grid cells. This allowed an individual stimulus feature vector to activate the associated pattern of place cells and arc length cells, and the arc length cell representation to activate the associated grid cell oscillation phase. This allowed cueing of retrieval from a specific stimulus or position along the trajectory. The examples of simulations using cued retrieval in the model are shown in figures 4.5, 4.6 and 4.7.

Acetylcholine Prevents Interference during Encoding

As described above, the episodic memory model must use separate dynamics during the encoding of a new trajectory versus the retrieval of a previously stored trajectory. On a more general level, the formation of accurate associations in many memory systems requires separate dynamics of encoding and retrieval, as described extensively in chapter 6. The functional need for this separation of encoding and retrieval is demonstrated with a simple example in this section. This simple example illustrates the basic function of associative memory models (Hasselmo et al., 1992; Hasselmo, 1993; Hasselmo and Schnell, 1994), demonstrates the problem of interference during encoding, and illustrates the need for different dynamics during encoding and retrieval. This example will show how selective presynaptic inhibition of glutamatergic release by acetylcholine can prevent interference during learning of overlapping associations in a heteroassociative network.

In this example, the application of the learning rule is after the activation rule has been applied for one step. The activation rule is

$$a_i(t+1) = A_i^{(k)}(t) + c(t)\sum_j W_{ij} A_j^{(k)}(t)$$

where A_i^k is the direct afferent input to the postsynaptic neurons, A_j^k is the direct afferent input to the presynaptic neurons, and $c(t)$ represents the selective presynaptic inhibition by acetylcholine of the modifiable synapses W (with no presynaptic inhibition of afferent input). The cholinergic modulation takes the value of 1 or 0 (where 0 represents the presence of cholinergic presynaptic inhibition).

The example will use the following learning rule:

$$W_{ij}(t+1) = W_{ij}(t) + a_i(t+1)A_j^{(k)}(t)$$

where W is the pattern of connections between the two regions, $a_i(t + 1)$ is the postsynaptic activation, and A_j is the presynaptic activation for input pattern k.

Initial Learning of an Association

Assume the synapses between the neurons start out at zero strength. The first association $k = 1$ is presented to the network at time $t = 1$, resulting in postsynaptic input $A_i^{k=1}(t=1) = [1 \quad 0 \quad 1]$ and presynaptic input $A_j^{k=1}(t=1) = [1 \quad 1 \quad 0]$. For this step, we will assume the absence of cholinergic suppression (therefore, $c = 1$). The activations of the neurons are computed after one time step:

$$a_i(t=2) = \begin{bmatrix}1\\0\\1\end{bmatrix} + (1)\begin{bmatrix}0&0&0\\0&0&0\\0&0&0\end{bmatrix}\begin{bmatrix}1\\1\\0\end{bmatrix} = \begin{bmatrix}1\\0\\1\end{bmatrix} + [1]\begin{bmatrix}0\\0\\0\end{bmatrix} = \begin{bmatrix}1\\0\\1\end{bmatrix}.$$

The learning rule is used to store the first association:

$$\Delta W_{ij} = a_i a_j = \begin{bmatrix}1\\0\\1\end{bmatrix}[1 \quad 1 \quad 0] = \begin{bmatrix}1&1&0\\0&0&0\\1&1&0\end{bmatrix}, \quad W_{ij}(t=2) = \begin{bmatrix}0&0&0\\0&0&0\\0&0&0\end{bmatrix} + \begin{bmatrix}1&1&0\\0&0&0\\1&1&0\end{bmatrix} = \begin{bmatrix}1&1&0\\0&0&0\\1&1&0\end{bmatrix}.$$

Interference during Encoding of Overlapping Association

Now consider another round of encoding on a later time step. For simplicity, this is designated as time step $t = 3$, but we assume all activity from the first steps has decayed to nothing. The network is presented with the second association by giving postsynaptic input $A_i^{k=2}(t=3) = [0 \quad 1 \quad 0]$ and presynaptic input $A_j^{k=2}(t=3) = [0 \quad 1 \quad 1]$ to the network (with no cholinergic suppression, so $c = 1$). The activation of the neurons after one time step is

$$a_i(t=3) = \begin{bmatrix}0\\1\\0\end{bmatrix} + (1)\begin{bmatrix}1&1&0\\0&0&0\\1&1&0\end{bmatrix}\begin{bmatrix}0\\1\\1\end{bmatrix} = \begin{bmatrix}0\\1\\0\end{bmatrix} + 1\begin{bmatrix}1\\0\\1\end{bmatrix} = \begin{bmatrix}1\\1\\1\end{bmatrix}.$$

The learning rule is used to store this pattern (again, with no cholinergic suppression).

$$\Delta W_{ij} = \begin{bmatrix} 1 \\ 1 \\ 1 \end{bmatrix} [0 \quad 1 \quad 1] = \begin{bmatrix} 0 & 1 & 1 \\ 0 & 1 & 1 \\ 0 & 1 & 1 \end{bmatrix}, \quad W_{ij}(t=4) = \begin{bmatrix} 1 & 1 & 0 \\ 0 & 0 & 0 \\ 1 & 1 & 0 \end{bmatrix} + \begin{bmatrix} 0 & 1 & 1 \\ 0 & 1 & 1 \\ 0 & 1 & 1 \end{bmatrix} = \begin{bmatrix} 1 & 2 & 1 \\ 0 & 1 & 1 \\ 1 & 2 & 1 \end{bmatrix}.$$

To test the retrieval of the network, the network is then presented with the following unambiguous (nonoverlapping) presynaptic input pattern as a retrieval cue $A_j^{cue2}(t=5) = [0 \quad 0 \quad 1]$. This cue only overlaps with the presynaptic activity of the second association and does not overlap with the first association, so it should not cause interference during retrieval. There is no postsynaptic input, and the weights are not updated:

$$a_i(t=5) = \begin{bmatrix} 0 \\ 0 \\ 0 \end{bmatrix} + (1) \begin{bmatrix} 1 & 2 & 1 \\ 0 & 1 & 1 \\ 1 & 2 & 1 \end{bmatrix} \begin{bmatrix} 0 \\ 0 \\ 1 \end{bmatrix} = \begin{bmatrix} 0 \\ 0 \\ 0 \end{bmatrix} + 1 \begin{bmatrix} 1 \\ 1 \\ 1 \end{bmatrix} = \begin{bmatrix} 1 \\ 1 \\ 1 \end{bmatrix}.$$

This shows that the retrieved postsynaptic activity in response to the cue for the second association contains elements of the postsynaptic activity from both of the stored associations. This demonstrates how interference during encoding can cause a complete breakdown of associative memory function, resulting in excessive synaptic modification forming spurious associations that impair retrieval.

Acetylcholine Reduces Interference

The encoding of the second association will be repeated with different dynamics, to show how acetylcholine can prevent interference during encoding through modulation of the presynaptic inhibition of glutamate release.

This example starts after the weights have been strengthened for learning of the first association (after $t = 2$). Replacing the step computed above for $t = 3$, the second association is presented, with postsynaptic input $A_i^{k=2}(t=3) = [0 \quad 1 \quad 0]$ and presynaptic input $A_j^{k=2}(t=3) = [0 \quad 1 \quad 1]$. However, this time the steps are performed with cholinergic presynaptic inhibition of glutamate release (i.e., setting c to 0). In this case, the activations of the postsynaptic neurons after one time step are as follows. Activity is driven only by the direct postsynaptic input, with no contribution from the spread across previously modified synapses:

$$a_i(t=3) = \begin{bmatrix} 0 \\ 1 \\ 0 \end{bmatrix} + (0) \begin{bmatrix} 1 & 1 & 0 \\ 0 & 0 & 0 \\ 1 & 1 & 0 \end{bmatrix} \begin{bmatrix} 0 \\ 1 \\ 1 \end{bmatrix} = \begin{bmatrix} 0 \\ 1 \\ 0 \end{bmatrix} + 0 \begin{bmatrix} 1 \\ 0 \\ 1 \end{bmatrix} = \begin{bmatrix} 0 \\ 1 \\ 0 \end{bmatrix}.$$

Applying the learning rule to store this pattern (with c still equal to 0) gives the following modification of the weights:

$$\Delta W_{ij} = \begin{bmatrix} 0 \\ 1 \\ 0 \end{bmatrix} [0 \quad 1 \quad 1] = \begin{bmatrix} 0 & 0 & 0 \\ 0 & 1 & 1 \\ 0 & 0 & 0 \end{bmatrix}, \quad W_{ij}(t=4) = \begin{bmatrix} 1 & 1 & 0 \\ 0 & 0 & 0 \\ 1 & 1 & 0 \end{bmatrix} + \begin{bmatrix} 0 & 0 & 0 \\ 0 & 1 & 1 \\ 0 & 0 & 0 \end{bmatrix} = \begin{bmatrix} 1 & 1 & 0 \\ 0 & 1 & 1 \\ 1 & 1 & 0 \end{bmatrix}.$$

Now the alternative learning dynamics can be tested by presenting the network with the unambiguous presynaptic retrieval cue $A_j^{cue2}(t=5) = [0 \quad 0 \quad 1]$, with no postsynaptic input and without cholinergic presynaptic inhibition:

$$a_i(t=5) = \begin{bmatrix} 0 \\ 0 \\ 0 \end{bmatrix} + (1) \begin{bmatrix} 1 & 1 & 0 \\ 0 & 1 & 1 \\ 1 & 1 & 0 \end{bmatrix} \begin{bmatrix} 0 \\ 0 \\ 1 \end{bmatrix} = \begin{bmatrix} 0 \\ 0 \\ 0 \end{bmatrix} + 1 \begin{bmatrix} 0 \\ 1 \\ 0 \end{bmatrix} = \begin{bmatrix} 0 \\ 1 \\ 0 \end{bmatrix}.$$

This shows that the retrieved postsynaptic activity matches the postsynaptic activity during encoding of the second association, indicating that the presence of acetylcholine prevents interference during encoding, allowing effective learning of associations with overlapping presynaptic activity.

Separation of Encoding and Retrieval on Phases of Theta Rhythm

The dynamics of theta rhythm oscillations also provide a mechanism for separating the dynamics of encoding and retrieval of associations within the hippocampal formation (Hasselmo et al., 2002). Notice that the use of theta rhythm for separating encoding and retrieval differs from the use of theta rhythm phase for encoding of location and time in episodic memory, but the two functions can be complementary. The example below demonstrates how changes in the magnitude of synaptic transmission and LTP induction during theta rhythm oscillations could prevent interference during encoding of overlapping patterns in a heteroassociative network, giving the mathematical result from the initial publication (Hasselmo et al., 2002).

Learning is applied during continuous trials in which the following activation rule is applied:

$$a_i(t) = \theta_{EC} A_{ECi}^{(k)}(t) + \theta_{CA3} \sum_j W_{ij} A_{CA3j}^{(k)}(t).$$

The activation equation incorporates the experimentally reported change in magnitude of synaptic transmission during different phases of the theta rhythm oscillation (Brankack et al., 1993; Wyble et al., 2000) by using the following functions that modulate synaptic transmission. These are functions of time.

$$\theta_{EC} = 1/2 * \sin(t + \varphi_{EC}) + 1/2$$

and

$$\theta_{CA3} = 1/2 * \sin(t + \varphi_{CA3}) + 1/2.$$

The activation is used during trials from t_m to t_{m+1} in the following learning rule:

$$W_{ij}(t_{m+1}) = W_{ij}(t_m) + \int_{m}^{m+1} \theta_{LTP} a_i(t) A_{CA3j}^{(k)}(t) dt.$$

In the learning rule, the data showing oscillatory shifts between induction of LTP and long-term depression is modeled with the oscillatory function:

$$\theta_{LTP} = \sin(t + \varphi_{LTP}).$$

This represents the experimental data on phasic changes in magnitude of synaptic modification during theta rhythm (Huerta and Lisman, 1995; Holscher et al., 1997; Hyman et al., 2003).

Assume that synapses between the neurons start out at zero strength. For simplicity, we will assume that storage of the first association between patterns $k = 1$ has already occurred. Therefore, we will just apply $A_{EC}^{k=1} = (1 \quad 0)$ to the neurons as the postsynaptic input and $A_{CA3}^{k=1} = (1 \quad 1 \quad 0)$ as the presynaptic input and will assume the weights grow to a value of 1:

$$W_{ij} = \begin{pmatrix} 1 & 1 & 0 \\ 0 & 0 & 0 \end{pmatrix}.$$

Then we present a second association as constant input, with postsynaptic input $A_{EC}^{k=2} = (0 \quad 1)$ and presynaptic input $A_{CA3}^{k=2} = (0 \quad 1 \quad 1)$. Applying the activation rule gives the following postsynaptic activity of neurons on each time step:

$$a_i(t) = \theta_{EC} \begin{pmatrix} 0 \\ 1 \end{pmatrix} + \theta_{CA3} \begin{pmatrix} 1 & 1 & 0 \\ 0 & 0 & 0 \end{pmatrix} \begin{pmatrix} 0 \\ 1 \\ 1 \end{pmatrix} = \theta_{EC} \begin{pmatrix} 0 \\ 1 \end{pmatrix} + \theta_{CA3} \begin{pmatrix} 1 \\ 0 \end{pmatrix} = \begin{pmatrix} \theta_{CA3} \\ \theta_{EC} \end{pmatrix}.$$

Now we use the learning rule to store the second associative pair:

$$\Delta W_{ij} = \theta_{LTP} \begin{pmatrix} \theta_{CA3} \\ \theta_{EC} \end{pmatrix} (0 \quad 1 \quad 1) = \begin{pmatrix} 0 & \int \theta_{LTP}\theta_{CA3} & \int \theta_{LTP}\theta_{CA3} \\ 0 & \int \theta_{LTP}\theta_{EC} & \int \theta_{LTP}\theta_{EC} \end{pmatrix}$$

$$W_{ij} = \begin{pmatrix} 1 & 1 & 0 \\ 0 & 0 & 0 \end{pmatrix} + \begin{pmatrix} 0 & \int \theta_{LTP}\theta_{CA3} & \int \theta_{LTP}\theta_{CA3} \\ 0 & \int \theta_{LTP}\theta_{EC} & \int \theta_{LTP}\theta_{EC} \end{pmatrix} = \begin{pmatrix} 1 & 1 + \int \theta_{LTP}\theta_{CA3} & \int \theta_{LTP}\theta_{CA3} \\ 0 & \int \theta_{LTP}\theta_{EC} & \int \theta_{LTP}\theta_{EC} \end{pmatrix}.$$

Now we can test retrieval with an unambiguous cue that does not overlap with the first association and should only retrieve the second pattern. We cue the retrieval with presynaptic activity $A_{CA3}^{cue2} = (0 \quad 0 \quad 1)$ and with no postsynaptic input, so $A_{EC}^{cue2} = (0 \quad 0)$. These inputs are constant during the trial.

$$a_i = \theta_{EC} \begin{pmatrix} 0 \\ 0 \end{pmatrix} + \theta_{CA3} \begin{pmatrix} 1 & 1 + \int \theta_{LTP}\theta_{CA3} & \int \theta_{LTP}\theta_{CA3} \\ 0 & \int \theta_{LTP}\theta_{EC} & \int \theta_{LTP}\theta_{EC} \end{pmatrix} \begin{pmatrix} 0 \\ 0 \\ 1 \end{pmatrix}.$$

$$= \theta_{EC} \begin{pmatrix} 0 \\ 0 \end{pmatrix} + \theta_{CA3} \begin{pmatrix} \int \theta_{LTP}\theta_{CA3} \\ \int \theta_{LTP}\theta_{EC} \end{pmatrix} = \begin{pmatrix} \theta_{CA3} \int \theta_{LTP}\theta_{CA3} \\ \theta_{CA3} \int \theta_{LTP}\theta_{EC} \end{pmatrix}$$

The retrieval performance of the network can be tested in response to the unambiguous cue. Because the retrieval cue is the nonoverlapping component of the second presynaptic pattern, this should evoke retrieval of only the second postsynaptic pattern. Thus, the desired pattern of postsynaptic activity is $A_{EC}^{k=2} = (0 \quad 1)$. We do not want to retrieve the other postsynaptic pattern, so the undesired pattern of output is $A_{EC}^{k=1} = (1 \quad 0)$. Therefore, the quality of retrieval performance can be measured with a performance measure that compares the pattern of postsynaptic activity with the desired and undesired patterns. This is done by taking the dot product of the desired output with postsynaptic activity and subtracting the dot product of the undesired output with postsynaptic activity. This is shown below with insertion of postsynaptic activity in the solution:

$$M = (0 \quad 1)\begin{pmatrix} a_1 \\ a_2 \end{pmatrix} - (1 \quad 0)\begin{pmatrix} a_1 \\ a_2 \end{pmatrix} = \theta_{CA3}\left(\int \theta_{LTP}\theta_{EC} - \int \theta_{LTP}\theta_{CA3}\right).$$

The sine wave function for LTP oscillates between a maximum value of 1 (at π/2 radians) and a minimum value of –1 (at 3*π/2 radians). The functions for EC and CA3 oscillate between a maximum of 1 (at π/2) and a minimum of 0 (at 3*π/2). If we only consider the time points when $\theta_{LTP} = 1$ (at π/2), then we are considering encoding. When LTP has that value, we want $\theta_{EC} = 1$ to make the positive part as large as possible (1) and we want $\theta_{CA3} = 0$ to make the negative part as small as possible (zero). Thus, we want EC be at its maximum when LTP is at its maximum (we want EC to be at phase π/2 when the LTP is at this phase, so they should have phase difference of zero). For maximum performance, at the time that LTP is at its maximum at π/2, then CA3 should be at its minimum with the value it has at 3*π/2. Thus, CA3 should be out of phase from EC and LTP by π radians (180 degrees).

If we only consider the time points when $\theta_{CA3} = 1$, then we are considering retrieval because we are focusing on the times when the spread of activity across modified connections is maximum. In this case, we want $\theta_{LTP} = -1$ to make the negative part become positive. Thus, LTD that mediates forgetting of retrieval will enhance the above performance measure because that measure is only concerned with retrieval of the second pattern. To relate this simple example to the mathematical description in the initial presentation of this theory (Hasselmo et al., 2002), just assume that initial learning grows to strength *K*, and use the ambiguous input pattern $A_{CA3}^{k=2} = (0 \quad 1 \quad 0)$ as the retrieval cue. This results in retrieval:

$$a_i = \left\{ \theta_{EC}\begin{pmatrix} 0 \\ 0 \end{pmatrix} + \theta_{CA3}\begin{pmatrix} K & K + \int \theta_{LTP}\theta_{CA3} & \int \theta_{LTP}\theta_{CA3} \\ 0 & \int \theta_{LTP}\theta_{EC} & \int \theta_{LTP}\theta_{EC} \end{pmatrix}\begin{pmatrix} 0 \\ 1 \\ 0 \end{pmatrix} \right\} = \theta_{CA3}\begin{pmatrix} (K + \int \theta_{LTP}\theta_{CA3}) \\ \int \theta_{LTP}\theta_{EC} \end{pmatrix}.$$

The performance measure then matches the previous article (Hasselmo et al., 2002):

$$M = (0 \quad 1)\begin{pmatrix} a_1 \\ a_2 \end{pmatrix} - (1 \quad 0)\begin{pmatrix} a_1 \\ a_2 \end{pmatrix} = \theta_{CA3}\left(\int \theta_{LTP}\theta_{EC} - K - \int \theta_{LTP}\theta_{CA3}\right).$$

Sequence Encoding and Goal Finding

The two examples of associative memory function above focus on individual pairs of discrete input patterns. The associative memory mechanism can be extended to encoding a sequence of discrete input patterns. The weight change from association of each pair of patterns $A(t)$ and $A(t + 1)$ can be summed over time: $W = \sum_{t=1}^{N} A(t+1)A(t)^T$. This process can use spike timing dependent plasticity if the sequence is compressed to occur within a brief time window (Hasselmo et al., 2002c; Gorchetchnikov and Hasselmo, 2005; Hasselmo, 2005b). The sequence can then be cued by the first pattern and can retrieve the chain of discrete patterns by iterative application of the activation rule: $a(t+1) = Wa(t)$. Note that this discrete mechanism differs from the more continuous mechanism used for encoding of spatiotemporal trajectories in the episodic memory model.

The sequence can be retrieved in reverse order if learning takes the opposite order: ${}_BW = \sum_{t=1}^{N} A(t)A(t+1)^T$. The reverse order of input can be obtained by reverse replay during sharp waves (Foster and Wilson, 2006; Davidson et al., 2009). If reverse sequence retrieval cued by a goal location interacts with a single step of forward retrieval, the first overlapping pattern of activity can allow selection of the next appropriate step toward a goal (Hasselmo et al., 2002c; Gorchetchnikov and Hasselmo, 2005; Hasselmo, 2005b). This is a highly simplified version of the gradient mechanism used for goal-directed behavior in chapter 7.

Simple Example of Attractor Dynamics

As described in chapter 5, attractor dynamics can arise from the interaction of feedback excitation and inhibition within cortical structures. This section presents equations for the example of a fixed-point attractor shown in figure 5.9. In hippocampal region CA3, excitatory neurons send recurrent excitatory connections to other pyramidal cells, and connect to inhibitory interneurons that make inhibitory connections on excitatory neurons. For certain parameters, subtractive inhibition can balance excitatory feedback (Wilson and Cowan, 1972; Hasselmo et al., 1995a), to allow the network to approach a stable equilibrium value without explosive growth. The term subtractive refers to the mathematical representation of inhibition with a reversal potential well below resting potential.

The network can be described by the following equations using linear algebra notation:

$$\frac{da}{dt} = A + W[a-\theta] + H[h-\theta_h] - \eta a$$

$$\frac{dh}{dt} = W'[a-\theta] - \eta h$$

In these equations a is the activation vector of the population of pyramidal cells, and h is the activation of the inhibitory interneurons, and η is the passive decay parameter. The square brackets represent a threshold linear input-output function, with threshold θ for excitatory pyramidal cells and θ_h for inhibitory interneurons. A is the vector of synaptic input to the pyramidal cells. W is the pattern of excitatory synaptic connectivity between pyramidal cells. W' is the excitatory connectivity from pyramidal cells to interneurons and H is the inhibitory synaptic connectivity from interneurons to pyramidal cells.

As shown in figure 5.9B and C and 5.10, this network can enter stable, fixed-point attractor states, in which pyramidal cells that are components of a stored memory vector show stable activity, and other pyramidal cells are inhibited and inactive. For the interaction of cells that are above threshold, the steady state level of pyramidal cell activity can be determined by setting $da/dt = 0$ and $dh/dt = 0$ and combining and rearranging the above equations to obtain:

$$a = \frac{A - W\theta + HW'\theta/\eta + H\theta_h}{\eta - W + HW'/\eta}.$$

This defines the activity level of the attractor states (Hasselmo et al., 1995a). For example, the figure shows results for: $A = 0.03$; $\theta = 0$, $W = 0.14/n$, $\theta h = 1.0$, $H = 0.1$, $W' = 0.14/n$, $\eta = 0.1$ (where n is the number of active excitatory neurons in a pattern). This results in a steady state activation in the attractors of $a(\infty) = 1.3$ with input and 1.0 without input. (Note that setting threshold to zero removes two terms in the numerator, and setting $W = W'$, and $H = \eta$ balances two terms in the denominator). The equation can describe the equilibrium activity of multiple neurons above threshold (figure 5.10).

Biophysical Simulations of Neuron Properties

The network models above commonly use highly simplified representations of neuron properties, but these can be linked to more detailed biophysical models that effectively simulate many of the changes in membrane potential of single neurons. The ultimate goal is to simulate the full network function using biophysically detailed neurons, so that the function of channels and receptors influencing membrane potential can be directly linked to the behavioral function of the network. Here, I will briefly review models of some of the biophysical properties of neurons relevant to the models described in this book.

As shown in figure A.16A, as an intracellular recording electrode enters a neuron it records a resting membrane potential around –70 mV relative to ground. Most

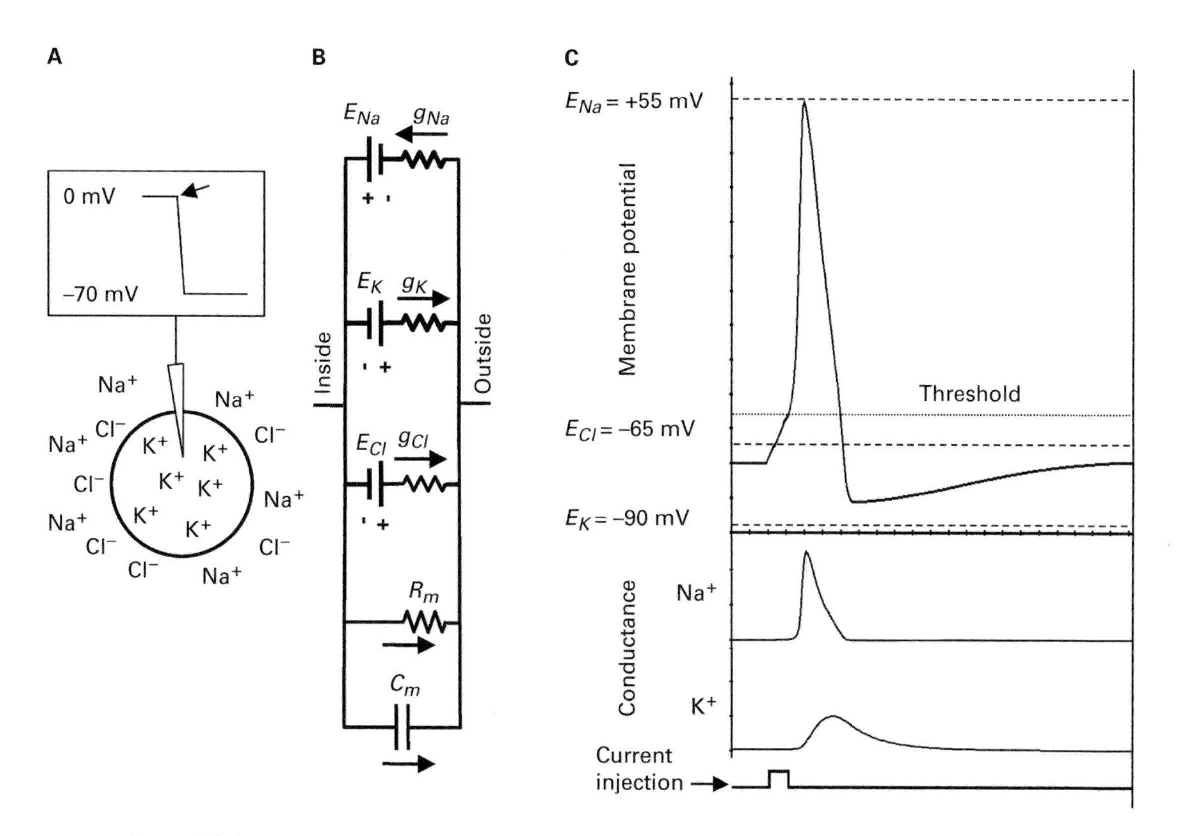

Figure A.16

(A) Top: After a recording electrode enters the cell, it displays a membrane potential that is about –70 mV in the absence of input. Bottom: The membrane potential arises from differences in concentration of ions inside and outside the cell. (B) Equivalent circuit model of membrane potential. Nernst potentials for different ions E_{Na}, E_K and E_{Cl} influence the membrane potential in proportion to specific conductances for g_{Na}, g_K and g_{Cl}. Models also usually include passive currents due to a leak current resistance R_m and capacitance C_m. (C) Graph showing how increased conductance for an ion drives the membrane potential toward the Nernst potential for that ion. Here, current injection depolarizes the cell to firing threshold, activating voltage-dependent Na+ conductances that drive the cell toward the Nernst potential of E_{Na} = +55 mV. Slower opening of voltage-dependent K^+ conductances then drive the cell toward the Nernst potential of E_K = –90 mV. As the K^+ conductance decreases, the cell returns to resting potential.

biophysical simulations of the changes in membrane potential use an equivalent circuit representation of membrane conductances and driving forces. This potential arises from driving forces due to concentration differences of ions, including sodium (Na^+) and chloride (Cl^-) at high concentrations outside of the cell, and potassium (K^+) at a high concentration inside the cell. Nernst potentials describe the potential that would be necessary to maintain the concentration difference. A Nernst potential of E_{Na} = +55 mV would be necessary to maintain the Na^+ concentration difference, whereas the K^+ concentrations result in a Nernst potential of E_K = –90 mV and Cl^- concentrations result in E_{Cl} = –65 mV.

The equivalent circuit in figure A.16B models these Nernst potentials as batteries driving current through variable conductances g_{Na}, g_K, and g_{Cl} corresponding to channels that allow these ions through the membrane. These conductances are in parallel with the membrane resistance R_m and capacitance C_m described at the start of the appendix. The influence of the Nernst potentials contribute to membrane potential in proportion to the conductance of each ion as follows:

$$C_m \frac{dV_m}{dt} = g_{Na}(E_{Na} - V_m) + g_K(E_K - V_m) + g_{Cl}(E_{Cl} - V_m) + \frac{(E_L - V_m)}{R_m} + I_{IN}$$

where V_m is membrane potential, I_{IN} is input current, and E_L is the leak potential driving force for passive conductance $g_L = 1/R_m$. The function of this equation can be best understood graphically in figure A.16C according to a simple rule: As the conductance for an ion increases, the membrane potential is driven toward the Nernst potential for that ion. In A.16C, a brief current injection depolarizes the cell to firing threshold. This increases the conductance of voltage-dependent sodium channels, driving the membrane potential strongly upward toward E_{Na} = +55. These channels then inactivate, causing the membrane potential to move away from +55. Meanwhile slower activation of voltage-dependent potassium channels drives the membrane potential down towards E_K = –90 mV. Then, as these channels close, the membrane potential returns to stable resting potential. Any conductance change will drive the potential toward a steady state (dV_m/dt = 0), that can be computed as $V_m = (g_{Na}E_{Na} + g_KE_K + g_{Cl}E_{Cl} + g_LE_L + I_{in})/(g_{Na}+g_K+g_{Cl}+g_L)$. Further equations describe the sensitivity of different conductances to membrane voltage or to the concentration of intracellular calcium and second messengers, or to extracellular transmitters. Additional properties of conductance changes can be included in the above equation in order to simulate the intrinsic properties described in the following sections. The dynamics of changes in conductances are commonly described using the structure of the Hodgkin-Huxley equations (Hodgkin and Huxley, 1952; Bower and Beeman, 2005).

Membrane Potential Oscillations

As described in chapter 2, entorhinal stellate cells in layer II show prominent membrane potential oscillations that appear to arise from the effect of the hyperpolarization activated cation current, or h current. The oscillations also appear to be amplified by the effects of the voltage-dependent persistent sodium current, or NaP current.

Erik Fransén created a model of these oscillations (Fransén et al., 2004) using detailed Hodgkin–Huxley representations of the h current conductances based on data from our collaboration with Angel Alonso and Clayton Dickson (Dickson et al., 2000). He used the GENESIS simulation package (Bower and Beeman, 1995) to create a multicompartmental model of an entorhinal stellate cell. The theta frequency oscillations arise in this model because the h current has a relatively slow time constant for activation. As shown in figure A.17, when the membrane potential is hyperpolarized, the h current conductance g_h turns on slowly, causing a depolarization. This depolarization then gradually decreases the conductance of the channel mediating the h current, resulting in a hyperpolarization. The hyperpolarization then turns on the h current slowly to start the next cycle of the ongoing oscillation.

In simulations of resonance, the h current causes a peak resonance frequency in response to the oscillating current of increasing frequency (see figures 2.6 and 2.15). The peak occurs when the hyperpolarizing part of the oscillating current activates the h current with a time course that matches the subsequent depolarizing part of the

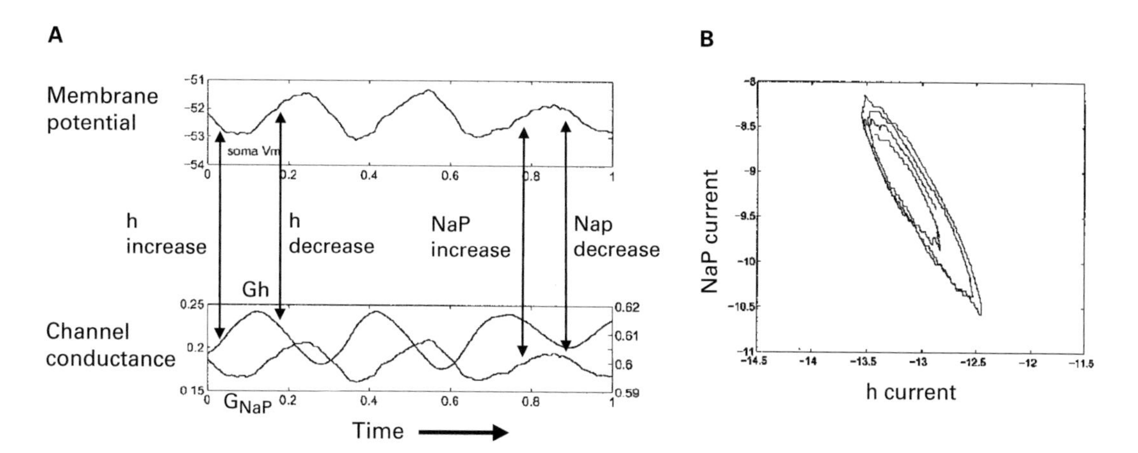

Figure A.17

(A) Top: Simulation of subthreshold membrane potential oscillations. Bottom: Conductances of h current (G_h) and NaP current (G_{NaP}) underlying membrane potential oscillations in the model. As the membrane potential hyperpolarizes, this causes a slow increase in the h current, resulting in a depolarization. As the membrane potential depolarizes, this causes a slow decrease in the h current, causing a hyperpolarization. The persistent sodium conductance NaP rapidly increases during depolarization to amplify the depolarization and then decreases during hyperpolarization. (B) Phase plot of NaP current versus h current showing the cyclical changes in both of these currents during membrane potential oscillations. Adapted from Fransén et al. (2004).

oscillating current, and this deactivates the h current with a time course matching the hyperpolarizing part of the oscillating current. The h current can be blocked by a selective blocker ZD7288 that also blocks the membrane potential oscillations and resonance of entorhinal layer II stellate cells (Dickson et al., 2000; Giocomo and Hasselmo, 2008b).

The channels mediating the h current are also sensitive to cyclic nucleotides, such as cAMP. Because of this, the molecular name for the subunits of the channels mediating the h current are HCN1 to HCN4 (for hyperpolarization activated, cyclic nucleotide gated channel). Subunits HCN1 and HCN2 appear to be the most common in the medial entorhinal cortex, and different combinations of these subunits create channels with different time constants of activation (Chen et al., 2001). Channels consisting of HCN1 subunits alone have faster time constants than channels combining HCN1 and HCN2 subunits which have faster time constants than channels consisting of HCN2 subunits alone.

The data on the time course of h current with different subunits led to the hypothesis that the dorsal to ventral gradient of differences in frequency of membrane potential oscillations and resonance in medial entorhinal cortex could arise from differences in the expression of the HCN1 subunit. This was supported by data showing a dorsal to ventral difference in the time constant of h current in rats (Giocomo and Hasselmo, 2008b) and a study in mice showing loss of the dorsal to ventral gradient of oscillation and resonance frequency with genetic knockout of the HCN1 subunit (Giocomo and Hasselmo, 2009). As an alternative, the frequency difference could result from a difference in the magnitude of the h current (Garden et al., 2008).

Mathematically, the phenomenon of subthreshold oscillations have been extensively analyzed and linked with the resonance properties of neurons (White et al., 1995; White et al., 1998; Pervouchine et al., 2006; Rotstein et al., 2006; Izhikevich, 2007). In the phase space analysis of these models, the dynamics of the neuron have the resting state as a stable focus. Any perturbations from the resting state cause damped oscillations back to the focus. In the model, inclusion of noise representing random opening and closings of membrane channels allows the oscillations to be observed for an extended period. The dynamics of the neuron are near a subcritical Andronov-Hopf bifurcation (Izhikevich, 2007), so the stable focus coexists with a large amplitude limit cycle that corresponds to spiking. Thus, the noise can cause the neuron to transition between periods of damped oscillations and periods of repetitive spike clusters on the limit cycle as shown in figure A.18. The spike clusters are observed both experimentally (Alonso and Llinas, 1989) and in detailed biophysical models (Fransén et al., 2004).

Simulations (Heys et al., 2010) using the NEURON software package (Carnevale and Hines, 2005) show that the dorsal to ventral decrease in frequency of subthreshold membrane potential oscillations and resonance can be replicated with biophysical

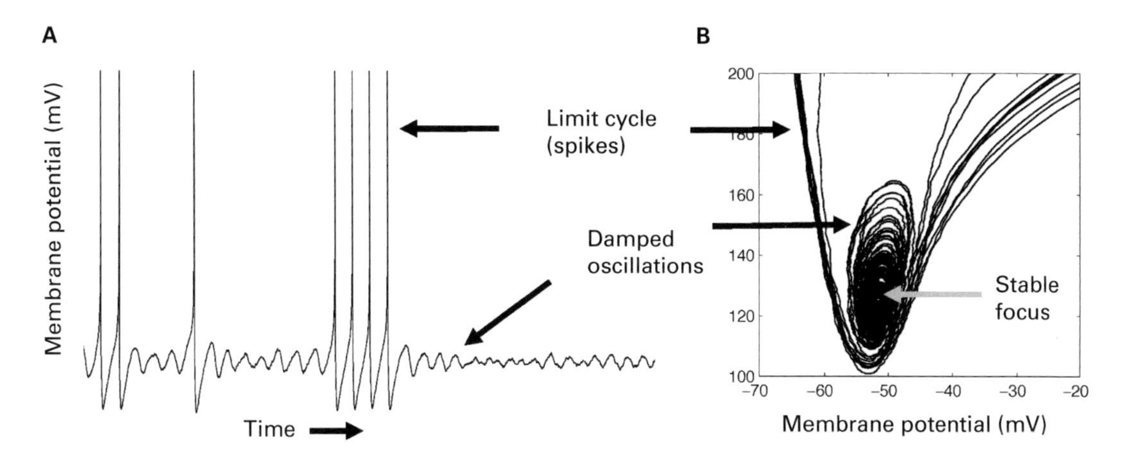

Figure A.18
(A) Simulation of subthreshold membrane potential oscillations in EC stellate cells replicating the simulation developed by Izhikevich (Izhikevich, 2007, p. 314), showing transition between clusters of spikes and damped oscillations sustained by noise. (B) Phase plot of the resonance variable in the simple model plotted against membrane potential to show how the subcritical Andronov-Hopf bifurcation allows the coexistence of damped oscillations (center) and limit cycle spiking (lines leaving top of plot).

simulations using the increase in time constants observed in voltage-clamp recording at different dorsal to ventral positions (Giocomo and Hasselmo, 2008b). Other cellular properties including the magnitude of h current and the input resistance of the cell can also influence oscillation and resonance frequency (Heys et al., 2010). The frequency of membrane potential oscillations and resonance in simulations also decreases with the decrease in magnitude of h current and the increase in input R_m observed along the dorsal to ventral axis in experiments (Garden et al., 2009).

Spike Frequency Accommodation

As shown in figure 2.3, many cortical neurons respond to input with spiking activity that shows a decrease in frequency known as spike frequency accommodation. The mechanism of spike frequency accommodation can also be modeled in detailed biophysical simulations (Barkai et al., 1994) using Hodgkin–Huxley representations of the M current and the calcium-activated potassium current or K(AHP). In the model, depolarization associated with spikes activates the M current, increasing potassium conductance. In addition, spikes cause calcium influx through voltage-gated calcium channels causing a buildup of intracellular calcium that activates the additional potassium conductance of the AHP current. Both these effects push the membrane potential down toward the Nernst potential for potassium and gradually slow down or stop the

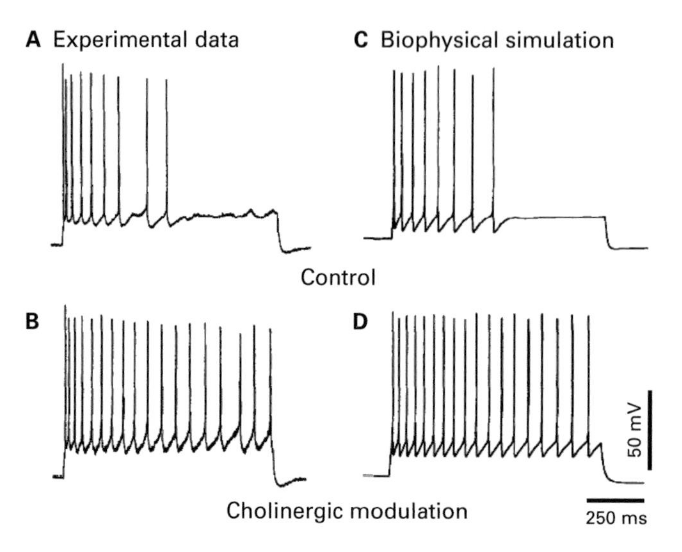

Figure A.19
(A) Experimental data showing spike frequency accommodation in control conditions. (B) Reduced spike frequency accommodation during cholinergic activation of muscarinic receptors that reduces the M current and AHP current. (C) Biophysical simulation of spike frequency accommodation in control conditions. (D) Simulation of reduction in spike frequency accommodation with reduced M current and AHP current. Figure from Barkai et al., 1994.

spiking, resulting in spike frequency accommodation as shown in figure A.19. This current can also balance the depolarizing current underlying persistent spiking.

Persistent Spiking

The cellular mechanisms of persistent spiking in the entorhinal cortex have also been simulated using detailed biophysical simulations (Fransén et al., 2002; Fransén et al., 2006) developed in GENESIS (Bower and Beeman, 1995). Persistent spiking appears to arise from activation of a calcium-sensitive nonspecific cation current, known as the CAN current or NCM current. The current appears to be increased by second messengers activated by metabotropic glutamate receptors and muscarinic acetylcholine receptors, resulting in greater likelihood of persistent spiking of individual neurons. Computational models demonstrate how persistent spiking could arise from the CAN current as summarized in figure A.20. The extensive details of the Hodgkin–Huxley models are available in the original publications (Fransén et al., 2002; Fransén et al., 2006). As shown in figure A.20B, during cholinergic modulation, the CAN current causes depolarization that induces spiking and increases calcium influx. This calcium activates the calcium-sensitive CAN current causing further depolarization in a regenerative cycle. The stable firing frequency of the persistent spiking model arises from

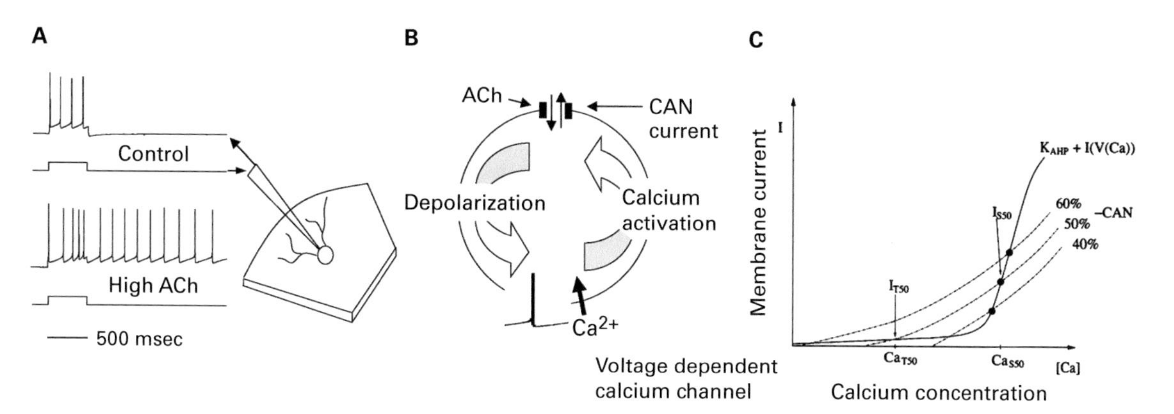

Figure A.20
(A) Simulation of persistent spiking in an entorhinal pyramidal cell. In control conditions (top), the neuron only spikes during a square pulse input of increased current injection and then stops when the square pulse ends. With acetylcholine (ACh) present to activate the calcium-activated, nonspecific cation current (CAN; bottom), a square pulse of current injection induces spiking activity that continues as persistent spiking after injection ends. (B) Mechanism of persistent spiking in the model. Activation of muscarinic receptors by acetylcholine activates the CAN current channel. During this period of activation, depolarization can cause calcium influx through voltage-dependent calcium channels. This calcium influx further activates the CAN current, causing further depolarization. The depolarization causes further calcium influx that further activates the CAN current, continuing the cycle to maintain persistent spiking. Figure based on material presented in Hasselmo and Stern (2006). (C) As calcium concentration varies, this changes the balance of the hyperpolarizing calcium-activated potassium current (AHP) relative to the CAN current, so that the firing rate and membrane current will balance where the calcium-sensitivity of the currents overlap (black dots). Figure from Fransén et al. (2006).

the balance of calcium-sensitive conductances in the model (Fransén et al., 2006), as shown in figure A.20C. In particular, calcium influx during spiking activates both the nonspecific cation current (the CAN current) that causes depolarization with higher calcium, and activates the calcium-dependent potassium current (the AHP current) that causes hyperpolarization with higher calcium (and causes spike frequency accommodation). These currents can balance each other because they have different functions of response to calcium that cross each other. The steady state firing rate will occur at the point where these two conductances balance each other (black dots in figure A.20C). If firing rate increases, the increase in calcium influx causes an increase in AHP relative to CAN that will cause greater hyperpolarization to push down the firing rate. If firing rate decreases, the decrease in calcium decreases the AHP relative to CAN and causes greater depolarization to increase the rate (Fransén et al., 2006). Persistent spiking can be terminated in some cases by sustained hyperpolarizing current input that sufficiently lowers calcium levels, or by strong depolarizing current input that causes sufficiently strong activation of the AHP to push the system out of the equilibrium state.

As shown in these figures, the detailed physiological properties of cortical neurons that provide intrinsic mechanisms for memory function can be simulated in terms of the dynamical properties of specific membrane currents mediated by specific membrane channels. Pharmacological work has shown how these membrane channels can be influenced by drugs, and molecular work demonstrates how these membrane channels change in function dependent upon alterations in their protein structure. This provides an important link to clinical research on the genetic and drug influences on cellular function. Ongoing research shows how the network properties such as grid cells and place cells could arise from these intrinsic properties. Thus, computational models provide an important link between the molecular mechanisms that can be influenced by clinical interventions and the behavioral function of the cortical circuits described here.

References

Abbott LF. 1991. Realistic synaptic inputs for model neural networks. *Network* 2: 245–258.

Acker CD, Kopell N, White JA. 2003. Synchronization of strongly coupled excitatory neurons: relating network behavior to biophysics. *J Comput Neurosci* 15: 71–90.

Acquas E, Wilson C, Fibiger HC. 1996. Conditioned and unconditioned stimuli increase frontal cortical and hippocampal acetylcholine release: effects of novelty, habituation, and fear. *J Neurosci* 16: 3089–3096.

Adams SV, Winterer J, Muller W. 2004. Muscarinic signaling is required for spike-pairing induction of long-term potentiation at rat Schaffer collateral-CA1 synapses. *Hippocampus* 14: 413–416.

Adey WR, Dunlop CW, Hendrix CE. 1960. Hippocampal slow waves: distribution and phase relationships in the course of approach learning. *Arch Neurol* 3: 74–90.

Agam Y, Bullock D, Sekuler R. 2005. Imitating unfamiliar sequences of connected linear motions. *J Neurophysiol* 94: 2832–2843.

Aggleton JP, Hunt PR, Rawlins JN. 1986. The effects of hippocampal lesions upon spatial and non-spatial tests of working memory. *Behav Brain Res* 19: 133–146.

Aggleton JP, Neave N, Nagle S, Hunt PR. 1995. A comparison of the effects of anterior thalamic, mamillary body and fornix lesions on reinforced spatial alternation. *Behav Brain Res* 68: 91–101.

Agster KL, Fortin NJ, Eichenbaum H. 2002. The hippocampus and disambiguation of overlapping sequences. *J Neurosci* 22: 5760–5768.

Aigner TG, Walker DL, Mishkin M. 1991. Comparison of the effects of scopolamine administered before and after acquisition in a test of visual recognition memory in monkeys. *Behav Neural Biol* 55: 61–67.

Ainge JA, van der Meer MA, Langston RF, Wood ER. 2007. Exploring the role of context-dependent hippocampal activity in spatial alternation behavior. *Hippocampus* 17: 988–1002.

Alkondon M, Albuquerque EX. 2001. Nicotinic acetylcholine receptor alpha7 and alpha4beta2 subtypes differentially control GABAergic input to CA1 neurons in rat hippocampus. *J Neurophysiol* 86: 3043–3055.

Alonso A, Garcia-Austt E. 1987. Neuronal sources of theta rhythm in the entorhinal cortex of the rat. I. Laminar distribution of theta field potentials. *Exp Brain Res* 67: 493–501.

Alonso A, Klink R. 1993. Differential electroresponsiveness of stellate and pyramidal-like cells of medial entorhinal cortex layer II. *J Neurophysiol* 70: 128–143.

Alonso A, Kohler C. 1984. A study of the reciprocal connections between the septum and the entorhinal area using anterograde and retrograde axonal transport methods in the rat brain. *J Comp Neurol* 225: 327–343.

Alonso A, Llinas RR. 1989. Subthreshold Na-dependent theta-like rhythmicity in stellate cells of entorhinal cortex layer II. *Nature* 342: 175–177.

Alvarez P, Squire LR. 1994. Memory consolidation and the medial temporal lobe: a simple network model. *Proc Natl Acad Sci USA* 91: 7041–7045.

Alvarez P, Zola-Morgan S, Squire LR. 1994. The animal model of human amnesia: long-term memory impaired and short-term memory intact. *Proc Natl Acad Sci USA* 91: 5637–5641.

Amaral DG, Witter MP. 1989. The 3-dimensional organization of the hippocampal formation—a review of anatomical data. *Neurosci* 31: 571–591.

Amaral DG, Ishizuka N, Claiborne B. 1990. Neurons, numbers and the hippocampal network. *Prog Brain Res* 83: 1–11.

Amit DJ. 1988. *Modeling Brain Function: The World of Attractor Neural Networks.* Cambridge, UK: Cambridge University Press.

Amit DJ, Brunel N. 1997. Model of global spontaneous activity and local structured activity during delay periods in the cerebral cortex. *Cereb Cortex* 7: 237–252.

Anderson JA. 1972. A simple neural network generating an interactive memory. *Math Biosci* 14: 197–220.

Anderson SJ, Conway MA. 1993. Investigating the structure of autobiographical memories. *J Exp Psychol Learn Mem Cogn* 19: 1178–1196.

Andrade R. 1991. Cell excitation enhances muscarinic cholinergic responses in rat association cortex. *Brain Res* 548: 81–93.

Arleo A, Gerstner W. 2000. Spatial cognition and neuro-mimetic navigation: a model of hippocampal place cell activity. *Biol Cybern* 83: 287–299.

Atri A, Sherman S, Norman KA, Kirchhoff BA, Nicolas MM, Greicius MD, Cramer SC, Breiter HC, Hasselmo ME, Stern CE. 2004. Blockade of central cholinergic receptors impairs new learning and increases proactive interference in a word paired-associate memory task. *Behav Neurosci* 118: 223–236.

Ault B, Nadler JV. 1982. Baclofen selectively inhibits transmission at synapses made by axons of CA3 pyramidal cells in the hippocampal slice. *J Pharmacol Exp Ther* 223: 291–297.

Baddeley A. 2001. The concept of episodic memory. In: *Episodic Memory: New Directions in Research* (Baddeley A, Conway M, Aggleton JP, eds), pp 1–10. Oxford, UK: Oxford University Press.

Baddeley AD, Hitch GJ. 1974. Working memory. In: *The Psychology of Learning and Motivation* (Bower GA, ed), pp 47–89. New York: Academic Press.

Baddeley AD, Warrington EK. 1970. Amnesia and the distinction between long- and short-term memory. *J Verbal Learn Verbal Behav* 9: 176–189.

Bakker A, Kirwan CB, Miller M, Stark CE. 2008. Pattern separation in the human hippocampal CA3 and dentate gyrus. *Science* 319: 1640–1642.

Bang SJ, Brown TH. 2009. Muscarinic receptors in perirhinal cortex control trace conditioning. *J Neurosci* 29: 4346–4350.

Bannerman DM, Yee BK, Lemaire M, Wilbrecht L, Jarrard L, Iversen SD, Rawlins JN, Good MA. 2001. The role of the entorhinal cortex in two forms of spatial learning and memory. *Exp Brain Res* 141: 281–303.

Barkai E, Hasselmo ME. 1994. Modulation of the input/output function of rat piriform cortex pyramidal cells. *J Neurophysiol* 72: 644–658.

Barkai E, Bergman RE, Horwitz G, Hasselmo ME. 1994. Modulation of associative memory function in a biophysical simulation of rat piriform cortex. *J Neurophysiol* 72: 659–677.

Barry C, Lever C, Hayman R, Hartley T, Burton S, O'Keefe J, Jeffery K, Burgess N. 2006. The boundary vector cell model of place cell firing and spatial memory. *Rev Neurosci* 17: 71–97.

Barry C, Hayman R, Burgess N, Jeffery KJ. 2007. Experience-dependent rescaling of entorhinal grids. *Nat Neurosci* 10: 682–684.

Barry C, Fleming SM, Jeewajee A, O'Keefe J, Burgess N. 2008. Effect of novelty on grid cell firing. *Proc ICCNS* 12: 35.

Becker JT, Walker JA, Olton DS. 1980. Neuroanatomical bases of spatial memory. *Brain Res* 200: 307–320.

Berry SD, Seager MA. 2001. Hippocampal theta oscillations and classical conditioning. *Neurobiol Learn Mem* 76: 298–313.

Berry SD, Thompson RF. 1978. Prediction of learning rate from the hippocampal electroencephalogram. *Science* 200: 1298–1300.

Bi G, Poo M. 2001. Synaptic modification by correlated activity: Hebb's postulate revisited. *Annu Rev Neurosci* 24: 139–166.

Bi GQ, Poo MM. 1998. Synaptic modifications in cultured hippocampal neurons: dependence on spike timing, synaptic strength, and postsynaptic cell type. *J Neurosci* 18: 10464–10472.

Bilkey DK, Goddard GV. 1985. Medial septal facilitation of hippocampal granule cell activity is mediated by inhibition of inhibitory interneurones. *Brain Res* 361: 99–106.

Blair HT, Cho J, Sharp PE. 1998. Role of the lateral mammillary nucleus in the rat head direction circuit: a combined single unit recording and lesion study. *Neuron* 21: 1387–1397.

Blair HT, Welday AC, Zhang K. 2007. Scale-invariant memory representations emerge from moiré interference between grid fields that produce theta oscillations: a computational model. *J Neurosci* 27: 3211–3229.

Blair HT, Gupta K, Zhang K. 2008. Conversion of a phase- to a rate-coded position signal by a three-stage model of theta cells, grid cells, and place cells. *Hippocampus* 18: 1239–1255.

Bland BH. 1986. The physiology and pharmacology of hippocampal-formation theta rhythms. *Prog Neurobiol* 26: 1–54.

Bland BH, Oddie SD. 2001. Theta band oscillation and synchrony in the hippocampal formation and associated structures: the case for its role in sensorimotor integration. *Behav Brain Res* 127: 119–136.

Bliss TV, Collingridge GL. 1993. A synaptic model of memory: long-term potentiation in the hippocampus. *Nature* 361: 31–39.

Bliss TV, Lømo T. 1973. Long-lasting potentiation of synaptic transmission in the dentate area of the anaesthetized rabbit following stimulation of the perforant path. *J Physiol* 232: 331–356.

Blokland A, Honig W, Raaijmakers WGM. 1992. Effects of intra-hippocampal scopolamine injections in a repeated spatial acquisition task in the rat. *Psychopharmacology (Berl)* 109: 373–376.

Blum KI, Abbott LF. 1996. A model of spatial map formation in the hippocampus of the rat. *Neural Comput* 8: 85–93.

Boccara CN, Sargolini F, Thoresen VH, Solstad T, Witter MP, Moser EI, Moser MB. 2010. Grid cells in pre- and parasubiculum. *Nat Neurosci* 13: 987–994.

Boehlen A, Heinemann U, Erchova I. 2010. The range of intrinsic frequencies represented by medial entorhinal cortex stellate cells extends with age. *J Neurosci* 30: 4585–4589.

Bogacz R, Brown MW. 2003. Comparison of computational models of familiarity discrimination in the perirhinal cortex. *Hippocampus* 13: 494–524.

Bohbot VD, Kalina M, Stepankova K, Spackova N, Petrides M, Nadel L. 1998. Spatial memory deficits in patients with lesions to the right hippocampus and to the right parahippocampal cortex. *Neuropsychologia* 36: 1217–1238.

Bolhuis JJ, Strijkstra AM, Kramers RJ. 1988. Effects of scopolamine on performance of rats in a delayed-response radial maze task. *Physiol Behav* 43: 403–409.

Borgers C, Kopell N. 2003. Synchronization in networks of excitatory and inhibitory neurons with sparse, random connectivity. *Neural Comput* 15: 509–538.

Bower JM, Beeman D. 1995. *The Book of GENESIS: Exploring Realistic Neural Models with the GEneral NEural SImulation System*. New York: Springer-Verlag.

Bower MR, Euston DR, McNaughton BL. 2005. Sequential-context-dependent hippocampal activity is not necessary to learn sequences with repeated elements. *J Neurosci* 25: 1313–1323.

Braak H, Braak E. 1991. Neuropathological stageing of Alzheimer-related changes. *Acta Neuropathol* 82: 239–259.

Brandon MP, Bogaard AR, Andrews C, Hasselmo ME. 2011a. Head direction cells in the postsubiculum do not show replay of prior waking sequences during sleep. *Hippocampus* in press.

Brandon MP, Bogaard AR, Libby CP, Connerney MA, Gupta K, Hasselmo ME. 2011b. Reduction of theta rhythm dissociates grid cell spatial periodicity from directional tuning. *Science* 332: 595–599.

Brandon MP, Bogaard AR, Hasselmo ME. 2011c. Grid cells, head direction cells, and theta oscillations: an analysis of theta cycle skipping and speed modulation. *Soc Neurosci Abstr* 37.

Brankack J, Stewart M, Fox SE. 1993. Current source density analysis of the hippocampal theta rhythm: associated sustained potentials and candidate synaptic generators. *Brain Res* 615: 310–327.

Brazhnik ES, Fox SE. 1999. Action potentials and relations to the theta rhythm of medial septal neurons in vivo. *Exp Brain Res* 127: 244–258.

Brazhnik ES, Muller RU, Fox SE. 2003. Muscarinic blockade slows and degrades the location-specific firing of hippocampal pyramidal cells. *J Neurosci* 23: 611–621.

Brewer JB, Zhao Z, Desmond JE, Glover GH, Gabrieli JD. 1998. Making memories: brain activity that predicts how well visual experience will be remembered. *Science* 281: 1185–1187.

Broks P, Preston GC, Traub M, Poppleton P, Ward C, Stahl SM. 1988. Modeling dementia: effects of scopolamine on memory and attention. *Neuropsychologia* 26: 685–700.

Brun VH, Otnass MK, Molden S, Steffenach HA, Witter MP, Moser MB, Moser EI 2002. Place cells and place recognition maintained by direct entorhinal-hippocampal circuitry. *Science* 296: 2243–2246.

Brun VH, Solstad T, Kjelstrup KB, Fyhn M, Witter MP, Moser EI, Moser MB. 2008. Progressive increase in grid scale from dorsal to ventral medial entorhinal cortex. *Hippocampus* 18: 1200–1212.

Buccafusco JJ, Letchworth SR, Bencherif M, Lippiello PM. 2005. Long-lasting cognitive improvement with nicotinic receptor agonists: mechanisms of pharmacokinetic-pharmacodynamic discordance. *Trends Pharmacol Sci* 26: 352–360.

Buckmaster CA, Eichenbaum H, Amaral DG, Suzuki WA, Rapp PR. 2004. Entorhinal cortex lesions disrupt the relational organization of memory in monkeys. *J Neurosci* 24: 9811–9825.

Buckner RL, Petersen SE, Ojemann JG, Miezin FM, Squire LR, Raichle ME. 1995. Functional anatomical studies of explicit and implicit memory retrieval tasks. *J Neurosci* 15: 12–29.

Bunce JG, Sabolek HR, Chrobak JJ. 2004. Intraseptal infusion of the cholinergic agonist carbachol impairs delayed-non-match-to-sample radial arm maze performance in the rat. *Hippocampus* 14: 450–459.

Burak Y, Fiete I. 2006. Do we understand the emergent dynamics of grid cell activity? *J Neurosci* 26: 9352–9354, discussion 9354.

Burak Y, Fiete IR. 2009. Accurate path integration in continuous attractor network models of grid cells. *PLOS Comput Biol* 5: e1000291.

Buresova O, Bolhuis JJ, Bures J. 1986. Differential effects of cholinergic blockade on performance of rats in the water tank navigation task and in a radial water maze. *Behav Neurosci* 100: 476–482.

Burgard EC, Sarvey JM. 1990. Muscarinic receptor activation facilitates the induction of long-term potentiation (LTP) in the rat dentate gyrus. *Neurosci Lett* 116: 34–39.

Burgess N. 2008. Grid cells and theta as oscillatory interference: theory and predictions. *Hippocampus* 18: 1157–1174.

Burgess N, Hitch G. 1999. Memory for serial order: a network model of the phonological loop and its timing. *Psychol Rev* 106: 551–581.

Burgess N, Hitch G. 2005. Computational models of working memory: putting long-term memory into context. *Trends Cogn Sci* 9: 535–541.

Burgess N, Donnett JG, Jeffery KJ, O'Keefe J. 1997. Robotic and neuronal simulation of the hippocampus and rat navigation. *Philos Trans R Soc Lond B Biol Sci* 352: 1535–1543.

Burgess N, Maguire EA, Spiers HJ, O'Keefe J. 2001. A temporoparietal and prefrontal network for retrieving the spatial context of lifelike events. *Neuroimage* 14: 439–453.

Burgess N, Barry C, Jeffery KJ, O'Keefe J. 2005. A grid and place cell model of path integration utilizing phase precession versus theta. In: Computational Cognitive Neuroscience Meeting. Computational Cognitive Neuroscience Meeting, Washington, DC.

Burgess N, Barry C, O'Keefe J. 2007. An oscillatory interference model of grid cell firing. *Hippocampus* 17: 801–812.

Burwell RD, Amaral DG. 1998. The cortical afferents of the perirhinal, postrhinal and entorhinal cortices of the rat. *J Comp Neurol* 398: 179–205.

Buzsaki G. 1989. Two-stage model of memory trace formation: a role for "noisy" brain states. *Neuroscience* 31: 551–570.

Buzsaki G. 2002. Theta oscillations in the hippocampus. *Neuron* 33: 325–340.

Buzsaki G. 2006. *Rhythms of the Brain.* New York: Oxford University Press.

Buzsaki G, Eidelberg E. 1983. Phase relations of hippocampal projection cells and interneurons to theta activity in the anesthetized rat. *Brain Res* 266: 334–339.

Buzsaki G, Grastyan E, Czopf J, Kellenyi L, Prohaska O. 1981. Changes in neuronal transmission in the rat hippocampus during behavior. *Brain Res* 225: 235–247.

Buzsaki G, Leung LW, Vanderwolf CH. 1983. Cellular bases of hippocampal EEG in the behaving rat. *Brain Res* 287: 139–171.

Buzsaki G, Czopf J, Kondakor I, Kellenyi L. 1986. Laminar distribution of hippocampal rhythmic slow activity (RSA) in the behaving rat: current-source density analysis, effects of urethane and atropine. *Brain Res* 365: 125–137.

Buzsaki G, Horvath Z, Urioste R, Hetke J, Wise K. 1992. High-frequency network oscillation in the hippocampus. *Science* 25: 1025–1027.

Byrne P, Becker S, Burgess N. 2007. Remembering the past and imagining the future: a neural model of spatial memory and imagery. *Psychol Rev* 114: 340–375.

Caballero-Bleda M, Witter MP. 1993. Regional and laminar organization of projections from the presubiculum and parasubiculum to the entorhinal cortex: an anterograde tracing study in the rat. *J Comp Neurol* 328: 115–129.

Caballero-Bleda M, Witter MP. 1994. Projections from the presubiculum and the parasubiculum to morphologically characterized entorhinal-hippocampal projection neurons in the rat. *Exp Brain Res* 101: 93–108.

Caine ED, Weingartner H, Ludlow CL, Cudahy EA, Wehry S. 1981. Qualitative analysis of scopolamine-induced amnesia. *Psychopharm* 74: 74–80.

Canto CB, Wouterlood FG, Witter MP. 2008. What does the anatomical organization of the entorhinal cortex tell us? *Neural Plast* 2008: 381243.

Caramanos Z, Shapiro ML. 1994. Spatial memory and N-methyl-D-aspartate receptor antagonists APV and MK-801: memory impairments depend on familiarity with the environment, drug dose, and training duration. *Behav Neurosci* 108: 30–43.

Carnevale NT, Hines ML. 2005. *The NEURON Book*. Cambridge, UK: Cambridge University Press.

Cassel JC, Kelche C. 1989. Scopolamine treatment and fimbria–fornix lesions: mimetic effects on radial maze performance. *Physiol Behav* 46: 347–353.

Cave CB, Squire LR. 1992. Intact verbal and nonverbal short-term memory following damage to the human hippocampus. *Hippocampus* 2: 151–163.

Chapman CA, Lacaille JC. 1999. Cholinergic induction of theta-frequency oscillations in hippocampal inhibitory interneurons and pacing of pyramidal cell firing. *J Neurosci* 19: 8637–8645.

Chen S, Wang J, Siegelbaum SA. 2001. Properties of hyperpolarization-activated pacemaker current defined by coassembly of HCN1 and HCN2 subunits and basal modulation by cyclic nucleotide. *J Gen Physiol* 117: 491–504.

Cheong MY, Yun SH, Mook-Jung I, Joo I, Huh K, Jung MW. 2001. Cholinergic modulation of synaptic physiology in deep layer entorhinal cortex of the rat. *J Neurosci Res* 66: 117–121.

Cho J, Sharp PE. 2001. Head direction, place, and movement correlates for cells in the rat retrosplenial cortex. *Behav Neurosci* 115: 3–25.

Chow CC, White JA, Ritt J, Kopell N. 1998. Frequency control in synchronized networks of inhibitory neurons. *J Comput Neurosci* 5: 407–420.

Chrobak JJ, Buzsaki G. 1994. Selective activation of deep layer (V-VI) retrohippocampal cortical neurons during hippocampal sharp waves in the behaving rat. *J Neurosci* 14: 6160–6170.

Chrobak JJ, Buzsaki G. 1996. High-frequency oscillations in the output networks of the hippocampal–entorhinal axis of the freely behaving rat. *J Neurosci* 16: 3056–3066.

Chrobak JJ, Buzsaki G. 1998. Gamma oscillations in the entorhinal cortex of the freely behaving rat. *J Neurosci* 18: 388–398.

Clayton NS, Bussey TJ, Dickinson A. 2003. Can animals recall the past and plan for the future? *Nat Rev Neurosci* 4: 685–691.

Cohen MA, Grossberg S. 1983. Absolute stability of global pattern formation and parallel memory storage by competitive neural networks. *IEEE Trans Syst Man Cybern* 13: 815–826.

Cohen NJ, Eichenbaum H. 1993. *Memory, Amnesia and the Hippocampal System.* Cambridge, MA: MIT Press.

Cohen NJ, Squire LR. 1980. Preserved learning and retention of pattern-analyzing skill in amnesia: dissociation of knowing how and knowing that. *Science* 210: 207–210.

Colbert CM, Levy W. 1992. Electrophysiological and pharmacological characterization of perforant path synapses in CA1: mediation by glutamate receptors. *J Neurophysiol* 68: 1–8.

Cole AE, Nicoll RA. 1984. Characterization of a slow cholinergic postsynaptic potential recorded in vitro from rat hippocampal pyramidal cells. *J Physiol* 352: 173–188.

Colgin LL, Denninger T, Fyhn M, Hafting T, Bonnevie T, Jensen O, Moser MB, Moser EI. 2009. Frequency of gamma oscillations routes flow of information in the hippocampus. *Nature* 462: 353–357.

Collins AM, Quillian MR. 1969. Retrieval time from semantic memory. *J Verbal Learn Verbal Behav* 8: 240–247.

Conway MA. 2009. Episodic memories. *Neuropsychologia* 47: 2305–2313.

Corkin S. 1984. Lasting consequences of bilateral medial temporal lobectomy: clinical course and experimental findings in H.M. *Semin Neurol* 4: 249–259.

Corkin S, Amaral DG, Gonzalez RG, Johnson KA, Hyman BT. 1997. H. M.'s medial temporal lobe lesion: findings from magnetic resonance imaging. *J Neurosci* 17: 3964–3979.

Correll RE, Scoville WB. 1965. Performance on delayed match following lesions of medial temporal lobe structures. *J Comp Physiol Psychol* 60: 360–367.

Coyle JT, Tsai G, Goff D. 2003. Converging evidence of NMDA receptor hypofunction in the pathophysiology of schizophrenia. *Ann NY Acad Sci* 1003: 318–327.

Csicsvari J, Hirase H, Czurko A, Mamiya A, Buzsaki G. 1999. Oscillatory coupling of hippocampal pyramidal cells and interneurons in the behaving rat. *J Neurosci* 19: 274–287.

Cutsuridis V, Hasselmo ME. 2011. Spatial memory sequence encoding and replay during theta and ripple oscillations. *Neural Netw* (in press).

Cutsuridis V, Cobb S, Graham BP. 2010. Encoding and retrieval in a model of the hippocampal CA1 microcircuit. *Hippocampus* 20: 423–446.

Dasari S, Gulledge AT. 2011. M1 and m4 receptors modulate hippocampal pyramidal neurons. *J Neurophysiol* 105: 779–792.

Davidson TJ, Kloosterman F, Wilson MA. 2009. Hippocampal replay of extended experience. *Neuron* 63: 497–507.

DeLuca J. 1993. Predicting neurobehavioral patterns following anterior communicating artery aneurysm. *Cortex* 29: 639–647.

De Rosa E, Hasselmo ME. 2000. Muscarinic cholinergic neuromodulation reduces proactive interference between stored odor memories during associative learning in rats. *Behav Neurosci* 114: 32–41.

De Rosa E, Hasselmo ME, Baxter MG. 2001. Contribution of the cholinergic basal forebrain to proactive interference from stored odor memories during associative learning in rats. *Behav Neurosci* 115: 314–327.

de Sevilla DF, Cabezas C, de Prada AN, Sanchez-Jimenez A, Buno W. 2002. Selective muscarinic regulation of functional glutamatergic Schaffer collateral synapses in rat CA1 pyramidal neurons. *J Physiol* 545: 51–63.

Denham MJ, Borisyuk RM. 2000. A model of theta rhythm production in the septal–hippocampal system and its modulation by ascending brain stem pathways. *Hippocampus* 10: 698–716.

Derdikman D, Whitlock JR, Tsao A, Fyhn M, Hafting T, Moser MB, Moser EI. 2009. Fragmentation of grid cell maps in a multicompartment environment. *Nat Neurosci* 12: 1325–1332.

Dere E, Kart-Teke E, Huston JP, De Souza Silva MA. 2006. The case for episodic memory in animals. *Neurosci Biobehav Rev* 30: 1206–1224.

Descarries L, Gisiger V, Steriade M. 1997. Diffuse transmission by acetylcholine in the CNS. *Prog Neurobiol* 53: 603–625.

Deshmukh SS, Yoganarasimha D, Voicu H, Knierim JJ. 2010. Theta modulation in the medial and the lateral entorhinal cortices. *J Neurophysiol* 104: 994–1006.

Dhillon A, Jones RS. 2000. Laminar differences in recurrent excitatory transmission in the rat entorhinal cortex in vitro. *Neuroscience* 99: 413–422.

Diba K, Buzsaki G. 2007. Forward and reverse hippocampal place-cell sequences during ripples. *Nat Neurosci* 10: 1241–1242.

Dickson CT, Mena AR, Alonso A. 1997. Electroresponsiveness of medial entorhinal cortex layer III neurons in vitro. *Neuroscience* 81: 937–950.

Dickson CT, Magistretti J, Shalinsky MH, Fransén E, Hasselmo ME, Alonso A. 2000. Properties and role of I(h) in the pacing of subthreshold oscillations in entorhinal cortex layer II neurons. *J Neurophysiol* 83: 2562–2579.

Doeller CF, Barry C, Burgess N. 2010. Evidence for grid cells in a human memory network. *Nature* 463: 657–661.

Dragoi G, Tonegawa S. 2011. Preplay of future place cell sequences by hippocampal cellular assemblies. *Nature* 469: 397–401.

Dragoi G, Carpi D, Recce M, Csicsvari J, Buzsaki G. 1999. Interactions between hippocampus and medial septum during sharp waves and theta oscillation in the behaving rat. *J Neurosci* 19: 6191–6199.

Drever BD, Anderson WG, Johnson H, O'Callaghan M, Seo S, Choi DY, Riedel G, Platt B. 2007. Memantine acts as a cholinergic stimulant in the mouse hippocampus. *J Alzheimers Dis* 12: 319–333.

Dudman JT, Nolan MF. 2009. Stochastically gating ion channels enable patterned spike firing through activity-dependent modulation of spike probability. *PLOS Comput Biol* 5: e1000290.

Durstewitz D, Seamans JK, Sejnowski TJ. 2000. Dopamine-mediated stabilization of delay-period activity in a network model of prefrontal cortex. *J Neurophysiol* 83: 1733–1750.

Eacott MJ, Norman G. 2004. Integrated memory for object, place, and context in rats: a possible model of episodic-like memory? *J Neurosci* 24: 1948–1953.

Eacott MJ, Gaffan D, Murray EA. 1994. Preserved recognition memory for small sets, and impaired stimulus identification for large sets, following rhinal cortex ablations in monkeys. *Eur J Neurosci* 6: 1466–1478.

Eacott MJ, Easton A, Zinkivskay A. 2005. Recollection in an episodic-like memory task in the rat. *Learn Mem* 12: 221–223.

Easton A, Zinkivskay A, Eacott MJ. 2009. Recollection is impaired, but familiarity remains intact in rats with lesions of the fornix. *Hippocampus* 19: 837–843.

Egorov AV, Hamam BN, Fransén E, Hasselmo ME, Alonso AA. 2002. Graded persistent activity in entorhinal cortex neurons. *Nature* 420: 173–178.

Eichenbaum H, Buckingham J. 1990. Studies on hippocampal processing: experiment, theory and model. In: *Learning and Computational Neuroscience: Foundation of Adaptive Networks* (Gabriel M, Moore J, eds), pp 171–231. Cambridge, MA: MIT Press.

Eichenbaum H, Cohen NJ. 2001. *From Conditioning to Conscious Recollection: Memory Systems of the Brain.* New York: Oxford University Press.

Eichenbaum H, Lipton PA. 2008. Towards a functional organization of the medial temporal lobe memory system: role of the parahippocampal and medial entorhinal cortical areas. *Hippocampus* 18: 1314–1324.

Eichenbaum H, Kuperstein M, Fagan A, Nagode J. 1987. Cue-sampling and goal-approach correlates of hippocampal unit-activity in rats performing an odor-discrimination task. *J Neurosci* 7: 716–732.

Eichenbaum H, Wiener SI, Shapiro ML, Cohen NJ. 1989. The organization of spatial coding in the hippocampus: a study of neural ensemble activity. *J Neurosci* 9: 2764–2775.

Eichenbaum H, Stewart C, Morris RG. 1990. Hippocampal representation in place learning. *J Neurosci* 10: 3531–3542.

Eichenbaum H, Dudchenko P, Wood E, Shapiro M, Tanila H. 1999. The hippocampus, memory, and place cells: is it spatial memory or a memory space? *Neuron* 23: 209–226.

Eichenbaum H, Yonelinas AP, Ranganath C. 2007. The medial temporal lobe and recognition memory. *Annu Rev Neurosci* 30: 123–152.

Elvander E, Schott PA, Sandin J, Bjelke B, Kehr J, Yoshitake T, Ogren SO. 2004. Intraseptal muscarinic ligands and galanin: influence on hippocampal acetylcholine and cognition. *Neuroscience* 126: 541–557.

Engel TA, Schimansky-Geier L, Herz AV, Schreiber S, Erchova I. 2008. Subthreshold membrane-potential resonances shape spike-train patterns in the entorhinal cortex. *J Neurophysiol* 100: 1576–1589.

Ennaceur A, Neave N, Aggleton JP. 1996. Neurotoxic lesions of the perirhinal cortex do not mimic the behavioral effects of fornix transection in the rat. *Behav Brain Res* 80: 9–25.

Epstein RA, Parker WE, Feiler AM. 2007a. Where am I now? Distinct roles for parahippocampal and retrosplenial cortices in place recognition. *J Neurosci* 27: 6141–6149.

Epstein RA, Higgins JS, Jablonski K, Feiler AM. 2007b. Visual scene processing in familiar and unfamiliar environments. *J Neurophysiol* 97: 3670–3683.

Erchova I, Kreck G, Heinemann U, Herz AV. 2004. Dynamics of rat entorhinal cortex layer II and III cells: characteristics of membrane potential resonance at rest predict oscillation properties near threshold. *J Physiol* 560: 89–110.

Erdem M, Hasselmo ME. 2010. A model of forward replay of grid cell activity for selection of goal-directed trajectories. *Soc Neurosci Abstr* 36: 101.24.

Erwin E, Obermayer K, Schulten K. 1995. Models of orientation and ocular dominance columns in the visual cortex: a critical comparison. *Neural Comput* 7: 425–468.

Esclassan F, Coutureau E, Di Scala G, Marchand AR. 2009. A cholinergic-dependent role for the entorhinal cortex in trace fear conditioning. *J Neurosci* 29: 8087–8093.

Fahy FL, Riches IP, Brown MW. 1993. Neuronal activity related to visual recognition memory: long-term memory and the encoding of recency and familiarity information in the primate anterior and medial inferior temporal and rhinal cortex. *Exp Brain Res* 96: 457–472.

Fantie BD, Goddard GV. 1982. Septal modulation of the population spike in the fascia dentata produced by perforant path stimulation in the rat. *Brain Res* 252: 227–237.

Fenton AA, Muller RU. 1998. Place cell discharge is extremely variable during individual passes of the rat through the firing field. *Proc Natl Acad Sci USA* 95: 3182–3187.

Fenton AA, Kao HY, Neymotin SA, Olypher A, Vayntrub Y, Lytton WW, Ludvig N. 2008. Unmasking the CA1 ensemble place code by exposures to small and large environments: more place cells and multiple, irregularly arranged, and expanded place fields in the larger space. *J Neurosci* 28: 11250–11262.

Ferbinteanu J, Shapiro ML. 2003. Prospective and retrospective memory coding in the hippocampus. *Neuron* 40: 1227–1239.

Fernandez G, Brewer JB, Zhao Z, Glover GH, Gabrieli JD. 1999. Level of sustained entorhinal activity at study correlates with subsequent cued-recall performance: a functional magnetic resonance imaging study with high acquisition rate. *Hippocampus* 9: 35–44.

Fiete IR, Burak Y, Brookings T. 2008. What grid cells convey about rat location. *J Neurosci* 28: 6858–6871.

Fletcher PC, Frith CD, Grasby PM, Shallice T, Frackowiak RS, Dolan RJ. 1995. Brain systems for encoding and retrieval of auditory-verbal memory. An in vivo study in humans. *Brain* 118(Pt 2): 401–416.

Flicker C, Serby M, Ferris SH. 1990. Scopolamine effects on memory, language visuospatial praxis and psychomotor speed. *Psychopharmacology* 100: 243–250.

Fortin NJ, Agster KL, Eichenbaum HB. 2002. Critical role of the hippocampus in memory for sequences of events. *Nat Neurosci* 5: 458–462.

Fortin NJ, Wright SP, Eichenbaum H. 2004. Recollection-like memory retrieval in rats is dependent on the hippocampus. *Nature* 431: 188–191.

Foster DJ, Wilson MA. 2006. Reverse replay of behavioral sequences in hippocampal place cells during the awake state. *Nature* 440: 680–683.

Foster DJ, Morris RG, Dayan P. 2000. A model of hippocampally dependent navigation, using the temporal difference learning rule. *Hippocampus* 10: 1–16.

Foster TC, Deadwyler SA. 1992. Acetylcholine modulates averaged sensory evoked responses and perforant path evoked field potentials in the rat dentate gyrus. *Brain Res* 587: 95–101.

Fox SE. 1989. Membrane potential and impedance changes in hippocampal pyramidal cells during theta rhythm. *Exp Brain Res* 77: 283–294.

Fox SE, Wolfson S, Ranck JB, Jr. 1986. Hippocampal theta rhythm and the firing of neurons in walking and urethane anesthetized rats. *Brain Res* 62: 495–508.

Frank LM, Brown EN, Wilson M. 2000. Trajectory encoding in the hippocampus and entorhinal cortex. *Neuron* 27: 169–178.

Fransén E, Lansner A. 1995. Low spiking rates in a population of mutually exciting pyramidal cells. *Network* 6: 271–288.

Fransén E, Alonso AA, Hasselmo ME. 2002. Simulations of the role of the muscarinic-activated calcium-sensitive nonspecific cation current INCM in entorhinal neuronal activity during delayed matching tasks. *J Neurosci* 22: 1081–1097.

Fransén E, Alonso AA, Dickson CT, Magistretti J, Hasselmo ME. 2004. Ionic mechanisms in the generation of subthreshold oscillations and action potential clustering in entorhinal layer II stellate neurons. *Hippocampus* 14: 368–384.

Fransén E, Tahvildari B, Egorov AV, Hasselmo ME, Alonso AA. 2006. Mechanism of graded persistent cellular activity of entorhinal cortex layer v neurons. *Neuron* 49: 735–746.

Freeman JH, Jr, Stanton ME. 1991. Fimbria–fornix transections disrupt the ontogeny of delayed alternation but not position discrimination in the rat. *Behav Neurosci* 105: 386–395.

Freund TF, Antal M. 1988. GABA-containing neurons in the septum control inhibitory interneurons in the hippocampus. *Nature* 336: 170–173.

Frith CD, Richardson JTE, Samuel M, Crow TJ, McKenna PJ. 1984. The effects of intravenous diazepam and hyoscine upon human memory. *Q J Exp Psychol* 36A: 133–144.

Fuhs MC, Touretzky DS. 2006. A spin glass model of path integration in rat medial entorhinal cortex. *J Neurosci* 26: 4266–4276.

Fujita Y, Sato T. 1964. Intracellular records from hippocampal pyramidal cells in rabbit during theta rhythm activity. *J Neurophysiol* 27: 1011–1025.

Fuster JM. 1995. *Memory in the Cerebral Cortex.* Cambridge, MA: MIT Press.

Fyhn M, Molden S, Witter MP, Moser EI, Moser MB. 2004. Spatial representation in the entorhinal cortex. *Science* 305: 1258–1264.

Fyhn M, Hafting T, Treves A, Moser MB, Moser EI. 2007. Hippocampal remapping and grid realignment in entorhinal cortex. *Nature* 446: 190–194.

Gabrieli JD, Brewer JB, Desmond JE, Glover GH. 1997. Separate neural bases of two fundamental memory processes in the human medial temporal lobe. *Science* 276: 264–266.

Gaffan D. 1974. Recognition impaired and association intact in the memory of monkeys after transection of the fornix. *J Comp Physiol Psychol* 86: 1100–1109.

Gaffan D, Harrison S. 1989. Place memory and scene memory: effects of fornix transection in the monkey. *Exp Brain Res* 74: 202–212.

Gaffan D, Murray EA. 1992. Monkeys (*Macaca fascicularis*) with rhinal cortex ablations succeed in object discrimination learning despite 24-hr intertrial intervals and fail at matching to sample despite double sample presentations. *Behav Neurosci* 106: 30–38.

Gaffan D, Saunders RC, Gaffan EA, Harrison S, Shields C, Owen MJ. 1984. Effects of fornix transection upon associative memory in monkeys: role of the hippocampus in learned action. *Q J Exp Psychol B* 36: 173–221.

Gais S, Born J. 2004. Low acetylcholine during slow-wave sleep is critical for declarative memory consolidation. *Proc Natl Acad Sci USA* 101: 2140–2144.

Garden DL, Dodson PD, O'Donnell C, White MD, Nolan MF. 2008. Tuning of synaptic integration in the medial entorhinal cortex to the organization of grid cell firing fields. *Neuron* 60: 875–889.

Gardner-Medwin AR. 1976. The recall of events through the learning of associations between their parts. *Proc R Soc Lond B Biol Sci* 194: 375–402.

Gasparini S, Magee JC. 2006. State-dependent dendritic computation in hippocampal CA1 pyramidal neurons. *J Neurosci* 26: 2088–2100.

Gelbard-Sagiv H, Mukamel R, Harel M, Malach R, Fried I. 2008. Internally generated reactivation of single neurons in human hippocampus during free recall. *Science* 322: 96–101.

Ghoneim MM, Mewaldt SP. 1975. Effects of diazepam and scopolamine on storage, retrieval and organizational processes in memory. *Psychopharmacology (Berl)* 44: 257–262.

Ghoneim MM, Mewaldt SP. 1977. Studies on human memory: the interactions of diazepam, scopolamine, and physostigmine. *Psychopharmacology (Berl)* 52: 1–6.

Gil Z, Conners BW, Amitai Y. 1997. Differential regulation of neocortical synapses by neuromodulators and activity. *Neuron* 19: 679–686.

Gilbert PE, Kesner RP, Lee I. 2001. Dissociating hippocampal subregions: double dissociation between dentate gyrus and CA1. *Hippocampus* 11: 626–636.

Gioanni Y, Rougeot C, Clarke PB, Lepouse C, Thierry AM, Vidal C. 1999. Nicotinic receptors in the rat prefrontal cortex: increase in glutamate release and facilitation of mediodorsal thalamocortical transmission. *Eur J Neurosci* 11: 18–30.

Giocomo LM, Hasselmo ME. 2005. Nicotinic modulation of glutamatergic synaptic transmission in region CA3 of the hippocampus. *Eur J Neurosci* 22: 1349–1356.

Giocomo LM, Hasselmo ME. 2008a. Computation by oscillations: implications of experimental data for theoretical models of grid cells. *Hippocampus* 18: 1186–1199.

Giocomo LM, Hasselmo ME. 2008b. Time constants of h current in layer II stellate cells differ along the dorsal to ventral axis of medial entorhinal cortex. *J Neurosci* 28: 9414–9425.

Giocomo LM, Hasselmo ME. 2009. Knock-out of HCN1 subunit flattens dorsal–ventral frequency gradient of medial entorhinal neurons in adult mice. *J Neurosci* 29: 7625–7630.

Giocomo LM, Hussaini SA, Zhang F, Kandel ER, Moser M-B, Moser EI. 2010. Scale of grid cells increases in HCN1 knockout mice. *Soc Neurosci Abstr* 36: 101.16.

Giocomo LM, Zilli EA, Fransén E, Hasselmo ME. 2007. Temporal frequency of subthreshold oscillations scales with entorhinal grid cell field spacing. *Science* 315: 1719–1722.

Givens B. 1996. Stimulus-evoked resetting of the dentate theta rhythm: relation to working memory. *Neuroreport* 8: 159–163.

Givens B, Olton DS. 1994. Local modulation of basal forebrain: effects on working and reference memory. *J Neurosci* 14: 3578–3587.

Givens BS, Olton DS. 1990. Cholinergic and GABAergic modulation of the medial septal area: effect on working memory. *Behav Neurosci* 104: 849–855.

Golding N, Staff N, Spruston N. 2002. Dendritic spikes as a mechanism for cooperative long-term potentiation. *Nature* 418: 326–331.

Goodridge JP, Taube JS. 1997. Interaction between the postsubiculum and anterior thalamus in the generation of head direction cell activity. *J Neurosci* 17: 9315–9330.

Gorchetchnikov A, Grossberg S. 2007. Space, time and learning in the hippocampus: how fine spatial and temporal scales are expanded into population codes for behavioral control. *Neural Netw* 20: 182–193.

Gorchetchnikov A, Hasselmo ME. 2005. A biophysical implementation of a bidirectional graph search algorithm to solve multiple goal navigation tasks. *Connect Sci* 17: 145–164.

Gorchetchnikov A, Versace M, Hasselmo ME. 2005. A model of STDP based on spatially and temporally local information: derivation and combination with gated decay. *Neural Netw* 18: 458–466.

Gothard KM, Skaggs WE, McNaughton BL. 1996. Dynamics of mismatch correction in the hippocampal ensemble code for space: interaction between path integration and environmental cues. *J Neurosci* 16: 8027–8040.

Grady CL, McIntosh AR, Horwitz B, Maisog JM, Ungerleider LG, Mentis MJ, Pietrini P, Schapiro MB, Haxby JV. 1995. Age-related reductions in human recognition memory due to impaired encoding. *Science* 269: 218–221.

Graf P, Squire LR, Mandler G. 1984. The information that amnesic patients do not forget. *J Exp Psychol Learn Mem Cogn* 10: 164–178.

Gray JA. 1982. *The Neuropsychology of Anxiety: An Enquiry into the Functions of the Septo–Hippocampal System.* Oxford, UK: Oxford University Press.

Green A, Ellis KA, Ellis J, Bartholomeusz CF, Ilic S, Croft RJ, Luan Phan K, Nathan PJ. 2005. Muscarinic and nicotinic receptor modulation of object and spatial n-back working memory in humans. *Pharmacol Biochem Behav* 81: 575–584.

Green JD, Arduini AA. 1954. Hippocampal electrical activity and arousal. *J Neurophysiol* 17: 533–557.

Greenstein-Messica A, Ruppin E. 1998. Synaptic runaway in associative networks and the pathogenesis of schizophrenia. *Neural Comput* 10: 451–465.

Greicius MD, Krasnow B, Boyett-Anderson JM, Eliez S, Schatzberg AF, Reiss AL, Menon V. 2003. Regional analysis of hippocampal activation during memory encoding and retrieval: fMRI study. *Hippocampus* 13: 164–174.

Griffin AL, Asaka Y, Darling RD, Berry SD. 2004. Theta-contingent trial presentation accelerates learning rate and enhances hippocampal plasticity during trace eyeblink conditioning. *Behav Neurosci* 118: 403–411.

Griffin AL, Eichenbaum H, Hasselmo ME. 2007. Spatial representations of hippocampal CA1 neurons are modulated by behavioral context in a hippocampus-dependent memory task. *J Neurosci* 27: 2416–2423.

Guanella A, Kiper D, Verschure P. 2007. A model of grid cells based on a twisted torus topology. *Int J Neural Syst* 17: 231–240.

Gustafsson B, Wigstrom H, Abraham WC, Huang YY. 1987. Long-term potentiation in the hippocampus using depolarizing current pulses as the conditioning stimulus to single volley synaptic potentials. *J Neurosci* 7: 774–780.

Guzowski JF, Knierim JJ, Moser EI. 2004. Ensemble dynamics of hippocampal regions CA3 and CA1. *Neuron* 44: 581–584.

Hafting T, Fyhn M, Molden S, Moser MB, Moser EI. 2005. Microstructure of a spatial map in the entorhinal cortex. *Nature* 436: 801–806.

Hafting T, Fyhn M, Bonnevie T, Moser MB, Moser EI. 2008. Hippocampus-independent phase precession in entorhinal grid cells. *Nature* 453: 1248–1252.

Haj-Dahmane S, Andrade R. 1996. Muscarinic activation of a voltage-dependent cation nonselective current in rat association cortex. *J Neurosci* 16: 3848–3861.

Haj-Dahmane S, Andrade R. 1998. Ionic mechanism of the slow afterdepolarization induced by muscarinic receptor activation in rat prefrontal cortex. *J Neurophysiol* 80: 1197–1210.

Halliwell JV. 1986. M-current in human neocortical neurones. *Neurosci Lett* 67: 1–6.

Halliwell JV. 1989. Cholinergic responses in human neocortical neurones. In: *Central Cholinergic Synaptic Transmission* (Frotscher M, Misgeld U, eds). Boston: Birkhauser. Pp. 138–149.

Hargreaves EL, Rao G, Lee I, Knierim JJ. 2005. Major dissociation between medial and lateral entorhinal input to dorsal hippocampus. *Science* 308: 1792–1794.

Harvey CD, Collman F, Dombeck DA, Tank DW. 2009. Intracellular dynamics of hippocampal place cells during virtual navigation. *Nature* 461): 941–946.

Hasler G, Drevets WC, Manji HK, Charney DS. 2004. Discovering endophenotypes for major depression. *Neuropsychopharmacology* 29: 1765-1781.

Hassabis D, Kumaran D, Vann SD, Maguire EA. 2007. Patients with hippocampal amnesia cannot imagine new experiences. *Proc Natl Acad Sci USA* 104: 1726–1731.

Hasselmo ME. 1993. Acetylcholine and learning in a cortical associative memory. *Neural Comput* 5: 32–44.

Hasselmo ME. 1994. Runaway synaptic modification in models of cortex: implications for Alzheimer's disease. *Neural Netw* 7: 13–40.

Hasselmo ME. 1995. Neuromodulation and cortical function: modeling the physiological basis of behavior. *Behav Brain Res* 67: 1–27.

Hasselmo ME. 1999a. Neuromodulation and the hippocampus: memory function and dysfunction in a network simulation. *Prog Brain Res* 121: 3–18.

Hasselmo ME. 1999b. Neuromodulation: acetylcholine and memory consolidation. *Trends Cogn Sci* 3: 351–359.

Hasselmo ME. 2005a. What is the function of hippocampal theta rhythm?—Linking behavioral data to phasic properties of field potential and unit recording data. *Hippocampus* 15: 936–949.

Hasselmo ME. 2005b. A model of prefrontal cortical mechanisms for goal-directed behavior. *J Cogn Neurosci* 17: 1115–1129.

Hasselmo ME. 2006. The role of acetylcholine in learning and memory. *Curr Opin Neurobiol* 16: 710–715.

Hasselmo ME. 2007. Arc length coding by interference of theta frequency oscillations may underlie context-dependent hippocampal unit data and episodic memory function. *Learn Mem* 14: 782–794.

Hasselmo ME. 2008a. Neuroscience: the scale of experience. *Science* 321: 46–47.

Hasselmo ME. 2008b. Grid cell mechanisms and function: contributions of entorhinal persistent spiking and phase resetting. *Hippocampus* 18: 1213–1229.

Hasselmo ME. 2008c. Temporally structured replay of neural activity in a model of entorhinal cortex, hippocampus and postsubiculum. *Eur J Neurosci* 28: 1301–1315.

Hasselmo ME. 2009. A model of episodic memory: mental time travel along encoded trajectories using grid cells. *Neurobiol Learn Mem* 92: 559–573.

Hasselmo ME, Barkai E. 1995. Cholinergic modulation of activity-dependent synaptic plasticity in the piriform cortex and associative memory function in a network biophysical simulation. *J Neurosci* 15: 6592–6604.

Hasselmo ME, Bower JM. 1992. Cholinergic suppression specific to intrinsic not afferent fiber synapses in rat piriform (olfactory) cortex. *J Neurophysiol* 67: 1222–1229.

Hasselmo ME, Bower JM. 1993. Acetylcholine and memory. Trends Neurosci 16: 218–222.

Hasselmo ME, Brandon MA. 2008. Linking cellular mechanisms to behavior: entorhinal persistent spiking and membrane potential oscillations may underlie path integration, grid cell firing and episodic memory. *Neural Plast* 2008: 658323.

Hasselmo ME, Cekic M. 1996. Suppression of synaptic transmission may allow combination of associative feedback and self-organizing feedforward connections in the neocortex. *Behav Brain Res* 79: 153–161.

Hasselmo ME, Eichenbaum H. 2005. Hippocampal mechanisms for the context-dependent retrieval of episodes. *Neural Netw* 18: 1172–1190.

Hasselmo ME, Fehlau BP. 2001. Differences in time course of ACh and GABA modulation of excitatory synaptic potentials in slices of rat hippocampus. J *Neurophysiol* 86: 1792–1802.

Hasselmo ME, McGaughy J. 2004. High acetylcholine levels set circuit dynamics for attention and encoding and low acetylcholine levels set dynamics for consolidation. *Prog Brain Res* 145:207–231.

Hasselmo ME, Schnell E. 1994. Laminar selectivity of the cholinergic suppression of synaptic transmission in rat hippocampal region CA1: computational modeling and brain slice physiology. *J Neurosci* 14: 3898–3914.

Hasselmo ME, Stern CE. 2006. Mechanisms underlying working memory for novel information. *Trends Cogn Sci* 10: 487–493.

Hasselmo ME, Wyble BP. 1997. Free recall and recognition in a network model of the hippocampus: simulating effects of scopolamine on human memory function. *Behav Brain Res* 89: 1–34.

Hasselmo ME, Rolls ET, Baylis GC. 1989. The role of expression and identity in the face-selective responses of neurons in the temporal visual cortex of the monkey. *Behav Brain Res* 32: 203–218.

Hasselmo ME, Anderson BP, Bower JM. 1992. Cholinergic modulation of cortical associative memory function. *J Neurophysiol* 67: 1230–1246.

Hasselmo ME, Schnell E, Barkai E. 1995a. Dynamics of learning and recall at excitatory recurrent synapses and cholinergic modulation in rat hippocampal region CA3. *J Neurosci* 15: 5249–5262.

Hasselmo ME, Schnell E, Berke J, Barkai E. 1995b. A model of the hippocampus combining self-organization and associative memory function. In: *Advances in Neural Information Processing Systems* (Tesauro G, Touretzky D, Leen T, eds), pp 77–84. Cambridge, MA: MIT Press.

Hasselmo ME, Wyble BP, Wallenstein GV. 1996. Encoding and retrieval of episodic memories: role of cholinergic and GABAergic modulation in the hippocampus. *Hippocampus* 6: 693–708.

Hasselmo ME, Bodelon C, Wyble BP. 2002a. A proposed function for hippocampal theta rhythm: separate phases of encoding and retrieval enhance reversal of prior learning. *Neural Comput* 14: 793–817.

Hasselmo ME, Wyble BP, Cannon RC. 2002b. From spike frequency to free recall: how neural circuits perform encoding and retrieval. In: *The Cognitive Neuroscience of Memory: Encoding and Retrieval* (Parker A, Bussey TJ, Wilding E, eds), pp 325–354. London: Psychology Press.

Hasselmo ME, Hay J, Ilyn M, Gorchetchnikov A. 2002c. Neuromodulation, theta rhythm and rat spatial navigation. *Neural Netw* 15: 689–707.

Hasselmo ME, Giocomo LM, Zilli EA. 2007. Grid cell firing may arise from interference of theta frequency membrane potential oscillations in single neurons. *Hippocampus* 17: 1252–1271.

Haxby JV, Ungerleider LG, Horwitz B, Maisog JM, Rapoport SI, Grady CL. 1996. Face encoding and recognition in the human brain. *Proc Natl Acad Sci USA* 93: 922–927.

Hebb DO. 1949. *The Organization of Behavior.* New York: Wiley.

Henson RN. 1998. Short-term memory for serial order: the Start–End Model. *Cognit Psychol* 36: 73–137.

Herreras O, Solis JM, Herranz AS, Martin del Rio R, Lerma J. 1988. Sensory modulation of hippocampal transmission. II. Evidence for a cholinergic locus of inhibition in the Schaffer-CA1 synapse. *Brain Res* 461: 303–313.

Herrero JL, Roberts MJ, Delicato LS, Gieselmann MA, Dayan P, Thiele A. 2008. Acetylcholine contributes through muscarinic receptors to attentional modulation in V1. *Nature* 454: 1110–1114.

Heys JG, Giocomo LM, Hasselmo ME. 2010. Cholinergic modulation of the resonance properties of stellate cells in layer II of medial entorhinal cortex. *J Neurophysiol* 104: 258–270.

Hodgkin AL, Huxley AF. 1952. A quantitative description of membrane current and its application to conduction and excitation in nerve. *J Physiol Lond* 117: 500–544.

Holmes WR, Levy WB. 1990. Insights into associative long-term potentiation from computational models of NMDA receptor-mediated calcium influx and intracellular calcium concentration changes. *J Neurophysiol* 63: 1148–1168.

Holmes WR, Levy WB. 1997. Quantifying the role of inhibition in associative long-term potentiation in dentate granule cells with computational models. *J Neurophysiol* 78: 103–116.

Holscher C, Anwyl R, Rowan MJ. 1997. Stimulation on the positive phase of hippocampal theta rhythm induces long-term potentiation that can be depotentiated by stimulation on the negative phase in area CA1 in vivo. *J Neurosci* 17: 6470–6477.

Hopfield JJ. 1982. Neural networks and physical systems with emergent selective computational abilities. *Proc Natl Acad Sci USA* 79: 2554–2559.

Hopfield JJ. 1984. *Neurons* with graded response have collective computational properties like those of two-state neurons. *Proc Natl Acad Sci USA* 81: 3088–3092.

Hounsgaard J. 1978. Presynaptic inhibitory action of acetylcholine in area CA1 of the hippocampus. *Exp Neurol* 62: 787–797.

Howard MW, Kahana MJ. 2002. A distributed representation of temporal context. *J Math Psychol* 46: 269–299.

Howard MW, Fotedar MS, Datey AV, Hasselmo ME. 2005. The temporal context model in spatial navigation and relational learning: toward a common explanation of medial temporal lobe function across domains. *Psychol Rev* 112: 75–116.

Hudon C, Dore FY, Goulet S. 2002. Spatial memory and choice behavior in the radial arm maze after fornix transection. *Prog Neuropsychopharmacol Biol Psychiatry* 26: 1113–1123.

Huerta PT, Lisman JE. 1995. Bidirectional synaptic plasticity induced by a single burst during cholinergic theta oscillation in CA1 in vitro. *Neuron* 15: 1053–1063.

Huxter J, Burgess N, O'Keefe J. 2003. Independent rate and temporal coding in hippocampal pyramidal cells. *Nature* 425: 828–832.

Hyman JM, Wyble BP, Goyal V, Rossi CA, Hasselmo ME. 2003. Stimulation in hippocampal region CA1 in behaving rats yields long-term potentiation when delivered to the peak of theta and long-term depression when delivered to the trough. *J Neurosci* 23: 11725–11731.

Insausti R, Herrero MT, Witter MP. 1997. Entorhinal cortex of the rat: cytoarchitectonic subdivisions and the origin and distribution of cortical efferents. *Hippocampus* 7: 146–183.

Izhikevich EM. 2007. *Dynamical Systems in Neuroscience: The Geometry of Excitability and Bursting.* Cambridge, MA: MIT Press.

Jeewajee A, Barry C, O'Keefe J, Burgess N. 2008a. Grid cells and theta as oscillatory interference: electrophysiological data from freely moving rats. *Hippocampus* 18: 1175–1185.

Jeewajee A, Lever C, Burton S, O'Keefe J, Burgess N. 2008b. Environmental novelty is signaled by reduction of the hippocampal theta frequency. *Hippocampus* 18: 340–348.

Jeffery KJ, Donnett JG, O'Keefe J. 1995. Medial septal control of theta-correlated unit firing in the entorhinal cortex of awake rats. *Neuroreport* 6: 2166–2170.

Jensen O, Lisman JE. 1996a. Theta/gamma networks with slow NMDA channels learn sequences and encode episodic memory: role of NMDA channels in recall. *Learn Mem* 3: 264–278.

Jensen O, Lisman JE. 1996b. Novel lists of 7 +/– 2 known items can be reliably stored in an oscillatory short-term memory network: interaction with long-term memory. *Learn Mem* 3: 257–263.

Jensen O, Lisman JE. 1996c. Hippocampal CA3 region predicts memory sequences: accounting for the phase precession of place cells. *Learn Mem* 3: 279–287.

Jensen O, Lisman JE. 2005. Hippocampal sequence-encoding driven by a cortical multi-item working memory buffer. *Trends Neurosci* 28: 67–72.

Ji D, Wilson MA. 2007. Coordinated memory replay in the visual cortex and hippocampus during sleep. *Nat Neurosci* 10: 100–107.

Johnson A, Redish AD. 2007. Neural ensembles in CA3 transiently encode paths forward of the animal at a decision point. *J Neurosci* 27: 12176–12189.

Johnson A, Seeland K, Redish AD. 2005. Reconstruction of the postsubiculum head direction signal from neural ensembles. *Hippocampus* 15: 86–96.

Kahana MJ, Sekuler R, Caplan JB, Kirschen M, Madsen JR. 1999. Human theta oscillations exhibit task dependence during virtual maze navigation. *Nature* 399: 781–784.

Kahle JS, Cotman CW. 1989. Carbachol depresses synaptic responses in the medial but not the lateral perforant path. *Brain Res* 482: 159–163.

Kamondi A, Acsady L, Wang XJ, Buzsaki G. 1998. Theta oscillations in somata and dendrites of hippocampal pyramidal cells in vivo: activity-dependent phase-precession of action potentials. *Hippocampus* 8: 244–261.

Karlsson MP, Frank LM. 2009. Awake replay of remote experiences in the hippocampus. *Nat Neurosci* 12: 913–918.

Katz Y, Kath WL, Spruston N, Hasselmo ME. 2007. Coincidence detection of place and temporal context in a network model of spiking hippocampal neurons. *PLOS Comput Biol* 3: e234.

Kelso SR, Ganong AH, Brown TH. 1986. Hebbian synapses in hippocampus. *Proc Natl Acad Sci USA* 83: 5326–5330.

Kesner RP, Gilbert PE, Barua LA. 2002. The role of the hippocampus in memory for the temporal order of a sequence of odors. *Behav Neurosci* 116: 286–290.

Kim JJ, Fanselow MS. 1992. Modality-specific retrograde amnesia of fear. *Science* 256: 675–677.

Kimura F. 2000. Cholinergic modulation of cortical function: a hypothetical role in shifting the dynamics in cortical network. *Neurosci Res* 38: 19–26.

Kimura F, Baughman RW. 1997. Distinct muscarinic receptor subtypes suppress excitatory and inhibitory synaptic responses in cortical neurons. *J Neurophysiol* 77: 709–716.

Kirchhoff BA, Wagner AD, Maril A, Stern CE. 2000. Prefrontal–temporal circuitry for episodic encoding and subsequent memory. *J Neurosci* 20: 6173–6180.

Kirwan CB, Bayley PJ, Galvan VV, Squire LR. 2008. Detailed recollection of remote autobiographical memory after damage to the medial temporal lobe. *Proc Natl Acad Sci USA* 105: 2676–2680.

Kjelstrup KB, Solstad T, Brun VH, Hafting T, Leutgeb S, Witter MP, Moser EI, Moser MB. 2008. Finite scale of spatial representation in the hippocampus. *Science* 321: 140–143.

Klausberger T, Somogyi P. 2008. Neuronal diversity and temporal dynamics: the unity of hippocampal circuit operations. *Science* 321: 53–57.

Klausberger T, Magill PJ, Marton LF, Roberts JD, Cobden PM, Buzsaki G, Somogyi P. 2003. Brain-state- and cell-type-specific firing of hippocampal interneurons in vivo. *Nature* 421: 844–848.

Kleinfeld D. 1986. Sequential state generation by model neural networks. *Proc Natl Acad Sci USA* 83: 9469–9473.

Klink R, Alonso A. 1997a. Muscarinic modulation of the oscillatory and repetitive firing properties of entorhinal cortex layer II neurons. *J Neurophysiol* 77: 1813–1828.

Klink R, Alonso A. 1997b. Ionic mechanisms of muscarinic depolarization in entorhinal cortex layer II neurons. *J Neurophysiol* 77: 1829–1843.

Knierim JJ, Kudrimoti HS, McNaughton BL. 1995. Place cells, head direction cells, and the learning of landmark stability. *J Neurosci* 15: 1648–1659.

Knierim JJ. 2002. Dynamic interactions between local surface cues, distal landmarks, and intrinsic circuitry in hippocampal place cells. *J Neurosci* 22: 6254–6264.

Koene RA, Hasselmo ME. 2005. An integrate-and-fire model of prefrontal cortex neuronal activity during performance of goal-directed decision making. *Cereb Cortex* 15: 1964–1981.

Koene RA, Hasselmo ME. 2007. First-in-first-out item replacement in a model of short-term memory based on persistent spiking. *Cereb Cortex* 17: 1766–1781.

Koene RA, Gorchetchnikov A, Cannon RC, Hasselmo ME. 2003. Modeling goal-directed spatial navigation in the rat based on physiological data from the hippocampal formation. *Neural Netw* 16: 577–584.

Koenig J, Linder AN, Leutgeb JK, Leutgeb S. 2010. The spatial periodicity of grid cells is not sustained during reduced theta oscillations. *Science* 332: 592–595.

Kohonen T. 1972. Correlation matrix memories. *IEEE Trans Comput* C-21: 353–359.

Komorowski RW, Manns JR, Eichenbaum H. 2009. Robust conjunctive item-place coding by hippocampal neurons parallels learning what happens where. *J Neurosci* 29: 9918–9929.

Konopacki J, MacIver MB, Bland BH, Roth SH. 1987. Carbachol-induced EEG "theta" activity in hippocampal brain slices. *Brain Res* 405: 196–198.

Kramis R, Vanderwolf CH, Bland BH. 1975. Two types of hippocampal rhythmical slow activity in both the rabbit and the rat: relations to behavior and effects of atropine, diethyl ether, urethane, and pentobarbital. *Exp Neurol* 49: 58–85.

Kraus BJ, Robinson RJ, Hasselmo ME, Eichenbaum H, White JA. 2010. Time and distance dependence of rat hippocampal neuron responses. *Soc Neurosci Abstr* 36: 100.16.

Kremin T, Hasselmo ME. 2007. Cholinergic suppression of glutamatergic synaptic transmission in hippocampal region CA3 exhibits laminar selectivity: implication for hippocampal network dynamics. *Neuroscience* 149: 760–767.

Kropff E, Treves A. 2008. The emergence of grid cells: intelligent design or just adaptation? *Hippocampus* 18: 1256–1269.

Kumar SS, Jin X, Buckmaster PS, Huguenard JR. 2007. Recurrent circuits in layer II of medial entorhinal cortex in a model of temporal lobe epilepsy. *J Neurosci* 27: 1239–1246.

Kunec S, Hasselmo ME, Kopell N. 2005. Encoding and retrieval in the CA3 region of the hippocampus: a model of theta-phase separation. *J Neurophysiol* 94: 70–82.

Kunitake A, Kunitake T, Stewart M. 2004. Differential modulation by carbachol of four separate excitatory afferent systems to the rat subiculum in vitro. *Hippocampus* 14: 986–999.

Langston RF, Ainge JA, Couey JJ, Canto CB, Bjerknes TL, Witter MP, Moser EI, Moser MB. 2010. Development of the spatial representation system in the rat. *Science* 328: 1576–1580.

Lansner A, Fransén E. 1992. Modeling Hebbian cell assemblies comprised of cortical neurons. *Network* 3: 105–119.

Lashley KS. 1951. The problem of serial order in behavior. In: *Cerebral Mechanisms in Behavior* (Jeffress LAE, ed), pp 112–131. New York: Wiley.

Lawrence JJ, Statland JM, Grinspan ZM, McBain CJ. 2006. Cell type-specific dependence of muscarinic signalling in mouse hippocampal stratum oriens interneurones. *J Physiol* 570: 595–610.

Lee AK, Wilson MA. 2002. Memory of sequential experience in the hippocampus during slow wave sleep. *Neuron* 36: 1183–1194.

Lee I, Yoganarasimha D, Rao G, Knierim JJ. 2004. Comparison of population coherence of place cells in hippocampal subfields CA1 and CA3. *Nature* 430: 456–459.

Lee I, Griffin AL, Zilli EA, Eichenbaum H, Hasselmo ME. 2006. Gradual translocation of spatial correlates of neuronal firing in the hippocampus toward prospective reward locations. *Neuron* 51: 639–650.

Lee MG, Chrobak JJ, Sik A, Wiley RG, Buzsaki G. 1994. Hippocampal theta activity following selective lesion of the septal cholinergic system. *Neuroscience* 62: 1033–1047.

Lengyel M, Szatmary Z, Erdi P. 2003. Dynamically detuned oscillations account for the coupled rate and temporal code of place cell firing. *Hippocampus* 13: 700–714.

Leonard BW, Amaral DG, Squire LR, Zola-Morgan S. 1995. Transient memory impairment in monkeys with bilateral lesions of the entorhinal cortex. *J Neurosci* 15: 5637–5659.

Lerma J, Garcia-Austt E. 1985. Hippocampal theta rhythm during paradoxical sleep: effects of afferent stimuli and phase relationships with phasic events. *Electroencephalogr Clin Neurophysiol* 60: 46–54.

Leung LS, Shen B, Rajakumar N, Ma J. 2003. Cholinergic activity enhances hippocampal long-term potentiation in CA1 during walking in rats. *J Neurosci* 23: 9297–9304.

Leung VL, Zhao Y, Brown TH. 2006. Graded persistent firing in neurons of rat perirhinal cortex. *Soc Neurosci Abstr* 32: 636.618.

Leutgeb S, Leutgeb JK, Treves A, Moser MB, Moser EI. 2004. Distinct ensemble codes in hippocampal areas CA3 and CA1. *Science* 305: 1295–1298.

Lever C, Wills T, Cacucci F, Burgess N, O'Keefe J. 2002. Long-term plasticity in hippocampal place-cell representation of environmental geometry. *Nature* 416: 90–94.

Lever C, Burton S, Jeewajee A, O'Keefe J, Burgess N. 2009. Boundary vector cells in the subiculum of the hippocampal formation. *J Neurosci* 29: 9771–9777.

Levin ED, McClernon FJ, Rezvani AH. 2006. Nicotinic effects on cognitive function: behavioral characterization, pharmacological specification, and anatomic localization. *Psychopharmacology (Berl)* 184: 523–539.

Levine B, Svoboda E, Hay JF, Winocur G, Moscovitch M. 2002. Aging and autobiographical memory: dissociating episodic from semantic retrieval. *Psychol Aging* 17: 677–689.

Levy WB. 1989. A computational approach to hippocampal function. In: *Computational Models of Learning in Simple Neural Systems* (Hawkins RD, Bower GH, eds), pp 243–305. Orlando, FL: Academic Press.

Levy WB. 1996. A sequence predicting CA3 is a flexible associator that learns and uses context to solve hippocampal-like tasks. *Hippocampus* 6: 579–590.

Levy WB, Steward O. 1979. Synapses as associative memory elements in the hippocampal formation. *Brain Res* 175: 233–245.

Levy WB, Steward O. 1983. Temporal contiguity requirements for long-term associative potentiation/depression in the hippocampus. *Neuroscience* 8: 791–797.

Liljenstrom H, Hasselmo ME. 1995. Cholinergic modulation of cortical oscillatory dynamics. *J Neurophysiol* 74: 288–297.

Liljequist R, Matilda MJ. 1979. Effects of physostigmine and scopolamine on the memory functioning of chess players. *Med Biol* 51: 402–405.

Linster C, Hasselmo ME. 2001. Neuromodulation and the functional dynamics of piriform cortex. *Chem Senses* 26: 585–594.

Linster C, Maloney M, Patil M, Hasselmo ME. 2003. Enhanced cholinergic suppression of previously strengthened synapses enables the formation of self-organized representations in olfactory cortex. *Neurobiol Learn Mem* 80: 302–314.

Lipton PA, Eichenbaum H. 2008. Complementary roles of hippocampus and medial entorhinal cortex in episodic memory. *Neural Plast* 2008: 258467.

Lipton PA, White JA, Eichenbaum H. 2007. Disambiguation of overlapping experiences by neurons in the medial entorhinal cortex. *J Neurosci* 27: 5787–5795.

Lisman JE. 1999. Relating hippocampal circuitry to function: recall of memory sequences by reciprocal dentate–CA3 interactions. *Neuron* 22: 233–242.

Lisman JE. 2009. The pre/post LTP debate. *Neuron* 63: 281–284.

Lisman JE, Idiart MA. 1995. Storage of 7 +/– 2 short-term memories in oscillatory subcycles. *Science* 267: 1512–1515.

Lisman JE, Fellous JM, Wang XJ. 1998. A role for NMDA-receptor channels in working memory. *Nat Neurosci* 1: 273–275.

Losonczy A, Zemelman BV, Vaziri A, Magee JC. 2010. Network mechanisms of theta related neuronal activity in hippocampal CA1 pyramidal neurons. *Nat Neurosci* 13: 967–972.

Louie K, Wilson MA. 2001. Temporally structured replay of awake hippocampal ensemble activity during rapid eye movement sleep. *Neuron* 29: 145–156.

M'Harzi M, Palacios A, Monmaur P, Willig F, Houcine O, Delacour J. 1987. Effects of selective lesions of fimbria–fornix on learning set in the rat. *Physiol Behav* 40: 181–188.

Macrides FH, Eichenbaum H, Forbes WB. 1982. Temporal relationship between sniffing and limbic theta rhythm during odor discrimination reversal learning. *J Neurosci* 2: 1705.

Madison DV, Nicoll RA. 1984. Control of the repetitive discharge of rat CA 1 pyramidal neurones in vitro. *J Physiol* 354: 319–331.

Magee JC. 2001. Dendritic mechanisms of phase precession in hippocampal CA1 pyramidal neurons. *J Neurophysiol* 86: 528–532.

Magistretti J, Ma L, Shalinsky MH, Lin W, Klink R, Alonso A. 2004. Spike patterning by Ca2+-dependent regulation of a muscarinic cation current in entorhinal cortex layer II neurons. *J Neurophysiol* 92: 1644–1657.

Manns JR, Hopkins RO, Reed JM, Kitchener EG, Squire LR. 2003. Recognition memory and the human hippocampus. *Neuron* 37: 171–180.

Manns JR, Zilli EA, Ong KC, Hasselmo ME, Eichenbaum H. 2007. Hippocampal CA1 spiking during encoding and retrieval: relation to theta phase. *Neurobiol Learn Mem* 87: 9–20.

Markowska AL, Olton DS, Murray EA, Gaffan D. 1989. A comparative analysis of the role of fornix and cingulate cortex in memory: rats. *Exp Brain Res* 74: 187–201.

Markram H, Segal M. 1990. Acetylcholine potentiates responses to *N*-methyl-D-aspartate in the rat hippocampus. *Neurosci Lett* 113: 62–65.

Markram H, Lubke J, Frotscher M, Sakmann B. 1997. Regulation of synaptic efficacy by coincidence of postsynaptic APs and EPSPs. *Science* 275: 213–215.

Markus EJ, Qin YL, Leonard B, Skaggs WE, McNaughton BL, Barnes CA. 1995. Interactions between location and task affect the spatial and directional firing of hippocampal neurons. *J Neurosci* 15: 7079–7094.

Marr D. 1971. Simple memory: a theory for archicortex. *Phil Trans Roy Soc B* B262: 23–81.

Marrosu F, Portas C, Mascia MS, Casu MA, Fa M, Giagheddu M, Imperato A, Gessa GL. 1995. Microdialysis measurement of cortical and hippocampal acetylcholine release during sleep–wake cycle in freely moving cats. *Brain Res* 671: 329–332.

Martin MM, Horn KL, Kusman KJ, Wallace DG. 2007. Medial septum lesions disrupt exploratory trip organization: evidence for septohippocampal involvement in dead reckoning. *Physiol Behav* 90: 412–424.

Maurer AP, Vanrhoads SR, Sutherland GR, Lipa P, McNaughton BL. 2005. Self-motion and the origin of differential spatial scaling along the septo–temporal axis of the hippocampus. *Hippocampus* 15: 841–852.

McCartney H, Johnson AD, Weil ZM, Givens B. 2004. Theta reset produces optimal conditions for long-term potentiation. *Hippocampus* 14: 684–687.

McClelland JL, Goddard NH. 1996. Considerations arising from a complementary learning systems perspective on hippocampus and neocortex. *Hippocampus* 6: 654–665.

McClelland JL, McNaughton BL, O'Reilly RC. 1995. Why there are complementary learning systems in the hippocampus and neocortex: insights from the successes and failures of connectionist models of learning and memory. *Psychol Rev* 102: 419–457.

McGaughy J, Koene RA, Eichenbaum H, Hasselmo ME. 2005. Cholinergic deafferentation of the entorhinal cortex in rats impairs encoding of novel but not familiar stimuli in a delayed non-match to sample task (DNMS). *J Neurosci* 25: 10273–10281.

McHugh TJ, Jones MW, Quinn JJ, Balthasar N, Coppari R, Elmquist JK, Lowell BB, Fanselow MS, Wilson MA, Tonegawa S. 2007. Dentate gyrus NMDA receptors mediate rapid pattern separation in the hippocampal network. *Science* 317: 94–99.

McLennan H, Miller JJ. 1974. Gamma-aminobutyric acid and inhibition in the septal nuclei of the rat. *J Physiol* 237: 625–633.

McNaughton BL, Morris RGM. 1987. Hippocampal synaptic enhancement and information storage within a distributed memory system. *Trends Neurosci* 10: 408–415.

McNaughton BL, Nadel L. 1990. Hebb-Marr networks and the neurobiological representation of action in space. In: *Neuroscience and Connectionist Theory* (Gluck MA, Rumelhart DE, eds), pp 1–64. Hillsdale, NJ: Lawrence Erlbaum Assoc.

McNaughton BL, Douglas RM, Goddard GV. 1978. Synaptic enhancement in fascia dentata: cooperativity among coactive afferents. *Brain Res* 157: 277–293.

McNaughton BL, Barnes CA, O'Keefe J. 1983. The contributions of position, direction, and velocity to single unit-activity in the hippocampus of freely-moving rats. *Exp Brain Res* 52: 41–49.

McNaughton BL, Battaglia FP, Jensen O, Moser EI, Moser MB. 2006. Path integration and the neural basis of the "cognitive map." *Nat Rev Neurosci* 7: 663–678.

McNaughton N, Coop CF. 1991. Neurochemically dissimilar anxiolytic drugs have common effects on hippocampal rhythmic slow activity. *Neuropharmacology* 30: 855–863.

McNaughton N, Kocsis B, Hajós M. 2007. Elicited hippocampal theta rhythm: a screen for anxiolytic and pro-cognitive drugs through changes in hippocampal function? *Behav Pharmacol* 18: 329–346.

McQuiston AR, Madison DV. 1999. Muscarinic receptor activity has multiple effects on the resting membrane potentials of CA1 hippocampal interneurons. *J Neurosci* 19: 5693–5702.

Mehta MR, Barnes CA, McNaughton BL. 1997. Experience-dependent, asymmetric expansion of hippocampal place fields. *Proc Natl Acad Sci USA* 94: 8918–8921.

Mehta MR, Lee AK, Wilson MA. 2002. Role of experience and oscillations in transforming a rate code into a temporal code. *Nature* 417: 741–746.

Menschik ED, Finkel LH. 1998. Neuromodulatory control of hippocampal function: towards a model of Alzheimer's disease. *Artif Intell Med* 13: 99–121.

Mesulam M-M, Mufson EJ, Wainer BH, Levey AI. 1983. Central cholinergic pathways in the rat: an overview based on an alternative nomenclature (Ch1–Ch6). *Neurosci* 10: 1185–1201.

Metherate R, Hsieh CY. 2004. Synaptic mechanisms and cholinergic regulation in auditory cortex. *Prog Brain Res* 145: 143–156.

Meunier M, Bachevalier J, Mishkin M, Murray EA. 1993. Effects on visual recognition of combined and separate ablations of the entorhinal and perirhinal cortex in rhesus monkeys. *J Neurosci* 13: 5418–5432.

Migliore M, Hoffman DA, Magee JC, Johnston D. 1999. Role of an A-type K+ conductance in the back-propagation of action potentials in the dendrites of hippocampal pyramidal neurons. *J Comput Neurosci* 7: 5–15.

Miller EK, Lin L, Desimone R. 1993. Activity of neurons in anterior inferior temporal cortex during a short-term memory task. *J Neurosci* 13: 1460–1478.

Miller EK, Erickson CA, Desimone R. 1996. Neural mechanisms of visual working memory in prefrontal cortex of the macaque. *J Neurosci* 16: 5154–5167.

Milner B, Corkin S, Teuber H-L. 1968. Further analysis of the hippocampal amnesic syndrome: 14-year follow-up study of H.M. *Neuropsychologia* 6: 215–234.

Minai AA, Levy WB. 1993. The dynamics of sparse random networks. *Biol Cybern* 70: 177–187.

Mishkin M. 1978. Memory in monkeys severely impaired by combined but not by separate removal of amygdala and hippocampus. *Nature* 273: 297–298.

Mishkin M, Appenzeller T. 1987. The anatomy of memory. *Sci Am* 256: 80–89.

Mishkin M, Delacour J. 1975. An analysis of short-term visual memory in the monkey. *J Exp Psychol Anim Behav Process* 1: 326–334.

Mitchell SJ, Ranck JB, Jr. 1980. Generation of theta rhythm in medial entorhinal cortex of freely moving rats. *Brain Res* 189: 49–66.

Mitchell SJ, Rawlins JN, Steward O, Olton DS. 1982. Medial septal area lesions disrupt theta rhythm and cholinergic staining in medial entorhinal cortex and produce impaired radial arm maze behavior in rats. *J Neurosci* 2: 292–302.

Mizumori SJY, Barnes CA, McNaughton BL. 1989. Reversible inactivation of the medial septum—selective effects on the spontaneous unit-activity of different hippocampal cell-types. *Brain Res* 500: 99–106.

Mizumori SJY, Perez GM, Alvarado MC, Barnes CA, Mcnaughton BL. 1990. Reversible inactivation of the medial septum differentially affects 2 forms of learning in rats. *Brain Res* 528: 12–20.

Molyneaux BJ, Hasselmo ME. 2002. GABA(B) presynaptic inhibition has an in vivo time constant sufficiently rapid to allow modulation at theta frequency. *J Neurophysiol* 87: 1196–1205.

Monmaur P, Collet A, Puma C, Frankel-Kohn L, Sharif A. 1997. Relations between acetylcholine release and electrophysiological characteristics of theta rhythm: a microdialysis study in the urethane-anesthetized rat hippocampus. *Brain Res Bull* 42: 141–146.

Morris RG, Frey U. 1997. Hippocampal synaptic plasticity: role in spatial learning or the automatic recording of attended experience? *Philos Trans R Soc Lond B Biol Sci* 352: 1489–1503.

Morris RG, Garrud P, Rawlins JN, O'Keefe J. 1982. Place navigation impaired in rats with hippocampal lesions. *Nature* 297: 681–683.

Morris RG, Anderson E, Lynch GS, Baudry M. 1986. Selective impairment of learning and blockade of long-term potentiation by an N-methyl-D-aspartate receptor antagonist, AP5. *Nature* 319: 774–776.

Moser EI, Moser MB. 2008. A metric for space. *Hippocampus* 18: 1142–1156.

Moser MB, Moser EI. 1998. Functional differentiation in the hippocampus. *Hippocampus* 8: 608–619.

Muller RU, Kubie JL. 1987. The effects of changes in the environment on the spatial firing of hippocampal complex-spike cells. *J Neurosci* 7: 1951–1968.

Muller RU, Kubie JL, Ranck JB, Jr. 1987. Spatial firing patterns of hippocampal complex-spike cells in a fixed environment. *J Neurosci* 7: 1935–1950.

Murray EA, Wise SP. 2010. What, if anything, can monkeys tell us about human amnesia when they can't say anything at all? *Neuropsychologia* 48: 2385–2405.

Naber PA, Witter MP. 1998. Subicular efferents are organized mostly as parallel projections: a double-labeling, retrograde-tracing study in the rat. *J Comp Neurol* 393: 284–297.

Naber PA, Lopes da Silva FH, Witter MP. 2001. Reciprocal connections between the entorhinal cortex and hippocampal fields CA1 and the subiculum are in register with the projections from CA1 to the subiculum. *Hippocampus* 11: 99–104.

Nadasdy Z, Hirase H, Czurko A, Csicsvari J, Buzsaki G. 1999. Replay and time compression of recurring spike sequences in the hippocampus. *J Neurosci* 19: 9497–9507.

Nakashiba T, Young JZ, McHugh TJ, Buhl DL, Tonegawa S. 2008. Transgenic inhibition of synaptic transmission reveals role of CA3 output in hippocampal learning. *Science* 319: 1260–1264.

Nakazawa K, Quirk MC, Chitwood RA, Watanabe M, Yeckel MF, Sun LD, Kato A, Carr CA, Johnston D, Wilson MA, Tonegawa S. 2002. Requirement for hippocampal CA3 NMDA receptors in associative memory recall. *Science* 297: 211–218.

Navratilova Z, Giocomo LM, Fellous J-M, Hasselmo ME, McNaughton BL. 2011. Phase precession and variable spatial scale in a periodic attractor map model of medial entorhinal grid cells with realistic after-spike dynamics. *Hippocampus* (in press).

Newman EL, Norman KA. 2010. Moderate excitation leads to weakening of perceptual representations. *Cereb Cortex* 20: 2760–2770.

Nigro G, Neisser U. 1983. Point of view in personal memories. *Cognit Psychol* 15: 467–482.

Nissen MJ, Knopman DS, Schacter DL. 1987. Neurochemical dissociation of memory systems. *Neurology* 37: 789–794.

Norman KA, O'Reilly RC. 2003. Modeling hippocampal and neocortical contributions to recognition memory: a complementary-learning-systems approach. *Psychol Rev* 110: 611–646.

Norman KA, Newman E, Detre G, Polyn S. 2006. How inhibitory oscillations can train neural networks and punish competitors. *Neural Comput* 18: 1577–1610.

Norman KA, Newman EL, Detre G. 2007. A neural network model of retrieval-induced forgetting. *Psychol Rev* 114: 887–953.

Numan R, Quaranta JR, Jr. 1990. Effects of medial septal lesions on operant delayed alternation in rats. *Brain Res* 531: 232–241.

O'Keefe J. 1976. Place units in the hippocampus of the freely moving rat. *Exp Neurol* 51: 78–109.

O'Keefe J, Burgess N. 1996. Geometric determinants of the place fields of hippocampal neurons. *Nature* 381: 425–428.

O'Keefe J, Burgess N. 2005. Dual phase and rate coding in hippocampal place cells: theoretical significance and relationship to entorhinal grid cells. *Hippocampus* 15: 853–866.

O'Keefe J, Conway DH. 1978. Hippocampal place units in the freely moving rat: why they fire where they fire. *Exp Brain Res* 31: 573–590.

O'Keefe J, Dostrovsky J. 1971. The hippocampus as a spatial map: preliminary evidence from unit activity in the freely-moving rat. *Brain Res* 34: 171–175.

O'Keefe J, Nadel L. 1978. *The Hippocampus as a Cognitive Map*. Oxford, UK: Oxford University Press.

O'Keefe J, Recce ML. 1993. Phase relationship between hippocampal place units and the EEG theta rhythm. *Hippocampus* 3: 317–330.

O'Keefe J, Burgess N, Donnett JG, Jeffery KJ, Maguire EA. 1998. Place cells, navigational accuracy, and the human hippocampus. *Philos Trans R Soc Lond B Biol Sci* 353: 1333–1340.

O'Reilly RC, McClelland JL. 1994. Hippocampal conjunctive encoding, storage, and recall: avoiding a trade-off. *Hippocampus* 4: 661–682.

O'Reilly RC, Norman KA, McClelland JL. 1998. A hippocampal model of recognition memory. In: *Advances in Neural Information Processing Systems 10* (Jordan MI, Solla SA, Kearns, MJ, eds). Cambridge, MA: MIT Press.

Olson IR, Sledge Moore K, Stark M, Chatterjee A. 2006. Visual working memory is impaired when the medial temporal lobe is damaged. *J Cogn Neurosci* 18: 1087–1097.

Olton DS, Becker JT, Handelmann GE. 1979. Hippocampus, space and memory. *Behav Brain Sci* 2: 313–365.

Orr G, Rao G, Houston FP, McNaughton BL, Barnes CA. 2001. Hippocampal synaptic plasticity is modulated by theta rhythm in the fascia dentata of adult and aged freely behaving rats. *Hippocampus* 11: 647–654.

Otto T, Eichenbaum H. 1992. Complementary roles of the orbital prefrontal cortex and the perirhinal entorhinal cortices in an odor-guided delayed-nonmatching- to-sample task. *Behav Neurosci* 106: 762–775.

Ovsepian SV, Anwyl R, Rowan MJ. 2004. Endogenous acetylcholine lowers the threshold for long-term potentiation induction in the CA1 area through muscarinic receptor activation: in vivo study. *Eur J Neurosci* 20: 1267–1275.

Pang KC, Nocera R. 1999. Interactions between 192-IgG saporin and intraseptal cholinergic and GABAergic drugs: role of cholinergic medial septal neurons in spatial working memory. *Behav Neurosci* 113: 265–275.

Pang KC, Nocera R, Secor AJ, Yoder RM. 2001. GABAergic septohippocampal neurons are not necessary for spatial memory. *Hippocampus* 11: 814–827.

Parikh V, Kozak R, Martinez V, Sarter M. 2007. Prefrontal acetylcholine release controls cue detection on multiple timescales. *Neuron* 56: 141–154.

Pastalkova E, Itskov V, Amarasingham A, Buzsaki G. 2008. Internally generated cell assembly sequences in the rat hippocampus. *Science* 321: 1322–1327.

Patil MM, Hasselmo ME. 1999. Modulation of inhibitory synaptic potentials in the piriform cortex. *J Neurophysiol* 81: 2103–2118.

Patil MM, Linster C, Lubenov E, Hasselmo ME. 1998. Cholinergic agonist carbachol enables associative long-term potentiation in piriform cortex slices. *J Neurophysiol* 80: 2467–2474.

Pavlides C, Winson J. 1989. Influences of hippocampal place cell firing in the awake state on the activity of these cells during subsequent sleep episodes. *J Neurosci* 8: 2907–2918.

Pavlides C, Greenstein YJ, Grudman M, Winson J. 1988. Long-term potentiation in the dentate gyrus is induced preferentially on the positive phase of theta-rhythm. *Brain Res* 439: 383–387.

Penfield W, Milner B. 1958. Memory deficit produced by bilateral lesions in the hippocampal zone. *Arch Neurol Psychiatry* 79: 475–497.

Perry EK, Gibson PH, Blessed G, Perry RH, Tomlinson BE. 1977. Neurotransmitter enzyme abnormalities in senile dementia. *J Neurol Sci* 34: 247–265.

Pervouchine DD, Netoff TI, Rotstein HG, White JA, Cunningham MO, Whittington MA, Kopell NJ. 2006. Low-dimensional maps encoding dynamics in entorhinal cortex and hippocampus. *Neural Comput* 18: 2617–2650.

Pitler TA, Alger BE. 1992. Cholinergic excitation of GABAergic interneurons in the rat hippocampal slice. *J Physiol* 450: 127–142.

Poucet B, Lenck-Santini PP, Hok V, Save E, Banquet JP, Gaussier P, Muller RU. 2004. Spatial navigation and hippocampal place cell firing: the problem of goal encoding. *Rev Neurosci* 15: 89–107.

Quiroga RQ, Reddy L, Kreiman G, Koch C, Fried I. 2005. Invariant visual representation by single neurons in the human brain. *Nature* 435: 1102–1107.

Radcliffe KA, Fisher JL, Gray R, Dani JA. 1999. Nicotinic modulation of glutamate and GABA synaptic transmission of hippocampal neurons. *Ann N Y Acad Sci* 30: 591–610.

Raghavachari S, Kahana MJ, Rizzuto DS, Caplan JB, Kirschen MP, Bourgeois B, Madsen JR, Lisman JE. 2001. Gating of human theta oscillations by a working memory task. *J Neurosci* 21: 3175–3183.

Ramon y Cajal S. 1911. *Histologie du Systeme Nerveux: De l'Homme et des Vertebres, II.* Paris: Maloin.

Ranck JBJ. 1984. Head-direction cells in the deep cell layers of dorsal presubiculum in freely moving rats. *Soc Neurosci Abstr* 10: 599.

Rasch BH, Born J, Gais S. 2006. Combined blockade of cholinergic receptors shifts the brain from stimulus encoding to memory consolidation. *J Cogn Neurosci* 18: 793–802.

Rawlins JN, Feldon J, Gray JA. 1979. Septo–hippocampal connections and the hippocampal theta rhythm. *Exp Brain Res* 37: 49–63.

Redish AD. 1999. *Beyond the Cognitive Map.* Cambridge, MA: MIT Press.

Redish AD, Touretzky DS. 1998. The role of the hippocampus in solving the Morris water maze. *Neural Comput* 10: 73–111.

Remme MW, Lengyel M, Gutkin BS. 2009. The role of ongoing dendritic oscillations in single-neuron dynamics. *PLOS Comput Biol* 5: e1000493.

Remme MW, Lengyel M, Gutkin BS. 2010. Democracy-independence trade-off in oscillating dendrites and its implications for grid cells. *Neuron* 66: 429–437.

Rempel-Clower NL, Zola SM, Squire LR, Amaral DG. 1996. Three cases of enduring memory impairment after bilateral damage limited to the hippocampal formation. *J Neurosci* 16: 5233–5255.

Revonsuo A. 2000. The reinterpretation of dreams: An evolutionary hypothesis of the function of dreaming. *Behav Brain Sci* 23: 988–901.

Richardson JTE, Frith CD, Scott E, Crow TJ, Cunningham-Owens D. 1984. The effects of intravenous diazepam and hyoscine on recognition memory. *Behav Brain Res* 14: 193–199.

Riches IP, Wilson FA, Brown MW. 1991. The effects of visual stimulation and memory on neurons of the hippocampal formation and the neighboring parahippocampal gyrus and inferior temporal cortex of the primate. *J Neurosci* 11: 1763–1779.

Rivas J, Gaztelu JM, Garcia-Austt E. 1996. Changes in hippocampal cell discharge patterns and theta rhythm spectral properties as a function of walking velocity in the guinea pig. *Exp Brain Res* 108: 113–118.

Rizzuto DS, Madsen JR, Bromfield EB, Schulze-Bonhage A, Seelig D, Aschenbrenner-Scheibe R, Kahana MJ. 2003. Reset of human neocortical oscillations during a working memory task. *Proc Natl Acad Sci USA* 100: 7931–7936.

Rizzuto DS, Madsen JR, Bromfield EB, Schulze-Bonhage A, Kahana MJ. 2006. Human neocortical oscillations exhibit theta phase differences between encoding and retrieval. *Neuroimage* 31: 1352–1358.

Robbins TW, Semple J, Kumar R, Truman MI, Shorter J, Ferraro A, Fox B, McKay G, Matthews K. 1997. Effects of scopolamine on delayed-matching-to-sample and paired associates tests of visual memory and learning in human subjects: comparison with diazepam and implications for dementia. *Psychopharmacology (Berl)* 134: 95–106.

Roberts MJ, Zinke W, Guo K, Robertson R, McDonald JS, Thiele A. 2005. Acetylcholine dynamically controls spatial integration in marmoset primary visual cortex. *J Neurophysiol* 93: 2062–2072.

Robertson RG, Rolls ET, Georges-Francois P, Panzeri S. 1999. Head direction cells in the primate pre-subiculum. *Hippocampus* 9: 206–219.

Robinson JA, Swanson KL. 1993. Field and observer modes of remembering. *Memory* 1: 169–184.

Rogers JL, Kesner RP. 2003. Cholinergic modulation of the hippocampus during encoding and retrieval. *Neurobiol Learn Mem* 80: 332–342.

Rolls ET, O'Mara SM. 1995. View-responsive neurons in the primate hippocampal complex. *Hippocampus* 5: 409–424.

Rolls ET, Baylis GC, Hasselmo ME, Nalwa V. 1989. The effect of learning on the face selective responses of neurons in the cortex in the superior temporal sulcus of the monkey. *Exp Brain Res* 76: 153–164.

Rolls ET, Robertson RG, Georges-Francois P. 1997a. Spatial view cells in the primate hippocampus. *Eur J Neurosci* 9: 1789–1794.

Rolls ET, Treves A, Foster D, Perez-Vicente C. 1997b. Simulation studies of the CA3 hippocampal subfield modeled as an attractor neural network. *Neural Netw* 10: 1559–1569.

Rolls ET, Treves A, Robertson RG, Georges-François P, Panzeri S. 1998. Information about spatial view in an ensemble of primate hippocampal cells. *J Neurophysiol* 79: 1797–813.

Rolls ET, Stringer SM, Elliot T. 2006. Entorhinal cortex grid cells can map to hippocampal place cells by competitive learning. *Network* 17: 447–465.

Rosenbaum DA, Cohen RG, Jax SA, Weiss DJ, van der Wel R. 2007. The problem of serial order in behavior: Lashley's legacy. *Hum Mov Sci* 26: 525–554.

Rotstein HG, Pervouchine DD, Acker CD, Gillies MJ, White JA, Buhl EH, Whittington MA, Kopell N. 2005. Slow and fast inhibition and an H-current interact to create a theta rhythm in a model of CA1 interneuron network. *J Neurophysiol* 94: 1509–1518.

Rotstein HG, Oppermann T, White JA, Kopell N. 2006. The dynamic structure underlying subthreshold oscillatory activity and the onset of spikes in a model of medial entorhinal cortex stellate cells. *J Comput Neurosci* 21: 271–292.

Rovira C, Ben-Ari Y, Cherubini E. 1983. Dual cholinergic modulation of hippocampal somatic and dendritic field potentials by the septo–hippocampal pathway. *Exp Brain Res* 49: 151–155.

Rubin JE, Gerkin RC, Bi GQ, Chow CC. 2005. Calcium time course as a signal for spike-timing-dependent plasticity. *J Neurophysiol* 93: 2600–2613.

Rudell AP, Fox SE, Ranck JB, Jr. 1984. Hippocampal excitability phase-locked to the theta rhythm in walking rats. *Exp Neurol* 68: 87–96.

Sabolek HR, Penley SC, Hinman JR, Bunce JG, Markus EJ, Escabi M, Chrobak JJ. 2009. Theta and gamma coherence along the septotemporal axis of the hippocampus. *J Neurophysiol* 101: 1192–1200.

Sainsbury RS, Harris JL, Rowland GL. 1987a. Sensitization and hippocampal type 2 theta in the rat. *Physiol Behav* 41: 489–493.

Sainsbury RS, Heynen A, Montoya CP. 1987b. Behavioral correlates of hippocampal type 2 theta in the rat. *Physiol Behav* 39: 513–519.

Samsonovich A, McNaughton BL. 1997. Path integration and cognitive mapping in a continuous attractor neural network model. *J Neurosci* 17: 5900–5920.

Sargolini F, Fyhn M, Hafting T, McNaughton BL, Witter MP, Moser MB, Moser EI. 2006. Conjunctive representation of position, direction, and velocity in entorhinal cortex. *Science* 312: 758–762.

Sarter M, Hasselmo ME, Bruno JP, Givens B. 2005. Unraveling the attentional functions of cortical cholinergic inputs: interactions between signal-driven and cognitive modulation of signal detection. *Brain Res Brain Res Rev* 48: 98–111.

Savelli F, Knierim JJ. 2010. Hebbian analysis of the transformation of medial entorhinal grid-cell inputs to hippocampal place fields. *J Neurophysiol* 103: 3167–3183.

Savelli F, Yoganarasimha D, Knierim JJ. 2008. Influence of boundary removal on the spatial representations of the medial entorhinal cortex. *Hippocampus* 18: 1270–1282.

Schacter DL, Addis DR. 2007. The cognitive neuroscience of constructive memory: remembering the past and imagining the future. *Philos Trans R Soc Lond B Biol Sci* 362: 773–786.

Schacter DL, Addis DR, Buckner RL. 2007. Remembering the past to imagine the future: the prospective brain. *Nat Rev Neurosci* 8: 657–661.

Schon K, Hasselmo ME, Lopresti ML, Tricarico MD, Stern CE. 2004. Persistence of parahippocampal representation in the absence of stimulus input enhances long-term encoding: a functional magnetic resonance imaging study of subsequent memory after a delayed match-to-sample task. *J Neurosci* 24: 11088–11097.

Schon K, Atri A, Hasselmo ME, Tricarico MD, LoPresti ML, Stern CE. 2005. Scopolamine reduces persistent activity related to long-term encoding in the parahippocampal gyrus during delayed matching in humans. *J Neurosci* 25: 9112–9123.

Schon K, Quiroz YT, Hasselmo ME, Stern CE. 2009. Greater working memory load results in greater medial temporal activity at retrieval. *Cereb Cortex* 19: 2561–2571.

Schwindt PC, Spain WJ, Foehring RC, Chubb MC, Crill WE. 1988. Slow conductances in neurons from cat sensorimotor cortex in vitro and their role in slow excitability changes. *J Neurophysiol* 59: 450–467.

Scoville WB, Milner B. 1957. Loss of recent memory after bilateral hippocampal lesions. *J Neurol Neurosurg Psychiatry* 20: 11–21.

Seager MA, Johnson LD, Chabot ES, Asaka Y, Berry SD. 2002. Oscillatory brain states and learning: impact of hippocampal theta-contingent training. *Proc Natl Acad Sci USA* 99: 1616–1620.

Seidenbecher T, Laxmi TR, Stork O, Pape HC. 2003. Amygdalar and hippocampal theta rhythm synchronization during fear memory retrieval. *Science* 301: 846–850.

Semba K, Komisaruk BR. 1984. Neural substrates of two different rhythmical vibrissal movements in the rat. *Neuroscience* 12: 761–774.

Shalinsky MH, Magistretti J, Ma L, Alonso AA. 2002. Muscarinic activation of a cation current and associated current noise in entorhinal-cortex layer-II neurons. *J Neurophysiol* 88: 1197–1211.

Shallice T, Warrington EK. 1970. Independent functioning of verbal memory stores: a neuropsychological study. *Q J Exp Psychol* 22: 261–273.

Shapiro ML, Simon DK, Olton DS, Gage FH, 3rd, Nilsson O, Bjorklund A. 1989. Intrahippocampal grafts of fetal basal forebrain tissue alter place fields in the hippocampus of rats with fimbria–fornix lesions. *Neuroscience* 32: 1–18.

Shapiro ML, Kennedy PJ, Ferbinteanu J. 2006. Representing episodes in the mammalian brain. *Curr Opin Neurobiol* 16: 701–709.

Sharp PE. 1991. Computer simulation of hippocampal place cells. *Psychobiology* 19: 103–115.

Sharp PE. 1996. Multiple spatial/behavioral correlates for cells in the rat postsubiculum: multiple regression analysis and comparison to other hippocampal areas. *Cereb Cortex* 6: 238–259.

Sharp PE, Turner-Williams S. 2005. Movement-related correlates of single-cell activity in the medial mammillary nucleus of the rat during a pellet-chasing task. *J Neurophysiol* 94: 1920–1927.

Sharp PE, Blair HT, Cho J. 2001. The anatomical and computational basis of the rat head-direction cell signal. *Trends Neurosci* 24: 289–294.

Sharp PE, Turner-Williams S, Tuttle S. 2006. Movement-related correlates of single cell activity in the interpeduncular nucleus and habenula of the rat during a pellet-chasing task. *Behav Brain Res* 166: 55–70.

Shay CF, Boardman IS, Hasselmo ME. 2010. Comparison between rat lateral and medial entorhinal cortex neuronal resonance and subthreshold membane potential oscillation properties in whole cell patch recordings in slices. *Soc Neurosci Abstr* 36: 101.23.

Shen B, McNaughton BL. 1996. Modeling the spontaneous reactivation of experience-specific hippocampal cell assemblies during sleep. *Hippocampus* 6: 685–692.

Sheynikhovich D, Chavarriaga R, Strosslin T, Arleo A, Gerstner W. 2009. Is there a geometric module for spatial orientation? Insights from a rodent navigation model. *Psychol Rev* 116: 540–566.

Shirey JK, Xiang Z, Orton D, Brady AE, Johnson KA, Williams R, Ayala JE, et al. 2008. An allosteric potentiator of M4 mAChR modulates hippocampal synaptic transmission. *Nat Chem Biol* 4(1): 42–50.

Shouval HZ, Wang SS, Wittenberg GM. 2010. Spike timing dependent plasticity: a consequence of more fundamental learning rules. *Front Comput Neurosci* 4:19.

Siegle GJ, Hasselmo ME. 2002. Using connectionist models to guide assessment of psychological disorder. *Psychol Assess* 14: 263–278.

Siekmeier PJ, Hasselmo ME, Howard MW, Coyle, J. 2006. Modeling of context-dependent retrieval in hippocampal region CA1: Implications for cognitive function in schizophrenia. *Schizophr Res* 89: 177–190.

Silver MA, Shenhav A, D'Esposito M. 2008. Cholinergic enhancement reduces spatial spread of visual responses in human early visual cortex. *Neuron* 60: 904–914.

Siok CJ, Rogers JA, Kocsis B, Hajos M. 2006. Activation of alpha7 acetylcholine receptors augments stimulation-induced hippocampal theta oscillation. *Eur J Neurosci* 23: 570–574.

Skaggs WE, McNaughton BL. 1996. Replay of neuronal firing sequences in rat hippocampus during sleep following spatial experience. *Science* 271: 1870–1873.

Skaggs WE, Knierim JJ, Kudrimoti HS, McNaughton BL. 1995. A model of the neural basis of the rat's sense of direction. *Adv Neural Inf Process Syst* 7: 173–180.

Skaggs WE, McNaughton BL, Wilson MA, Barnes CA. 1996. Theta phase precession in hippocampal neuronal populations and the compression of temporal sequences. *Hippocampus* 6: 149–172.

Smith DM, Mizumori SJ. 2006a. Learning-related development of context-specific neuronal responses to places and events: the hippocampal role in context processing. *J Neurosci* 26: 3154–3163.

Smith DM, Mizumori SJ. 2006b. Hippocampal place cells, context, and episodic memory. *Hippocampus* 16: 716–729.

Sohal VS, Hasselmo ME. 1998a. GABA(B) modulation improves sequence disambiguation in computational models of hippocampal region CA3. *Hippocampus* 8: 171–193.

Sohal VS, Hasselmo ME. 1998b. Changes in GABAB modulation during a theta cycle may be analogous to the fall of temperature during annealing. *Neural Comput* 10: 869–882.

Sohal VS, Hasselmo ME. 2000. A model for experience-dependent changes in the responses of inferotemporal neurons. *Network* 11: 169–190.

Solstad T, Moser EI, Einevoll GT. 2006. From grid cells to place cells: a mathematical model. *Hippocampus* 16: 1026–1031.

Solstad T, Boccara CN, Kropff E, Moser MB, Moser EI. 2008. Representation of geometric borders in the entorhinal cortex. *Science* 322: 1865–1868.

Squire LR, Cohen N. 1979. Memory and amnesia: resistance to disruption develops for years after learning. *Behav Neural Biol* 25: 115–125.

Squire LR, Ojemann JG, Miezin FM, Petersen SE, Videen TO, Raichle ME. 1992. Activation of the hippocampus in normal humans: a functional anatomical study of memory. *Proc Natl Acad Sci USA* 89: 1837–1841.

Stackman RW, Taube JS. 1997. Firing properties of head direction cells in the rat anterior thalamic nucleus: dependence on vestibular input. *J Neurosci* 17: 4349–4358.

Stackman RW, Taube JS. 1998. Firing properties of rat lateral mammillary single units: head direction, head pitch, and angular head velocity. *J Neurosci* 18: 9020–9037.

Stanton ME, Thomas GJ, Brito GN. 1984. Posterodorsal septal lesions impair performance on both shift and stay working memory tasks. *Behav Neurosci* 98: 405–415.

Steele RJ, Morris RG. 1999. Delay-dependent impairment of a matching-to-place task with chronic and intrahippocampal infusion of the NMDA-antagonist D-AP5. *Hippocampus* 9: 118–136.

Steffenach HA, Witter M, Moser MB, Moser EI. 2005. Spatial memory in the rat requires the dorsolateral band of the entorhinal cortex. *Neuron* 45: 301–313.

Steinvorth S, Levine B, Corkin S. 2005. Medial temporal lobe structures are needed to re-experience remote autobiographical memories: evidence from H.M. and W.R. *Neuropsychologia* 43: 479–496.

Steinvorth S, Corkin S, Halgren E. 2006. Ecphory of autobiographical memories: an fMRI study of recent and remote memory retrieval. *Neuroimage* 30: 285–298.

Stensland H, Kirkesola T, Moser M-B, Moser EI. 2010. Orientational geometry of entorhinal grid cells. *Soc Neurosci Abstr* 36: 101.14.

Stepankova K, Fenton AA, Pastalkova E, Kalina M, Bohbot VD. 2004. Object-location memory impairment in patients with thermal lesions to the right or left hippocampus. *Neuropsychologia* 42: 1017–1028.

Stern CE, Corkin S, Gonzalez RG, Guimaraes AR, Baker JR, Jennings PJ, Carr CA, Sugiura RM, Vedantham V, Rosen BR. 1996. The hippocampal formation participates in novel picture encoding: evidence from functional magnetic resonance imaging. *Proc Natl Acad Sci USA* 93: 8660–8665.

Stern CE, Sherman SJ, Kirchhoff BA, Hasselmo ME. 2001. Medial temporal and prefrontal contributions to working memory tasks with novel and familiar stimuli. *Hippocampus* 11: 337–346.

Storm JF. 1990. Potassium currents in hippocampal pyramidal cells. *Prog Brain Res* 83: 161–187.

Strang G. 1988. *Linear Algebra and Its Applications.* San Diego: Harcourt, Brace, Jovanovich.

Sutton RS, Barto AG. 1998. *Reinforcement Learning (Adaptive Computation and Machine Learning).* Cambridge, MA: MIT Press.

Suzuki WA, Miller EK, Desimone R. 1997. Object and place memory in the macaque entorhinal cortex. *J Neurophysiol* 78: 1062–1081.

Swanson LW, Wyss JM, Cowan WM. 1978. An autoradiographic study of the organization of intrahippocampal association pathways in the rat. *J Comp Neurol* 181: 681–716.

Tahvildari B, Alonso A. 2005. Morphological and electrophysiological properties of lateral entorhinal cortex layers II and III principal neurons. *J Comp Neurol* 491: 123–140.

Tahvildari B, Fransén E, Alonso AA, Hasselmo ME. 2007. Switching between "On" and "Off" states of persistent activity in lateral entorhinal layer III neurons. *Hippocampus* 17: 257–263.

Tamamaki N, Nojyo Y. 1995. Preservation of topography in the connections between the subiculum, field CA1, and the entorhinal cortex in rats. *J Comp Neurol* 353: 379–390.

Tang AC, Hasselmo ME. 1994. Selective suppression of intrinsic but not afferent fiber synaptic transmission by baclofen in the piriform (olfactory) cortex. *Brain Res* 659: 75–81.

Tang Y, Mishkin M, Aigner TG. 1997. Effects of muscarinic blockade in perirhinal cortex during visual recognition. *Proc Natl Acad Sci USA* 94: 12667–12669.

Taube JS. 1995. Head direction cells recorded in the anterior thalamic nuclei of freely moving rats. *J Neurosci* 15: 70–86.

Taube JS. 1998. Head direction cells and the neurophysiological basis for a sense of direction. *Prog Neurobiol* 55: 225–256.

Taube JS, Bassett JP. 2003. Persistent neural activity in head direction cells. *Cereb Cortex* 13: 1162–1172.

Taube JS, Burton HL. 1995. Head direction cell activity monitored in a novel environment and during a cue conflict situation. *J Neurophysiol* 74: 1953–1971.

Taube JS, Muller RU, Ranck JB, Jr. 1990a. Head-direction cells recorded from the postsubiculum in freely moving rats. II. Effects of environmental manipulations. *J Neurosci* 10: 436–447.

Taube JS, Muller RU, Ranck JB, Jr. 1990b. Head-direction cells recorded from the postsubiculum in freely moving rats. I. Description and quantitative analysis. *J Neurosci* 10: 420–435.

Taube JS, Kesslak JP, Cotman CW. 1992. Lesions of the rat postsubiculum impair performance on spatial tasks. *Behav Neural Biol* 57: 131–143.

Taube JS, Goodridge JP, Golob EJ, Dudchenko PA, Stackman RW. 1996. Processing the head direction cell signal: a review and commentary. *Brain Res Bull* 40: 477–484.

Terrace HS. 2005. The simultaneous chain: a new approach to serial learning. *Trends Cogn Sci* 9: 202–210.

Thurley K, Leibold C, Gundlfinger A, Schmitz D, Kempter R. 2008. Phase precession through synaptic facilitation. *Neural Comput* 20: 1285–1324.

Toth K, Freund TF, Miles R. 1997. Disinhibition of rat hippocampal pyramidal cells by GABAergic afferent from the septum. *J Physiol* 500: 463–474.

Touretzky DS, Redish AD. 1996. Theory of rodent navigation based on interacting representations of space. *Hippocampus* 6: 247–270.

Toussaint M. 2006. A sensorimotor map: modulating lateral interactions for anticipation and planning. *Neural Comput* 18: 1132–1155.

Traub R, Miles R, Buzsaki G. 1992. Computer simulation of carbachol-driven rhythmic population oscillations in the CA3 region of the in vitro rat hippocampus. *J Physiol* 451: 653–672.

Traub RD, Wong RK, Miles R, Michelson H. 1991. A model of a CA3 hippocampal pyramidal neuron incorporating voltage-clamp data on intrinsic conductances. *J Neurophysiol* 66: 635–650.

Treves A. 2004. Computational constraints between retrieving the past and predicting the future, and the CA3–CA1 differentiation. *Hippocampus* 14: 539–556.

Treves A, Rolls ET. 1992. Computational constraints suggest the need for two distinct input systems to the hippocampal CA3 network. *Hippocampus* 2: 189–199.

Treves A, Rolls ET. 1994. Computational analysis of the role of the hippocampus in memory. *Hippocampus* 4: 374–391.

Tsodyks MV, Skaggs WE, Sejnowski TJ, McNaughton BL. 1996. Population dynamics and theta rhythm phase precession of hippocampal place cell firing: a spiking neuron model. *Hippocampus* 6: 271–280.

Tulving E. 1972. Episodic and semantic memory. In: *Organization of Memory* (Tulving E, Donaldson W, eds), pp 381–403. New York: Academic Press.

Tulving E. 1983. *Elements of Episodic Memory.* Oxford, UK: Oxford University Press.

Tulving E. 1984. Precis of *Elements of Episodic Memory. Behav Brain Sci* 7: 223–268.

Tulving E. 2001. Episodic memory and common sense: how far apart? *Philos Trans R Soc Lond B Biol Sci* 356: 1505–1515.

Tulving E. 2002. Episodic memory: from mind to brain. *Annu Rev Psychol* 53: 1–25.

Turchi J, Saunders RC, Mishkin M. 2005. Effects of cholinergic deafferentation of the rhinal cortex on visual recognition memory in monkeys. *Proc Natl Acad Sci USA* 102: 2158–2161.

Umbriaco D, Garcia S, Beaulieu C, Descarries L. 1995. Relational features of acetylcholine, noradrenaline, serotonin and GABA axon terminals in the stratum radiatum of adult rat hippocampus (CA1). *Hippocampus* 5: 605–620.

Valentino RJ, Dingledine R. 1981. Presynaptic inhibitory effect of acetylcholine in the hippocampus. *J Neurosci* 1: 784–792.

van Groen T, Wyss JM. 1990. The postsubicular cortex in the rat: characterization of the fourth region of the subicular cortex and its connections. *Brain Res* 529: 165–177.

van Hoesen GW. 1982. The parahippocampal gyrus. New observations regarding its cortical connections in the monkey. *Trends Neurosci* 5: 345–350.

van Turennout M, Bielamowicz L, Martin A. 2003. Modulation of neural activity during object naming: effects of time and practice. *Cereb Cortex* 13: 381–391.

van Turennout M, Ellmore T, Martin A. 2000. Long-lasting cortical plasticity in the object naming system. *Nat Neurosci* 3: 1329–1334.

Vanderwolf CH. 1969. Hippocampal electrical activity and voluntary movement in the rat. *Electroencephalogr Clin Neurophysiol* 26: 407–418.

Vann SD, Aggleton JP. 2004. The mammillary bodies: two memory systems in one? *Nat Rev Neurosci* 5: 35–44.

Vazdarjanova A, Guzowski JF. 2004. Differences in hippocampal neuronal population responses to modifications of an environmental context: evidence for distinct, yet complementary, functions of CA3 and CA1 ensembles. *J Neurosci* 24: 6489–6496.

Vertes RP, Kocsis B. 1997. Brainstem-diencephalo-septohippocampal systems controlling the theta rhythm of the hippocampus. *Neuroscience* 81: 893–926.

Villarreal DM, Gross AL, Derrick BE. 2007. Modulation of CA3 afferent inputs by novelty and theta rhythm. *J Neurosci* 27: 13457–13467.

Vinogradova OS. 2001. *Hippocampus* as comparator: role of the two input and two output systems of the hippocampus in selection and registration of information. *Hippocampus* 11: 578–598.

Vogt KE, Regehr WG. 2001. Cholinergic modulation of excitatory synaptic transmission in the CA3 area of the hippocampus. *J Neurosci* 21: 75–83.

Voicu H, Schmajuk N. 2002. Latent learning, shortcuts and detours: a computational model. *Behav Processes* 59(2): 67.

Wagner AD, Schacter DL, Rotte M, Koutstaal W, Maril A, Dale AM, Rosen BR, Buckner RL. 1998. Building memories: remembering and forgetting of verbal experiences as predicted by brain activity. *Science* 281: 1188–1191.

Wallenstein GV, Hasselmo ME. 1997a. Functional transitions between epileptiform-like activity and associative memory in hippocampal region CA3. *Brain Res Bull* 43: 485–493.

Wallenstein GV, Hasselmo ME. 1997b. GABAergic modulation of hippocampal population activity: sequence learning, place field development, and the phase precession effect. *J Neurophysiol* 78: 393–408.

Warrington EK, Shallice T. 1969. The selective impairment of auditory verbal short-term memory. *Brain* 92: 885–896.

Weinberger NM. 2003. The nucleus basalis and memory codes: auditory cortical plasticity and the induction of specific, associative behavioral memory. *Neurobiol Learn Mem* 80: 268–284.

Welinder PE, Burak Y, Fiete IR. 2008. Grid cells: the position code, neural network models of activity, and the problem of learning. *Hippocampus* 18: 1283–1300.

Whishaw IQ. 1972. Hippocampal electroencephalographic activity in the Mongolian gerbil during natural behaviors and wheel running and in the rat during wheel running and conditioned immobility. *Can J Psychol* 26: 219–239.

Whishaw IQ. 1985. Cholinergic receptor blockade in the rat impairs locale but not taxon strategies for place navigation in a swimming pool. *Behav Neurosci* 99: 979–1005.

Whishaw IQ, Vanderwolf CH. 1973. Hippocampal EEG and behavior: changes in amplitude and frequency of RSA (theta rhythm) associated with spontaneous and learned movement patterns in rats and cats. *Behav Biol* 8: 461–484.

White JA, Budde T, Kay AR. 1995. A bifurcation analysis of neuronal subthreshold oscillations. *Biophys J* 69: 1203–1217.

White JA, Klink R, Alonso A, Kay AR. 1998a. Noise from voltage-gated ion channels may influence neuronal dynamics in the entorhinal cortex. *J Neurophysiol* 80: 262–269.

White JA, Chow CC, Ritt J, Soto-Trevino C, Kopell N. 1998b. Synchronization and oscillatory dynamics in heterogeneous, mutually inhibited neurons. *J Comput Neurosci* 5: 5–16.

White NM, McDonald RJ. 2002. Multiple parallel memory systems in the brain of the rat. *Neurobiol Learn Mem* 77: 125–184.

Wiener SI, Paul CA, Eichenbaum H. 1989. Spatial and behavioral correlates of hippocampal neuronal activity. *J Neurosci* 9: 2737–2763.

Wigstrom H, Gustafsson B, Huang Y-Y, Abraham WC. 1986. Hippocampal long-term potentiation is induced by pairing single afferent volleys with intracellularly injected depolarizing current pulses. *Acta Physiol Scand* 126: 317–319.

Williams HL, Conway MA, Baddeley AD. 2008. The boundaries of episodic memories. In: *Understanding Events: From Perception to Action* (Shipley TF, Zacks JM, eds), pp 39–52. New York: Oxford University Press.

Wills TJ, Lever C, Cacucci F, Burgess N, O'Keefe J. 2005. Attractor dynamics in the hippocampal representation of the local environment. *Science* 308: 873–876.

Wills TJ, Cacucci F, Burgess N, O'Keefe J. 2010. Development of the hippocampal cognitive map in preweanling rats. *Science* 328: 1573–1576.

Wilson HR, Cowan JD. 1972. Excitatory and inhibitory interactions in localized populations of model neurons. *Biophys J* 12: 1–24.

Wilson MA, McNaughton BL. 1994. Reactivation of hippocampal ensemble memories during sleep. *Science* 265: 676–679.

Winson J. 1978. Loss of hippocampal theta rhythm results in spatial memory deficit in the rat. *Science* 201: 160–163.

Winters BD, Bussey TJ. 2005. Removal of cholinergic input to perirhinal cortex disrupts object recognition but not spatial working memory in the rat. *Eur J Neurosci* 21: 2263–2270.

Wirth S, Yanike M, Frank LM, Smith AC, Brown EN, Suzuki WA. 2003. Single neurons in the monkey hippocampus and learning of new associations. *Science* 300: 1578–1581.

Wirth S, Avsar E, Chiu CC, Sharma V, Smith AC, Brown E, Suzuki WA. 2009. Trial outcome and associative learning signals in the monkey hippocampus. *Neuron* 61: 930–940.

Witter MP, Moser EI. 2006. Spatial representation and the architecture of the entorhinal cortex. *Trends Neurosci* 29: 671–678.

Witter MP, Wouterlood FG, Naber PA, Van Haeften T. 2000. Anatomical organization of the parahippocampal–hippocampal network. *Ann N Y Acad Sci* 911: 1–24.

Wixted JT, Stretch V. 2004. In defense of the signal detection interpretation of remember/know judgments. *Psychon Bull Rev* 11: 616–641.

Wood ER, Dudchenko PA, Robitsek RJ, Eichenbaum H. 2000. Hippocampal neurons encode information about different types of memory episodes occurring in the same location. *Neuron* 27: 623–633.

Wyble BP, Linster C, Hasselmo ME. 2000. Size of CA1-evoked synaptic potentials is related to theta rhythm phase in rat hippocampus. *J Neurophysiol* 83: 2138–2144.

Wyble BP, Hyman JM, Rossi CA, Hasselmo ME. 2004. Analysis of theta power in hippocampal EEG during bar pressing and running behavior in rats during distinct behavioral contexts. *Hippocampus* 14: 662–674.

Yamamoto C, Kawai N. 1967. Presynaptic action of acetylcholine in thin sections from the guinea pig dentate gyrus in vitro. *Exp Neurol* 19: 176–187.

Yoder RM, Pang KC. 2005. Involvement of GABAergic and cholinergic medial septal neurons in hippocampal theta rhythm. *Hippocampus* 15: 381–392.

Yonelinas AP. 2001. Receiver-operating characteristics in recognition memory: evidence for a dual-process model. *J Exp Psychol Hum Learn* 20: 1341–1354.

Yonelinas AP, Otten LJ, Shaw KN, Rugg MD. 2005. Separating the brain regions involved in recollection and familiarity in recognition memory. *J Neurosci* 25: 3002–3008.

Yoshida M, Alonso A. 2007. Cell-type specific modulation of intrinsic firing properties and subthreshold membrane oscillations by the m(kv7)-current in neurons of the entorhinal cortex. *J Neurophysiol* 98: 2779–2794.

Yoshida M, Hasselmo ME. 2009a. Persistent firing supported by an intrinsic cellular mechanism in a component of the head direction system. *J Neurosci* 29: 4945–4952.

Yoshida M, Hasselmo ME. 2009b. Differences in persistent firing properties dependent upon anatomical location of neurons in rat medial entorhinal cortex. *Soc Neurosci Abstr* 35: 193.19.

Yoshida M, Boardman I, Hasselmo ME. 2010. Analysis of the frequency of subthreshold oscillations at different membrane potential voltages in neurons at different anatomical positions on the dorso-ventral axis in the rat medial entorhinal cortex. *Soc Neurosci Abstr* 36:101.21.

Yoshida M, Fransén E, Hasselmo ME. 2008. mGluR-dependent persistent firing in entorhinal cortex layer III neurons. *Eur J Neurosci* 28: 1116–1126.

Young BJ, Otto T, Fox GD, Eichenbaum H. 1997. Memory representation within the parahippocampal region. *J Neurosci* 17: 5183–5195.

Zacks JM, Tversky B. 2001. Event structure in perception and conception. *Psychol Bull* 127: 3–21.

Zacks JM, Tversky B, Iyer G. 2001. Perceiving, remembering, and communicating structure in events. *J Exp Psychol Gen* 130: 29–58.

Zalutsky RA, Nicoll RA. 1990. Comparison of two forms of long-term potentiation in single hippocampal neurons. *Science* 248: 1619–1624.

Zhang K. 1996. Representation of spatial orientation by the intrinsic dynamics of the head-direction cell ensemble: a theory. *J Neurosci* 16: 2112–2126.

Zilli E, Hasselmo ME. 2010. Coupled noisy spiking neurons as velocity-controlled oscillators in a model of grid cell spatial firing. *J Neurosci* 30: 13850–13860.

Zilli EA, Hasselmo ME. 2008a. Modeling the role of working memory and episodic memory in behavioral tasks. *Hippocampus* 18: 193–209.

Zilli EA, Hasselmo ME. 2008b. Analyses of Markov decision process structure regarding the possible strategic use of interacting memory systems. *Front Comput Neurosci* 2: 6.

Zilli EA, Hasselmo ME. 2008c. The influence of Markov decision process structure on the possible strategic use of working memory and episodic memory. *PLoS ONE* 3(7): e2756.

Zilli EA, Yoshida M, Tahvildari B, Giocomo LM, Hasselmo ME. 2009. Evaluation of the oscillatory interference model of grid cell firing through analysis and measured period variance of some biological oscillators. *PLOS Comput Biol* 5: e1000573.

Zola-Morgan S, Squire LR. 1986. Memory impairment in monkeys following lesions limited to the hippocampus. *Behav Neurosci* 100: 155–160.

Zola-Morgan S, Squire LR, Amaral DG. 1986. Human amnesia and the medial temporal region: enduring memory impairment following a bilateral lesion limited to field CA1 of the hippocampus. *J Neurosci* 6: 2950–2967.

Zola-Morgan S, Squire LR, Clower RP, Rempel NL. 1993. Damage to the perirhinal cortex exacerbates memory impairment following lesions to the hippocampal formation. *J Neurosci* 13: 251–265.

Zola-Morgan SM, Squire LR. 1990. The primate hippocampal formation: evidence for a time-limited role in memory storage. *Science* 250: 288–290.

Author Index

Subject Index

Printed in the United States
by Baker & Taylor Publisher Services